T0310239

Multifunctional Antennas and Arrays for Wireless Communication Systems

Multifunctional Antennas and Arrays for Wireless Communication Systems

Edited by

Satish K. Sharma and Jia-Chi S. Chieh
San Diego State University
San Diego, CA, USA

Registered Office
John Wiley & Sons, Inc., 111 River Street, Hoboken, NJ 07030, USA

Editorial Office
111 River Street, Hoboken, NJ 07030, USA

For details of our global editorial offices, customer services, and more information about Wiley products visit us at www.wiley.com.

Library of Congress Cataloging-in-Publication data applied for
ISBN: 9781119535058

Cover design by Wiley
Cover image: © Andrey Suslov/iStock.com

Set in 9.5/12.5pt STIXTwoText by SPi Global, Pondicherry, India

SKY10026104_040621

To our parents, teachers, and family members

Contents

List of Contributors

Behrouz Babakhani
Antenna and Microwave Lab (AML),
Department of Electrical and
Computer Engineering, San Diego
State University, San Diego, CA, USA

Sonika P. Biswal
Antenna and Microwave Lab (AML),
Department of Electrical and
Computer Engineering, San Diego
State University, San Diego, CA, USA

Jia-Chi S. Chieh
Antenna and Microwave Lab (AML),
Department of Electrical and
Computer Engineering, San Diego
State University, San Diego, CA, USA

Kumud R. Jha
Antenna and Microwave Lab (AML),
Department of Electrical and
Computer Engineering, San Diego
State University, San Diego, CA, USA
Department of Electronics and
Communication Engineering, Shri
Mata Vaishno Devi University, SMVD
University, Katra, India

Saeed I. Latif
Department of Electrical and
Computer Engineering, University of
South Alabama, Mobile, AL, USA

Sima Noghanian
Antenna and Microwave Lab (AML),
Department of Electrical and
Computer Engineering, San Diego
State University, San Diego, CA, USA

Satish K. Sharma
Antenna and Microwave Lab (AML),
Department of Electrical and
Computer Engineering, San Diego
State University, San Diego, CA, USA

Preface

Multifunctional antennas and arrays are the new trend in the field of antennas for diversified applications such as wireless and satellite communications as well as for radar applications. Reconfigurable antennas starting from frequency reconfiguration, pattern reconfiguration to polarization reconfiguration and their combinations make these antennas not only multifunctional but also reduce space requirements on the host communication devices. In the last two decades there has been great efforts to design and realize these reconfigurable antennas and we anticipate even more efforts to come in the near future. A wide range of sub-topics as they apply to multifunction antennas and arrays include the design and development of the reconfigurable multiple-input-multiple-output (MIMO) antennas, liquid metal antennas, piezoelectric antennas, radio frequency (RF) micro-electro-mechanical-systems (MEMS) based reconfigurable antennas, multifunctional antennas for 4G/5G communications and MIMO applications, metamaterials reconfigurable antennas, multifunctional antennas for user equipment (EUs), reconfigurable antennas for the defense applications and phased array antennas using 5G silicon RFICs.

The purpose of this book is to present in-depth theory, as well as design and development insight of these various multifunctional antennas and arrays. The book is aimed for use by practicing antenna engineers and researchers in the industry and academia. This book starts with an introduction to the antennas in Chapter 1, which discusses the importance of antennas. It also provides an introduction to antenna performance parameters, antenna types, multifunctional antennas, reconfigurable antennas, and antenna measurements. Next in Chapter 2, frequency reconfigurable antennas (FRAs) are detailed. This chapter starts with discussion of the mechanism of frequency reconfigurability, types of the FRAs using various switches and tunable components, FRAs by employing mechanical changes such as ground plane membrane deflection, and FRAs by using special materials and special shapes. Chapter 3 presents discussion on the pattern reconfigurable antennas which includes the following: pattern reconfiguration by electronically changing antenna elements and feeding networks, mechanically controlled pattern reconfigurable antennas, pattern reconfigurable arrays and optimizations, and reconfigurable wearable and implanted antennas. In Chapter 4, we discuss the polarization reconfigurable antennas with emphasis on the polarization reconfiguration mechanism using RF switches, polarization reconfigurable antennas using solid-state RF switches, mechanical and micro-electro-mechanical-system (MEMS) RF switches, switchable feed networks, usage of metasurfaces, as well as other methods. These chapters describe the three main types of reconfigurable antennas and arrays as described in the introduction.

Reconfigurable antennas using the liquid metal, piezoelectric and RF MEMS are discussed in Chapter 5. This chapter specifically includes discussion on the liquid metal based frequency, pattern, and directivity reconfigurable antennas, piezoelectric based pattern reconfigurable arrays, and RF MEMS based frequency and pattern reconfigurable antennas. Compact reconfigurable antennas are discussed in Chapter 6 with the main focus on the reconfigurable pixel antennas, and reconfigurable antennas using fluidic, ferrite and magnetic materials, metamaterials and metasurfaces.

Reconfigurable MIMO antennas are presented in Chapter 7, which discusses the following: reconfigurable antennas for MIMO applications, isolation techniques in MIMO antennas, pattern diversity scheme, reconfigurable polarization MIMO antennas, MIMO antenna performance parameters, and finally some reconfigurable MIMO antenna examples. Chapter 8 offers discussion on the MIMO antennas in multifunctional systems, MIMO antennas in Radar systems, MIMO antennas in communication systems, MIMO antennas for sensing applications, MIMO antennas for 5G systems, massive MIMO arrays, dielectric lens for millimeter wave MIMO, beamforming in massive MIMO, MIMO in imaging systems, and MIMO antenna in medical applications. Use of metamaterials in reconfigurable antennas have been addressed in Chapter 9. This chapter focuses the discussion on metamaterials in antenna reconfigurability, metamaterial-inspired reconfigurable antennas, and metasurface-inspired reconfigurable antennas.

Chapter 10 provides detailed discussion on the multifunctional antennas for user equipments (UEs) with emphasis on the lower/sub-6 GHz 5G band antennas, 5G *mm*-wave antenna arrays, collocated sub-6 GHz and *mm*-Wave 5G array antennas, and RF and electromagnetic fields (EMF) exposure limits. The department of defense (DoD) related reconfigurable antennas are presented in Chapter 11 with a focus on the tactical air navigation system (TACAN) antennas, sea-based X-Band Radar 1 (SBX-1) antennas, the advanced multifunction RF concept (AMRFC) antennas, integrated topside (InTop) antennas, the Defense Advanced Research Projects Agency (DARPA) arrays of commercial timescales (ACT), and the Air Force Research Laboratory (AFRL) transformational element level array (TELA). Finally, Chapter 12 discusses 5G silicon RFICs-based phased array antennas, which introduces silicon beamformer technology. It includes a short discussion of three phase shifting topologies using local oscillator (LO) based phase shifting, intermediate frequency (IF) based phase shifting and RF based phase shifting for beam steering array antennas. Several flat panel phased array antenna examples using the silicon beamforming chipsets both at Ku- and Ka-band with linear and circular polarizations are also presented.

We would like to mention that the slight overlap between the content in couple of chapters is acknowledged. We have done this intentionally so that discussion is complete in the respective chapters. While the contributors and authors have made great effort to present details for each topic area, they are by no means complete as the body of work in this field is large. They do represent the interpretations of each chapter's contributors. As time progresses, further improvements and innovations in the state-of-the-art technologies in reconfigurable antennas is anticipated. Therefore, it is expected that interested readers should continually refresh their knowledge to follow the growth of communication technologies.

1 February 2021   *Professor Satish K. Sharma, PhD*
San Diego, CA, USA   *Jia-Chi S. Chieh, PhD*

Acknowledgements

We would like to offer our sincere thanks to the chapter coauthors for their valuable contributions, patience and timely support throughout the development of this book. We would also like to thank the Wiley team members especially, Brett Kurzman, Victoria Bradshaw, Sarah Lemore, Sukhwinder Singh and most importantly S. M. Amudhapriya for their immense help throughout the completion of this book.

Professor Satish K. Sharma will like to take this opportunity to thank his research collaborators, past and present graduate students, post-doctoral fellows, visiting scholars, and undergraduate students at San Diego State University (SDSU) who have been the continuous source for his research growth. He thanks Dr. Jia-Chi S. Chieh for agreeing to work on this book. He also thanks the funding agencies: National Science Foundation (NSF) for the prestigious CAREER award, the Office of Naval Research (ONR), the Naval Information Warfare Center-Pacific (NIWC-PAC), the Space and Naval Warfare Systems Command (SPAWAR)-San Diego, and the SBIR/STTR Phase I and II research grants subcontracted through the local industries, which have helped him pursue his research work. Finally, he thanks his spouse Mamta Sharma (Author and Artist) and daughters Shiva Shree Sharma (Doctoral Student in Material Science Engineering at University of California, Riverside, California) and Shruti Shree Sharma (Undergraduate Student in Electrical Engineering at University of California, Irvine, California) who spared their valuable time to let him work on this book and offered their unconditional love and support as always. He also thanks his pet dog and cat Charlie Sharma and Razzle Sharma, respectively, for their unconditional love to him. Lastly, he is grateful to his parents (Mr. Rama Naresh Sharma and Mrs. Taravati Sharma), elders in his extended family, research advisors (Professors L. Shafai, the University of Manitoba and B. R. Vishvakarma, Indian Institute of Technology, Banaras Hindu University), teachers, colleagues, friends and the almighty God for bestowing continuous blessings on him.

Dr. Jia-Chi S. Chieh is grateful to his research group at the Naval Information Warfare Center in San Diego for their tireless efforts in the development of low-cost phased array antennas over the last decade. He is also grateful for the research collaboration opportunities he has had with Prof. Satish K. Sharma from San Diego State University (SDSU), as well as his mentorship and friendship over the years. He is thankful to his family for their love and support, and who have allowed him to complete this work including his wife Kristine, and his two daughters Joanna and Audrey. Lastly, he is grateful to his parents (Dr. Shih-Huang Chieh and Mrs. Dolly Chieh), who taught him the importance of learning and to never stop.

1 February 2021   *Professor Satish K. Sharma, PhD*
San Diego, CA, USA   *Jia-Chi S. Chieh, PhD*

1

Introduction

Satish K. Sharma and Jia-Chi S. Chieh

1.1 Introduction

In this chapter, we provide basic discussion about an antenna and its importance, type of antennas, and introductory information about the reconfigurable antenna, frequency agile antenna, multifunctional antenna, and antenna measurements.

1.2 Antenna: an Integral Component of Wireless Communications

An antenna is described as a device that radiates or receives transverse electromagnetic waves (TEM) from its surface, or structure. It is an integral component of all the wireless communication systems. As shown in Figure 1.1, the transmitter block which usually consists of the signal generator, modulator, and power amplifiers is terminated with an antenna to radiate the power in free space. A poor choice and design of antenna will result in the power being reflected to the source and cause waste of power, which is undesirable. Efficient power utilization becomes critical in applications such as onboard circuits in satellite communications. To emphasize the importance of antennas for the receiver circuitry, maximum power should be obtained from the incident wave to relax the burden on the succeeding blocks such as low noise amplifiers to maintain the required signal-to-noise-ratio (SNR) for satisfactory wireless links. Different communication application demands different minimum required SNR for a satisfactory link and efficient antenna design plays a big role in achieving this goal.

Multifunctional Antennas and Arrays for Wireless Communication Systems, First Edition.
Edited by Satish K. Sharma and Jia-Chi S. Chieh.
© 2021 John Wiley & Sons, Inc. Published 2021 by John Wiley & Sons, Inc.

Figure 1.1 The importance of antenna in a wireless communication system.

1.3 Antenna Performance Parameters

Antenna performance parameters can be categorized into two groups: circuit parameters and radiation parameters. Circuit parameters refer to the impedance matching properties such as reflection coefficient magnitudes ($|S_{ii}|$) and isolation ($|S_{ij}|$) between the antenna ports. Antenna radiation parameters refer to radiation patterns, gain, directivity, antenna efficiency, polarization, effective length and effective aperture, antenna temperature, etc. Readers should refer to the well-known text book by C. A. Balanis, Antenna Theory: Analysis and Design (Fourth Revised edition), Wiley publications [1] for detailed discussion and learning about these antenna performance parameters.

1.4 Antenna Types

Various antennas that find use in wireless communication systems can be classified in many ways. Antenna geometry of four selected antennas is shown in Figure 1.2. Figure 1.2a shows a well-known bow-tie planar antenna that is known for wideband operation with omnidirectional radiation pattern performance. A planar inverted F-antenna (PIFA) is shown in Figure 1.2b which has been known to provide single and multi-band operation based on suitable dimension of the radiating structures and feeding mechanism. It also offers omnidirectional radiation patterns. Figure 1.2c shows a quasi-Yagi planar antenna which offers end-fire directional radiation patterns. Similarly, Figure 1.2d shows a stepped Vivaldi planar antenna which is known for its extremely wideband antenna performance.

The antenna performance can be characterized using impedance matching and radiation patterns. One example is shown in Figure 1.3. Figure 1.3a shows reflection coefficient magnitude versus frequency and Figure 1.3b shows 3D gain radiation patterns of an antenna. There are numerous full-wave analysis tools, also called Maxwell Solvers, which provide accurate simulation and analysis results for an antenna. One such tool is Ansys high-frequency structure simulator (HFSS) which has been used to generate these impedance matching and radiation pattern.

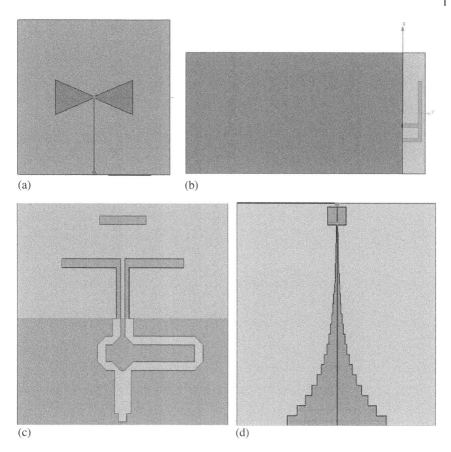

(a) (b)

(c) (d)

Figure 1.2 Some antenna types generated through Antenna Design Kit in Ansys Electronic Package.

1.5 Multifunctional Antennas

Multifunctional antennas can have features of frequency reconfiguration, polarization reconfiguration, beam steering, flexible radiation patterns, and radiation pattern reconfiguration in a single antenna structure. Combination of a couple of these features makes these antennas "multifunctional." These antennas can meet multiple wireless communication standards and hence can provide multiple communication applications. Such antennas can also be multiband in nature and can have multiple-input-multiple-output (MIMO) implementations. Also, multiple

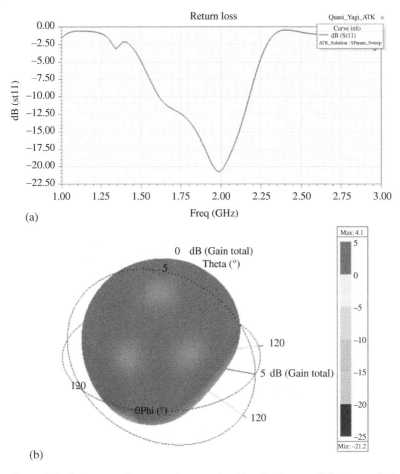

Figure 1.3 Antenna performance shown using (a) reflection coefficient magnitude (S_{11}, dB) and (b) 3D gain radiation pattern.

communication antennas on a common small size host platform, such as cellular phone size ground plane, can be categorized as "multifunctional" antenna.

The full-polarization reconfigurable antenna can switch between the vertical and horizontal linear polarizations, right-hand circular polarization (RHCP), and left-hand circular polarization (LHCP) depending on the communication system requirements. These antennas offer advantages of reduced antenna hardware, low weight, and low cost. Such antennas are very attractive to emerging wireless communications such as 5G communication systems. One such antenna

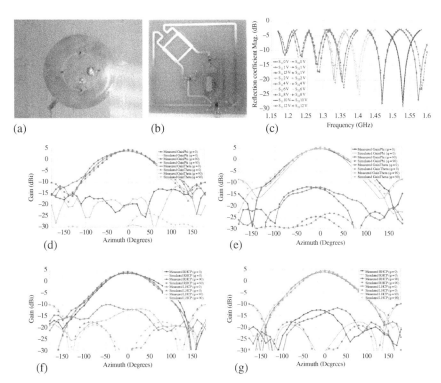

Figure 1.4 (a) Top view photograph of a frequency tunable concentric circular microstrip patch antenna along with varactor diode placement locations, (b) photograph of the fabricated control feed network, (c) measured frequency tunable response for both feed ports, and (d–g) comparison between the measured and simulated gain radiation patterns for 4 V bias voltage which corresponds to 1.36 GHz tunable band for horizontal linear, vertical linear, RHCP, and LHCP, respectively. *Source*: Babakhani and Sharma [2].

is shown in Figure 1.4, which offers both frequency tunability and polarization reconfiguration [2].

Tunable antennas, mostly, frequency tunable can be designed by incorporating variable capacitors in a radiating element in suitable placement arrangements. Figure 1.4a shows photograph of a prototype frequency tunable concentric circular patch antenna where four varactors (Skyworks SMV 1234) are placed between the central patch and the outer ring patch. Two feed ports are selected so that both linear and circular polarizations can be obtained by suitably exciting feed points. Figure 1.4b shows photograph of the control feed network which provides polarization reconfiguration along with simultaneously frequency tunability. The control feed network uses single pole double through (SPDT) and single pole 4 through (SP4T) RF switches along with quadrature power divider, dual in package

(DIP) switch, and control lines for biasing varactor diodes and RF switches. A DIP switch is included to apply DC voltage to control the state of the switches. To bias the varactors for tunability, the central patch is connected to the positive pole through the feed network circuit. For the radiation pattern measurements in the anechoic chamber, varactor diodes were biased with a Jameco Electronics DC–DC Boost Converter.

The frequency response with bias voltage variation is shown in Figure 1.4c. With 0 V bias, the antenna resonates at the lowest frequency (1.17 GHz). Similarly, with 12 V bias, the antenna operates at 1.58 GHz. Thus, this antenna provides frequency tunability between 1.17 and 1.58 GHz which corresponds to 30% tunability. By applying bias voltage to two varactors each along with one feed point, vertical and horizontal linear polarizations are achieved. Similarly, applying the same bias voltage to all the varactors and exciting both feed points in $\pm 90°$ time phase difference, right hand (RH) or left hand (LH) circular polarization is obtained. Measured and simulated radiation patterns are shown in Figure 1.4d–g for 1.36 GHz (4 V bias) for all four polarization cases. Similar was the pattern response for all other frequency tunable bands but are not shown for the sake of brevity.

1.6 Reconfigurable Antennas

Reconfigurable antennas can be frequency reconfigurable, pattern reconfigurable, or polarization reconfigurable or combination of these types. Once again, reconfigurable antennas can be categorized as a multifunctional antenna in general like the one discussed in the Section 1.5. Reconfiguration is obtained by varying the antenna structure with the help of RF or optical switches. One such antenna demonstrating frequency reconfiguration is shown in Figure 1.5.

The geometry of the proposed PIFA element is shown in Figure 1.5a and b. This miniaturized antenna is able to achieve consistent high band coverage while reconfigurable lower frequency bands are maintained. The antenna is matched across all the 4G/LTE lower reconfigurable bands and simultaneous higher consistent wireless bands. The antenna miniaturization and impedance matching are possible by exploiting the meandering and mutual coupling between the different parts of the antenna structure. It employs ground plane edge effect for its optimum performance.

The reconfigurable antenna element (Figure 1.5a, b) is designed by using Ansys HFSS on a tablet size ground plane of $L_4 = 180$ mm and $W_4 = 150$ mm [3, 4]. The corresponding antenna dimensions are: length arm, $L_1 = 73$ mm and width, $W_1 = 2.3$ mm to the first PIN diode switch, and second ground extension, $L_2 = 20$ mm with end width, $W_2 = 7$ mm, coupling grounding gap, $g = 5.6$ mm and $g_1 = 3.8$ mm designed on a FR4 ($\varepsilon_r = 4.4$, tan $\delta = 0.021$) substrate with thickness,

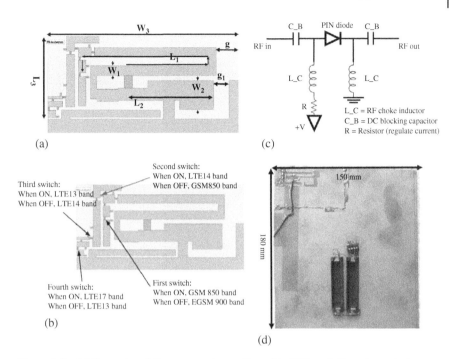

(a)

(c)

(b)

(d)

Second switch:
When ON, LTE14 band
When OFF, GSM850 band

Third switch:
When ON, LTE13 band
When OFF, LTE14 band

Fourth switch:
When ON, LTE17 band
When OFF, LTE13 band

First switch:
When ON, GSM 850 band
When OFF, EGSM 900 band

L_C = RF choke inductor
C_B = DC blocking capacitor
R = Resistor (regulate current)

Figure 1.5 (a) Geometry of the proposed reconfigurable PIFA radiating element along with design parameters, (b) location of the PIN diodes on the radiating element, (c) bias network used for PIN diodes, and (d) photograph of the fabricated radiating element on the tablet size ground plane.

$h = 0.8\,mm$, $W_4 = 150\,mm$, $L_4 = 180\,mm$. The antenna clearance and area required are $W_3 = 48.5\,mm$ and $L_3 = 20\,mm$. Four PIN diodes are used to generate frequency switching for the 4G/LTE lower communication bands (704–960 MHz) while it maintains simultaneous consistent higher frequency bands between 1710 and 2690 MHz. The bias network (Figure 1.5c) was used for each PIN diode (Microsemi MPP4203) to prevent DC and AC signal getting influenced by the power supply. Also, it prevents the power supply line from becoming part of the antenna. The 47 nH inductor choke and 560 pF DC blocking capacitor were used as a bias network to prevent RF signal distortion. Photograph of the fabricated antenna on the tablet size ground is shown in Figure 1.5d.

The designed antenna can operate in LTE bands 13, 14, 17, EGSM, GSM (lower frequency) and LTE bands 4, 7, DCS, PCS (higher frequency) with near omnidirectional radiation patterns for each band. There are a total of five switching states to reconfigure lower frequency as given in Table 1.1. The first state is when all the PIN diode switches are in the OFF state. The second state occurs when the first PIN

Table 1.1 PIN diode states for reconfiguring the lower frequency bands while the higher band is consistently maintained.

	Switch table list				
	LTE 17 (0.704– 0.746 GHz)	LTE 13 (0.746– 0.787 GHz)	LTE 14 (0.758– 0.798 GHz)	GSM 850 (0.824– 0.894 GHz)	EGSM (0.880– 0.960 GHz)
First switch	ON	ON	ON	ON	OFF
Second switch	ON	ON	ON	OFF	OFF
Third switch	ON	ON	OFF	OFF	OFF
Fourth switch	ON	OFF	OFF	OFF	OFF

diode switch is activated in the ON state while the other diodes are in the OFF state, which allows tuning of the center frequency from 930 to 850 MHz. The third state (780 MHz) occurs when the first and second PIN diodes are switched in the ON state while the remaining diodes are in the OFF state. The fourth state (750 MHz) occurs when the first, second, and third PIN diodes are switched in the ON state while the fourth PIN diode is in the OFF state. The fifth state (720 MHz) is obtained when all the PIN diode switches are in the ON state. While we manage these PIN diodes from the first state to the fifth state, we simultaneously maintain the higher frequency bands between 1710 and 2690 MHz.

Current distribution (Figure 1.6) shows current flow and hotspot for radiating element at the lower frequency end (780 MHz, Figure 1.6a) and the upper frequency end (1880 MHz, Figure 1.6b) for the proposed reconfigurable PIFA. Low and high bands have individual hotspots which can tune each band separately. Different sections of the radiator employ mutual coupling to obtain better matching and bandwidth.

Simulated and measured reflection coefficient magnitudes for the lower reconfigurable bands are compared in Figure 1.7a. It can be observed that the antenna is matched well below −7 dB between 704 and 960 MHz which includes the reconfigurable states of LTE 17, 13, 14, GSM, and EGSM. Similarly, Figure 1.7b shows the simulated and measured matching for the high-frequency band. Once again, it can be observed that the antenna consistently maintains the matching level between 1710 and 2690 MHz better than −7 dB for all the switch states.

The 3D radiation patterns and total antenna efficiency were measured using the Satimo chamber. The total efficiency for the reconfigurable stages at the lower

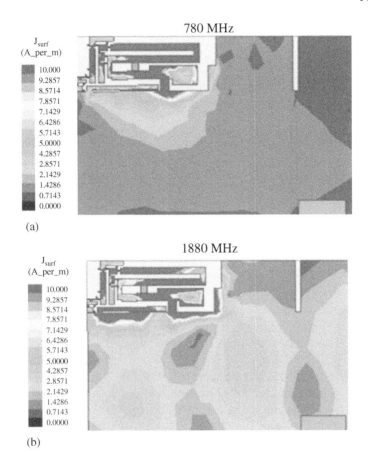

Figure 1.6 Surface current distribution for (a) 780 MHz in lower frequency band and (b) 1880 MHz in upper frequency band.

frequency bands and the consistent higher band is shown in Figure 1.8a and b, respectively. This efficiency takes care of all the possible losses in the antenna such as mismatch loss, Ohmic and dielectric losses, and losses due to the PIN diodes and bias components. The antenna efficiency is above 50% for all the switching states in the lower band and the higher band. The simulated and measured efficiencies agree reasonably well except toward edges of the bands.

The simulated (Figure 1.9a, c, e, g) and measured (Figure 1.9b, d, f, h) 3D radiation pattern for the lower reconfigurable bands for the selected 720, 770, 850, and 910 MHz is shown in Figure 1.9. Similarly, simulated (Figure 1.10a, c, e, g) and measured (Figure 1.10b, d, f, h) 3D radiation pattern for the higher

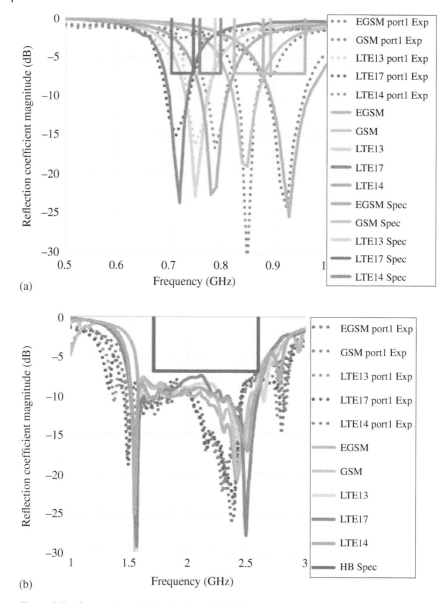

(a)

(b)

Figure 1.7 Comparison of the simulated (solid lines) and measured (dash lines) reflection coefficient magnitudes against specifications of the (a) lower 4G LTE frequency reconfigurable bands and (b) consistent higher frequency bands.

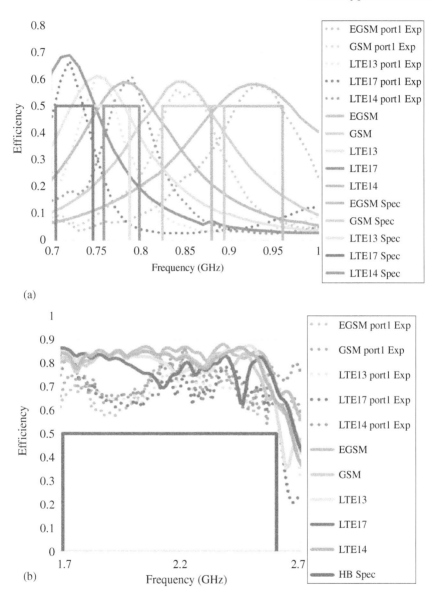

Figure 1.8 The simulated (solid lines) and measured (dash lines) total antenna efficiencies for (a) the reconfigurable states of the lower 4G bands and (b) the consistent higher frequency band.

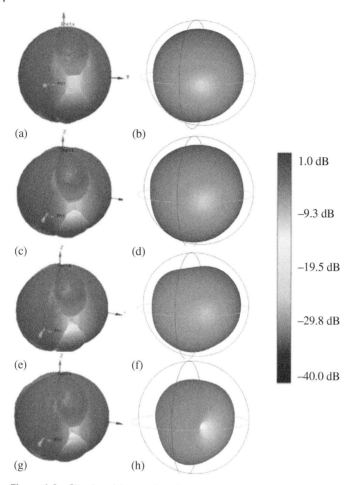

(a) (b)

(c) (d)

(e) (f)

(g) (h)

1.0 dB

−9.3 dB

−19.5 dB

−29.8 dB

−40.0 dB

Figure 1.9 Simulated (a, c, e, g) and measured (b, d, f, h) near-omnidirectional 3D radiation pattern at (a, b) 720 MHz, (c, d) 770 MHz, (e, f) 850 MHz, and (g, h) 910 MHz.

consistent bands for the selected 1710, 1880, 2100, and 2600 MHz is shown in Figure 1.10. From these figures, it can be observed that the patterns are near omnidirectional except toward the higher frequency end, where patterns show slight multilobes and nulls. In all the cases, realized peak gain is positive and around 2 dBi for both simulated and measured cases. Further, the patterns toward the higher end show some directionality due to the ground plane effect which tends to push the radiated energy in a direction causing diversity that is helpful when the MIMO antenna system is implemented.

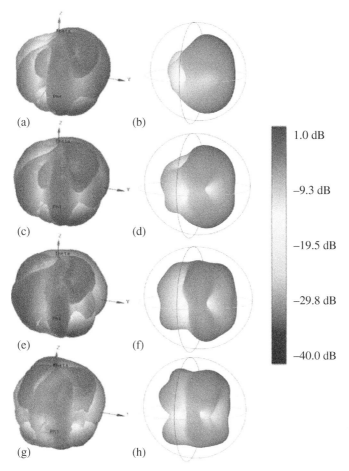

1.0 dB

−9.3 dB

−19.5 dB

−29.8 dB

−40.0 dB

Figure 1.10 Simulated (a, c, e, g) and measured (b, d, f, h) near-omnidirectional 3D radiation pattern at (a, b) 1710 MHz, (c, d) 1880 MHz, (e, f) 2100 MHz, and (g, h) 2600 MHz.

1.7 Frequency Agile/Tunable Antenna

A frequency agile or tunable antenna can be realized by integrating tunable RF components such as varactor diodes or RF micro-electro-mechanical-systems (MEMS) variable capacitors in an antenna structure. Like a reconfigurable antenna, such antennas also require necessary bias networks such as those implemented in [2] and shown in Figure 1.4a–c and hence is not repeated again. The antenna impedance and radiation mechanism depends on the variation in effective permittivity which in turn is related to the change in capacitance variation. Frequency (f) of an antenna can be described as follows:

$$f = \frac{1}{2\pi\sqrt{LC}}$$ (1.1)

where L and C are the inductance and capacitance of an antenna.

To demonstrate mechanism of such an antenna, an example is now discussed from [5]. The antenna utilizes a meandered planar portion on the right-hand side, which includes a varactor diode and its biasing network and a rectangular-shaped cutout portion on the left-hand side on the central feed arm (Figure 1.11). The varactor capacitance variation alters the electrical length of the meandered portion of the antenna, which in turn varies the resonance frequency of the antenna. In addition to the tunable meandered portion of the antenna, the design also includes a rectangular-shaped cutout geometry for realizing higher band. Thus, this antenna follows a two-step design approach.

In the first design step, the meandered line was chosen as the location for placement of the varactor diode. Since the meandered line enables the antenna to resonate at the lower frequency band, adding a varactor to this section allows the antenna to become tunable. To provide a wider tunable band between 690 and 970 MHz, the meandered line's electrical length is essentially increased by adding the varactor diode (SMV1281). After optimizing the antenna with the varactor diode set in place, Figure 1.12 shows the meandered line length as $M_L = 137.5$ mm, which is slightly below $\lambda/2$ at 970 MHz and the width of the meandered line is $M_W = 1.2$ mm.

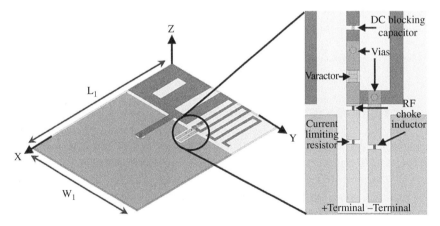

Figure 1.11 Antenna geometry with varactor diode and biasing network. *Source*: Damman et al. [5].

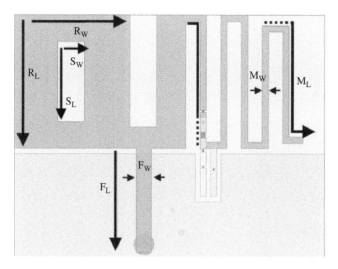

Figure 1.12 Design parameters for the low (right) and high bands (left). *Source:* Damman et al. [5].

The second design step requires adding a higher band antenna structure to the design such that, while the tuning of the lower band is obtained, the higher band is consistently present. The higher band antenna was chosen as a simple rectangular-shaped structure with a rectangular-shaped slot which was cutout within the middle of the structure (Figure 1.12). Adding this cutout improved the matching of the higher band and subsequently lowered the effect of frequency tuning in the higher band when the lower band was tuned. In such an antenna, current distributions on the two portions of the antenna should be as independent as possible, which is shown in Figure 1.13a and b for the lower and upper bands, respectively, for capacitance value of 1.6 pF. When the lower band meandering structure is radiating, it has little effect on the upper band antenna. The outer rectangular length and width are $R_L = 25$ mm and $R_W = 22.25$ mm. The rectangular slot cutout (slot length of $S_L = 15$ mm and a slot width of $S_W = 5.125$ mm) dimensions are noted. The centerline feed has a length and width of $F_L = 19.5$ mm and $F_W = 3$ mm. Photograph of this antenna is shown in Figure 1.14 with an overall substrate size of length $L_1 = 80$ mm and width $W_1 = 60$ mm. The material used was FR-4 ($\varepsilon_r = 4.4$) with a thickness of t = 0.762 mm. Response of this antenna can be found in [5] and hence is not repeated here.

As a reconfigurable antenna, the lower band meandered antenna also includes a biasing network. The DC biasing lines are placed on the same side as the ground plane (Figure 1.12), so as to limit the effect of coupling. The DC biasing lines include a RF choke inductor to block any high-frequency signal from entering the DC power supply. A current limiting resistor is also added to the positive voltage

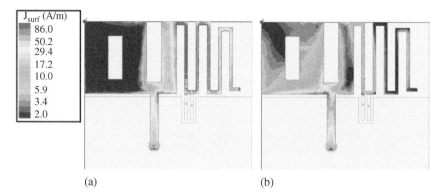

(a)   (b)

Figure 1.13 Surface current distribution with capacitance of 1.6 pF: (a) 810 MHz and (b) 1.68 GHz. *Source*: Damman et al. [5].

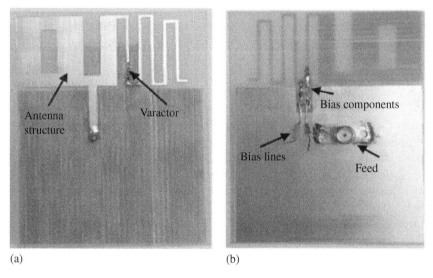

(a)   (b)

Figure 1.14 Photograph of the fabricated single-feed dual band antenna: (a) top view and (b) bottom view. *Source*: Damman et al. [5].

terminal to protect the varactor diode. On the top side of the antenna where the radiating structure is located, a single DC blocking capacitor is placed above the varactor diode such that the potentially damaging DC supply voltage does not enter the RF signal. For the varactor diode, the reverse biasing voltage is varied from 0 to 20 V which varies the capacitance from 13.3 to 0.69 pF, respectively.

Although mechanisms to achieve frequency reconfigurable and frequency agile/tunable antenna are different, for our discussion, we refer both antennas under the "Reconfigurable Antenna," category.

1.8 Antenna Measurements

Antenna performance parameters measurement is important for verifying computation, simulation, and analysis results. For measuring circuit parameters such as scattering parameters (S_{ii}/S_{ij}), where S_{ii} and S_{ij} refer to self-port reflection coefficient and coupling port transmission coefficient, vector network analyzers are preferred after proper calibration [6]. For example, Figure 1.15 shows photographs of vector network analyzers from Anritsu and Keysight, both of which are available at the Antenna and Microwave Laboratory (AML), San Diego State University.

For measuring radiation patterns, we can use far-field, near-field, and compact antenna test range (CATR) chambers. Figure 1.16 shows photographs of the far-field and CATR anechoic chambers at the AML, San Diego State University. The first anechoic chamber is shown in Figure 1.16a, which is capable of far-field radiation measurements. It can cover a frequency range from 800 to 40 GHz. The chamber comes with ORBIT/FR 959 acquisition measurement software and provides measurement results for 2D/3D radiation pattern, realized gain, and polarization with sense of rotation.

The Mini-Compact Antenna Test Range (M-CATR) from Microwave Vision Group (MVG) for millimeter-wave antenna measurement covers frequencies between 26.5 and 110 GHz, as shown in Figure 1.16b. Keysight N5225A Power Network Analyzer (PNA) serves its signal power generator that ranges from 10 to 50 GHz. The frequency is extended up to 110 GHz using proper external frequency extenders: V-band (50–75 GHz) and W-band (75–110 GHz). This chamber is capable of measuring realized gain, 2D and 3D radiation patterns, and polarization of the antenna with the sense of rotation using the ORBIT/FR 959 acquisition measurement software.

Interested readers should review text books on the theory behind antenna radiation pattern measurements such as [1].

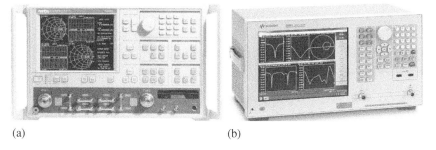

(a)   (b)

Figure 1.15 Vector network analyzer (VNA) can be used for measurement of the scattering parameter: (a) Anritsu's VNA and (b) Keysight's VNA.

(a)   (b)

Figure 1.16 Antenna testing in anechoic chamber at the Antenna and Microwave Laboratory (AML), San Diego State University: (a) far-field anechoic chamber covering 800–40 GHz and (b) Mini-Compact Antenna Test Range (M-CATR) system covering 26.5–110 GHz.

1.9 Conclusion

This chapter introduced basics of antennas as well as an introduction to reconfigurable, multifunctional, frequency agile/tunable, and antenna measurements. In the coming chapters, we dive into a more detailed discussion.

References

1 Balanis, C.A. (2016). *Antenna Theory: Analysis and Design*, 4e. Wiley.

2 Babakhani, B. and Sharma, S.K. (2015). Wideband frequency tunable concentric circular microstrip patch antenna with simultaneous polarization reconfiguration. *IEEE Antennas and Propagation Magazine* 57 (2): 203–216.

3 Sharma, S.K. and Wang, A. (2018). Two elements MIMO antenna for tablet size ground plane with reconfigurable lower bands and consistent high band radiating elements. *2018 IEEE International Symposium on Antennas and Propagation & USNC/URSI National Radio Science Meeting*, Boston (8–13 July 2018).

4 Wang, A. (2017). Four elements compact MIMO antenna with reconfigurable lower band and consistent high band for tablet applications. MS (Electrical Engineering) Thesis. San Diego State University.

5 Damman, R., Mishra, G., Sharma, S.K., and Babakhani, B. (2017). A single feed planar antenna with 4G tunable bands and consistent upper LTE bands between 1.29 GHz–2.05 GHz. *Microwave and Optical Technology Letters* 59 (8): 2070–2075.

6 Pozar, D.M. (2011). *Microwave Engineering*, 4e. Wiley.

2

Frequency Reconfigurable Antennas

Saeed I. Latif and Satish K. Sharma

2.1 Introduction

Reconfigurable antennas have been in use for many years in various forms as communication systems and have become more diverse and multifunctional. Here, we refer both frequency reconfigurable and frequency agile/tunable antennas under the "Reconfigurable Antenna," category. Reconfigurability enables a single antenna system to be used for multiple purposes in the communication system. With the rapid proliferation of wireless communications systems, more services are accommodated in the limited electromagnetic spectrum. Now, wireless devices have to access multiple services, which is achieved by using multiband, wideband, or reconfigurable antennas. In cognitive radio systems, these devices have to be intelligent and have to adapt to changes based on feedback sensed by wireless channels, and therefore, change operating features including operating frequency, direction of radiation, and modulation schemes. Reconfigurability of an antenna can be implemented to modify antenna's operating frequency, polarization, or radiation characteristics dynamically [1–6]. This chapter focuses on frequency reconfigurable antennas (FRAs). Multiband [7–10] and wideband antennas [11–15] are commonly used in various wireless devices when it comes to access multiple services operating at different frequencies. FRAs can be used to replace multiband antennas and, in some occasions, wideband antennas, if their functionality can offset the complexity of the control mechanism for reconfiguration.

The modification in characteristics of an antenna, i.e. change of the operating frequency, is achieved by redistribution of current in antenna. There are many techniques by which the antenna current can be redistributed, either by altering the antenna geometry or by changing the electrical properties of the antenna. To achieve this, switches, varactors, or tunable materials can be used. These concepts of reconfigurability can significantly decrease the complexity of hardware by

Multifunctional Antennas and Arrays for Wireless Communication Systems, First Edition.
Edited by Satish K. Sharma and Jia-Chi S. Chieh.
© 2021 John Wiley & Sons, Inc. Published 2021 by John Wiley & Sons, Inc.

reducing the number of components, and improve the performance of the wireless system by saving energy and space.

Frequency reconfigurability of antenna can be achieved by using various switches, electrical, optical, or micro-electro-mechanical-system (MEMS)-based switches, by introducing mechanical changes in the antenna, or by using tunable material. Recently, reconfigurable antennas using special shapes adopting the concept of origami have emerged. In this chapter, various FRAs will be discussed which are categorized based on how reconfiguration is achieved. In Section 2.2, the mechanism of frequency reconfigurability is briefly discussed. In Section 2.3, FRAs are described where reconfigurability is achieved using switches. In this section, other methods of frequency reconfigurability such as by mechanical changes or material changes are also discussed. Origami and other special shape-based FRAs will be addressed. In Section 2.4, the use of FRAs in emerging applications, such as cubesats and software-defined radios, will be presented. Conclusion will be drawn in Section 2.5.

2.2 Mechanism of Frequency Reconfigurability

There are several ways to achieve frequency reconfigurability: modifying the current distribution on the antenna to change its effective length by deforming the antenna element, changing the dielectric or diamagnetic properties of antenna elements, using external components, such as diodes, transistors, MEMS devices, etc. While some of these mechanisms rely on mechanical alteration of the structure forming the antenna, other rely on the reconfigurability of the external power supply or excitation of antenna elements. The use of external switches, such as diodes and varactors, allows the change of the size and/or the shape of the radiating element by introducing short circuits or slots, thanks to the advancement in electronics and MEMS. While switches provide the opportunity to modify the antenna element to control the reconfigurability discretely, varactors offer continuous reconfigurability. However, the addition of these external components requires significant real estate and, in many occasions, added layers and biasing networks, increasing the manufacturing complexity and cost. Semiconductor-based switching elements for reconfigurability, such as PIN diodes, gallium arsenide field-effect transistors (GaAs FETs), and varactors require lower operating voltage, have lower cost and higher reliability than those in MEMS-based devices. Nevertheless, MEMS devices are very low loss compared to semiconductor-based devices. Microwave RF switches, which have low insertion loss and high power handling capability with low DC power consumption, are also conveniently used for frequency reconfigurability [16, 17].

A mechanism to obtain frequency reconfigurability without the use of additional components is using "agile" or "smart" materials, whose dielectric properties (permittivity and/or permeability) can be changed by applying an external electric and/or magnetic field. These materials are used as the antenna metal or the substrate layer for the antenna. Liquid crystals, graphene, etc., are such materials, which have seen widespread applications for frequency reconfigurability along with other types of reconfigurability, e.g. pattern or polarization reconfigurability. Integrating these materials along with control structures in the antenna increases the fabrication complexities. Moreover, if the material properties degrade over time, the reconfigurability operation or tunability is ultimately affected. Altering the antenna geometry using mechanical movement is another method to obtain frequency reconfigurability. The use of motors or actuators for mechanical adjustment poses extra complexities in manufacturing and operation.

Other reconfiguration techniques are based on photoconductive [18–26] and thermal switches [27], and special shape-based designs, such as origami shapes [28–34], fractal shapes [35], etc.

2.3 Types of FRAs

The operating frequency of reconfigurable antennas can be changed to handle several wireless services over a wide frequency spectrum. Modern mobile devices are required to access multiple wireless standards such of various cellular bands, wireless local area networks (WLANs), Wi-Fi, Bluetooth, Global Positioning System (GPS), etc. If a single band is to be accessed at a given time, FRAs are very useful in a sense that the antenna topology can be modified to operate in that band. Although multiband and wideband antennas are the other choices, reconfigurable antennas can provide a more compact solution with an added advantage of high noise reduction for the unwanted bands. FRAs can be classified based on how they are made reconfigurable. A very popular and common method is to use switches to introduce the variation in current distribution and change in the aperture. Electrical, optical, and MEMS switches are used to obtain frequency reconfigurability. When it comes to mechanical changes, motors and actuators are widely used. A variation in current or a topology change can also be achieved using liquid crystals and plasmonics. Another group of FRAs is the ones using special shapes, e.g. origami based or fractal shape-based. These various groups of FRAs are discussed in this section.

2.3.1 Frequency Reconfigurability by Switches/ Tunable Components

Frequency reconfigurability by switches leads to the design of either discrete or continuous tuning of frequency. Electrical switches that are popularly in use are PIN diodes providing discrete tuning and varactors providing continuous tuning.

They provide fast switching and are compact in size, although require a biasing network. Optical switches use laser beam from laser diodes to become conductive, and when integrated in the antenna structure offer reconfigurability. Electrical, RF MEMS, and optical switch-based FRAs will be discussed here.

2.3.1.1 Electrical Switches

Basic type of frequency reconfigurability can be achieved by using RF/microwave switches, which are electromechanical relay switches. They are usually larger assemblies compared to solid-state and MEMS switches, since they incorporate a series of coils and mechanical contacts. These switches have low insertion loss and high isolation. Switch speed is in the milliseconds. Other benefits are: large operating frequency bands – DC up to millimeter wave frequencies (50 GHz or more) and high power handling capability.

A five-pointed spirograph planar monopole antenna (SPMA) operating as FRA is shown in Figure 2.1a [36], where current distribution control is shown on the radiating element as RF switches are turned ON or OFF. The 50 Ω microstrip feed line (with end-launch SMA connector) and radiating element are on the top layer of 0.508 mm-thick Rogers RT/Duroid 5880 (ε_r = 2.2, tan δ = 0.0009) substrate and the ground plane is on the bottom. The maximum diameter of the antenna element is 51.8 mm (0.54λ at 3.1 GHz) and the average ring thickness is 13.3 mm. The ground plane and substrate dimensions are as follows: r_{curv} = 4.0 cm, G_W = 59.26 mm, G_L = 59.26 mm, and S_L = 11.5 cm. The antenna is segmented by slicing a 1.85 mm gap at the angles of: 3°, 108°, 252°, and 324°. This gap size was selected so that two 0402 package size components could fit into this region. For the two gaps closest to the feed line (switches 1 and 2), a DC blocking capacitor is

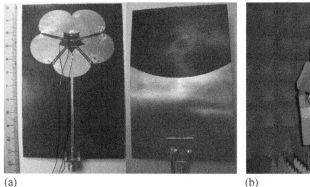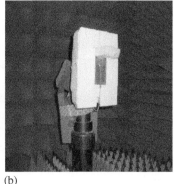

(a)   (b)

Figure 2.1 (a) Top and bottom views of the frequency reconfigurable Spirograph Planar Monopole Antenna (SPMA) and (b) radiation pattern measurement setup. *Source:* Rayno and Sharma [36] and Rayno [37].

in series with a PIN diode, and the positive voltage DC bias lines (bias line 1) have a RF choke inductor close to the PIN diode and a 0402 size resistor close to the beginning of the bias line (near the center of the SPMA). The middle DC bias line (bias line 2) is for ground. The second set of gaps (switches 3 and 4) contain a PIN diode each, and the final DC bias line (bias line 3) is for a positive voltage. The PIN diodes are located at the outer edge of the antenna since the current is most concentrated along these edges. The PIN diodes modeled are the Skyworks SMP1340. The HFSS model values for the ON state are: L = 0.45 nH and R = 1.2 Ω, and the model values for the OFF state are: L = 0.45 nH, R = 5 MΩ, and C = 0.14 pF. The RF choke inductors are 2.2 nH, the DC blocking capacitors are 15 pF, and the bias line current regulating resistors are 47 Ω (selected by assuming a voltage of +1.5 V). This antenna was tested for the impedance matching and radiation patterns in a far-field anechoic chamber (Figure 2.1b) at the Antenna and Microwave Lab (AML) at San Diego State University.

There are three possible frequency reconfiguration cases. Case 1 is when both sets of PIN diodes are OFF (all bias lines connected to ground), so only the first segment of the antenna is connected to the feed line. Case 2 is when the first set of PIN diodes (switches 1 and 2) is ON and the second set (switches 3 and 4) is OFF (first bias line connected to positive voltage and the other two grounded), so three segments of the antenna are connected to the feed line. Case 3 is when all PIN diodes are ON (bias lines 1 and 3 are connected to positive voltage and bias line 2 is grounded), so all segments of the antenna are connected to the feed line. The antenna topology and associated PIN diode biasing arrangements are shown in Figure 2.2. The antenna operates in three modes to cover a 1.07–3.36 GHz (UHF band) wide bandwidth. When operating in Case 1, all PIN diodes are OFF and a small portion of the total radiating patch takes a part in the radiation and operates over 2.17–3.36 GHz frequency band. In another case, when diodes close to feed lines are ON, a larger section of the patch radiates and the operating frequency shifts downward to 1.21–2.50 GHz due to the change in the effective aperture. Finally, when all switches are ON, the entire patch radiates and an operating bandwidth of 1.07–3.36 GHz is achieved. During this operation, surface current distribution of the radiating patch also changes (Figure 2.3), which demonstrates how the reconfiguration affects the current distribution on the edge of the patch which attributes to the desired radiation patterns of the antenna. For Case 1, although only the first segment is connected to the feed line, some current is still electromagnetically coupled to the other segments. For Case 2, the current plot indicate switches 1 and 2 are ON. For Case 3, the current plot indicates all switches are ON.

The simulated and measured voltage standing wave ratios (VSWRs) are shown in Figure 2.4. For each case, the DC battery is connected to the appropriate bias wires, which are held straight directly in front of the antenna using foam. For the Case 1, the battery is not connected to anything. For the Case 2, bias line 1 is connected to

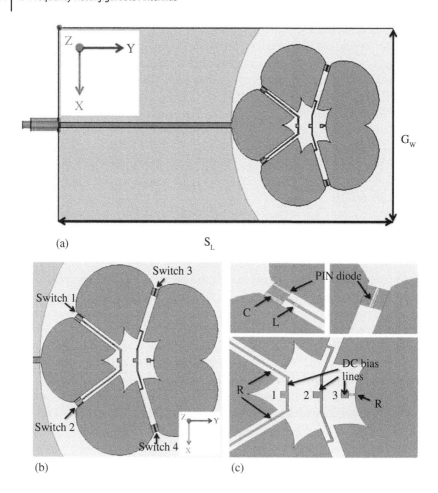

Figure 2.2 Ansys HFSS model of the SPMA frequency reconfigurable antenna: (a) top view, (b) zoomed-in top view, and (c) close-up of bias circuitry. *Source:* Rayno and Sharma [36] and Rayno [37].

a positive voltage, and bias line 2 is connected to ground (bias line 3 is not connected to anything). For the Case 3, bias lines 1 and 3 are connected to a positive
voltage, while bias line 2 is connected to ground. The measured results show close
agreement to the simulated for all three cases. Assuming a VSWR = 2 criterion, the
following measured bands are achieved: Case 1, 2.17–3.36 GHz (43.0%); Case 2,
1.21–2.50 GHz (69.5%); and Case 3, 1.07–2.92 GHz (92.7%). For comparison, the
corresponding simulated results were achieved: Case 1, 2.27–3.08 GHz (30.3%);
Case 2, 1.11–2.41 GHz (73.9%); and Case 3, 0.98–2.81 GHz (96.6%).

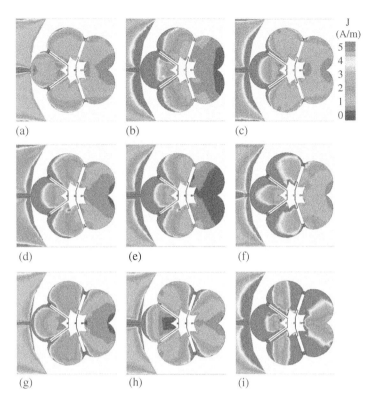

Figure 2.3 Surface current distribution for (a) Case 1 at 2.3 GHz; (b) Case 1 at 2.7 GHz; (c) Case 1 at 3.1 GHz; (d) Case 2 at 1.2 GHz; (e) Case 2 at 1.7 GHz; (f) Case 2 at 2.4 GHz; (g) Case 3 at 1 GHz; (h) Case 3 at 2 GHz; (i) Case 3 at 2.8 GHz. *Source:* Rayno and Sharma [36].

A comparison of the simulated and measured radiation patterns at the start, middle, and stop frequency of each band is shown in Figure 2.5 for the XZ-plane and Figure 2.6 for the YZ-plane, indicating an omnidirectional radiation pattern for all three cases. A reasonable agreement of measured and simulated results is achieved, considering the multiple unknowns of this antenna design (bias wires, battery, PIN diode model).

Another implementation of frequency reconfigurability with RF switches is demonstrated in [16]. A small size $(55 \times 5 \times 3 \, \text{mm}^3)$ internal loop antenna with frequency agility is designed, where a single-pole-four-throw RF switch (model RF-1604) is used to excite four modes so that the antenna works at five operating frequency bands: GSM850/900, DCS1800, PCS1900, UMTS2100, and LTE2300/2500. The antenna comprises a folded loop antenna that provides four resonant modes and two other branches to cover higher frequencies, as shown in Figure 2.7. The terminal end of the loop is connected to the RF switch, whose four

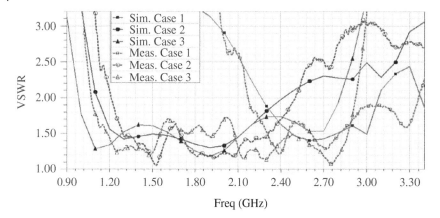

Figure 2.4 Comparison of measured and simulated VSWR for the frequency-reconfigurable SPMA for Cases 1–3. *Source:* Rayno and Sharma [36].

RF ports are linked to four lumped components. Four states of the RF switch are controlled by two bias voltages. Two AA batteries were used to supply 3-V for the RF switch biasing. At State 1, the antenna operates at 860–970, 1670–1755, and 1890–2380 MHz. At State 2, operating bands are: 780–900, 1810–2380, and 2590–2740 MHz. It works in 1020–1095 and 1795–2690 MHz bands at State 3, and in 900–990 and 1710–2365 MHz at State 4, covering all desirable operating bands. Measured reflection coefficients at these states are shown in Figure 2.8.

The most popular electrical reconfiguration techniques are the one with PIN diodes, which have a fast switching speed in the range of 1–100 ns [38]. These provide discrete tuning by physically disconnecting any portion of the antenna from the rest [39, 40]. An example is presented in Figure 2.9, where a unidirectional printed monopole antenna with frequency and polarization reconfigurable properties is presented [41]. By electronically controlling the state of PIN diodes between two slots etched on the ground plane of the printed monopole antenna, the reconfigurability is achieved in frequency and polarization. The antenna consists of a monopole structure with a partial ground plane, two switch circuits, and a reflector. The antenna is modeled and fabricated on a square FR4 substrate. Two slots are introduced on the ground plane, and two metal vias are added besides the slots. BAR50–02 V RF PIN diodes are used in this antenna. Each diode is in ON state by forward bias with a DC voltage which can provide 100 mA bias current, while it is in OFF state if left unbiased. These PIN diodes connect two metal vias and RF choke inductances which are used to keep the RF signal out of the DC lines. DC blocking capacitances are also used to keep the DC lines out of the circuit. The DC current is obtained by a resistance of value 50 Ω.

Figure 2.5 (Left column): Comparison of simulated and measured co-pol and cross-pol realized gain (dBi) radiation patterns in the XZ-plane for (a) Case 1 at 2.3 GHz; (b) Case 1 at 2.7 GHz; (c) Case 1 at 3.1 GHz; (d) Case 2 at 1.2 GHz; (e) Case 2 at 1.7 GHz; (f) Case 2 at 2.4 GHz; (g) Case 3 at 1 GHz; (h) Case 3 at 2 GHz; (i) Case 3 at 2.8 GHz.

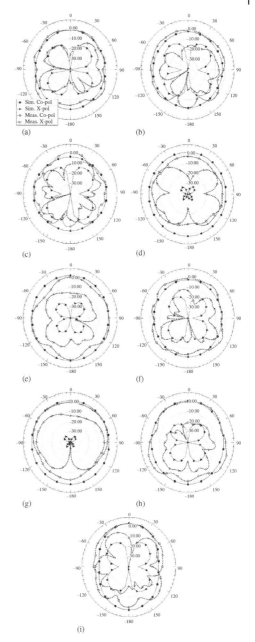

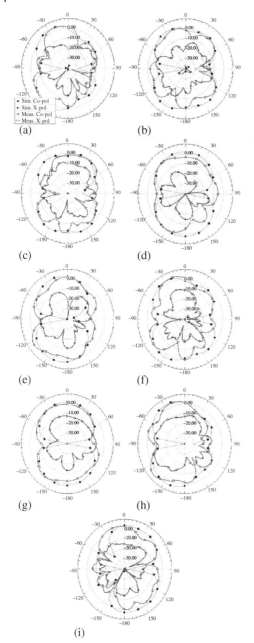

(a)   (b)   (c)   (d)   (e)   (f)   (g)   (h)   (i)

Figure 2.6 (Right column): Comparison of simulated and measured co-pol and cross-pol realized gain (dBi) radiation patterns in the YZ-plane for (a) Case 1 at 2.3 GHz; (b) Case 1 at 2.7 GHz; (c) Case 1 at 3.1 GHz; (d) Case 2 at 1.2 GHz; (e) Case 2 at 1.7 GHz; (f) Case 2 at 2.4 GHz; (g) Case 3 at 1 GHz; (h) Case 3 at 2 GHz; (i) Case 3 at 2.8 GHz.

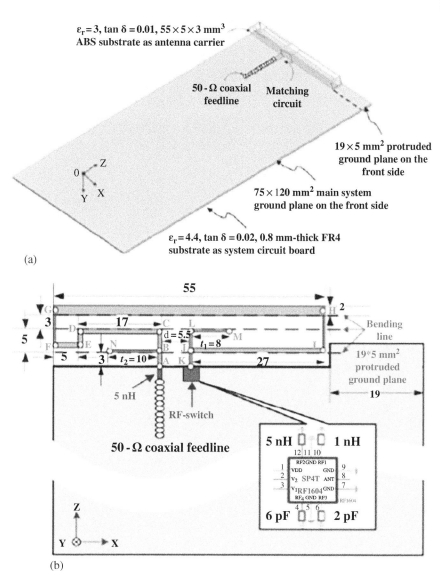

(a)

(b)

Figure 2.7 (a) Geometry of the frequency reconfigurable loop antenna with a RF switch, (b) antenna dimensions in detail (unit: mm), and photographs of (c) the fabricated antenna and (d) the RF switch. *Source:* Wang et al. [16]. © 2016, IEEE.

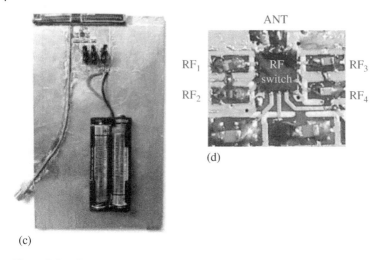

(c)

ANT

(d)

Figure 2.7 *(Continued)*

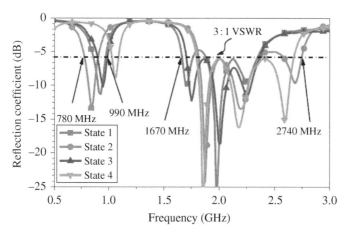

Figure 2.8 Measured reflection coefficients of the frequency reconfigurable antenna in Figure 2.7 operated by a RF switch. *Source:* Wang et al. [16]. © 2016, IEEE.

When both diodes are ON, the antenna operates at 2.35 GHz with a bandwidth of 23.6% (from 2.02 to 2.56 GHz). When both diodes are OFF, the operating frequency of the antenna shifts to 2.62 GHz due to the effect of slots on the ground plane. The bandwidth of the antenna at this state is 23.9% (from 2.32 to 2.95 GHz). The simulated and measured S_{11} versus frequency in these states are shown in Figure 2.10a. In both cases, the antenna exhibits linear polarization with peak gains

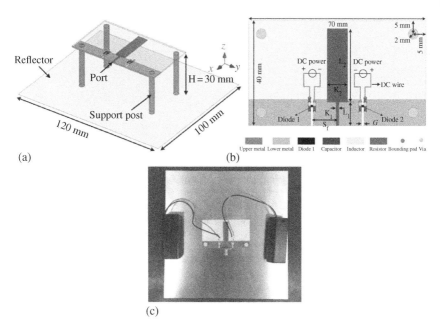

(a) (b)

(c)

Figure 2.9 The antenna geometry for frequency and polarization reconfigurability from a printed monopole antenna using PIN diodes: (a) 3D view and (b) top view. Some antenna parameters: K_1 = 1.0 mm, K_2 = 8.0 mm, L_1 = 9.0 mm, L_2 = 28.0 mm, S_f = 10.0 mm, G = 1.0 mm. Substrate dimension: $70 \times 40 \times 1.0 \, mm^3$. (c) Picture of the fabricated printed monopole antenna. *Source:* Liu et al. [41]. © 2018, IEEE.

of 7.9 and 8 dBi at States 1 and 2, respectively. However, when either of the diodes is ON, the antenna radiates circularly polarized signal operating at 2.22 GHz with a bandwidth of about 34%. The measured and simulated reflection coefficient plots at these states are shown in Figure 2.10b. The measured 3 dB axial ratio bandwidth of this antenna is 14.3% (2.08–2.4 GHz), as can be seen in Figure 2.10c. The gain patterns for the linear case (States 1 and 2) are shown in Figure 2.11a and b. When diode 1 is ON, and the other is OFF, the antenna is left-hand circularly polarized (LHCP, State 3), whereas, when diode 1 is OFF and diode 2 is ON, it is right-hand circularly polarized (RHCP, State 4). Measured and simulated circularly polarized gain patterns at State 3 are shown in Figure 2.11c. Peak gain reaches 7.7 dBic having the difference between the measured RHCP and LHCP gains at +z direction to be about 21.0 dB.

2.3.1.2 Varactor Diodes

If a continuous tuning is necessary for an application, varactors are employed, which have a dynamic reconfiguration capability [42, 43]. By changing the bias voltages of varactor diodes, the effective electrical size of the patch can be varied

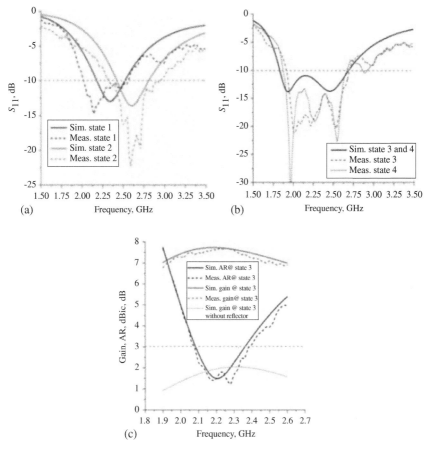

Figure 2.10 Simulated and measured S_{11} versus frequency of the printed monopole antenna in Figure 2.9: (a) when both diodes are ON (state 1) or OFF (state 2), (b) when diode 1 is ON and diode 2 is OFF (state 3) or when diode 1 is OFF and diode 2 is ON (state 4), and (c) antenna gain and axial ratio at states 3 and 4 in the circularly polarized case. *Source:* Liu et al. [41]. © 2018, IEEE.

to obtain smooth frequency tuning. Both microstrip patch [44] and microstrip slot antennas [45] can be used where varactor diodes can be used to change their effective lengths. Slot antennas offer wider tuning ratio, however, with lower gain and efficiency.

To achieve agility or tunability, a varactor diode is integrated to an antenna. The FRA operates over a wide range of frequency instead of toggling over the different stage. A conventional antenna operates at its resonance frequency in the dominant mode where the resonance frequency is governed by the dimensions of the

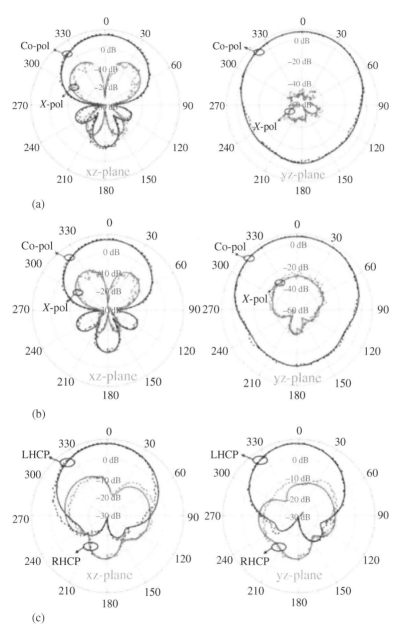

Figure 2.11 Measured and simulated radiation patterns in the linear polarization case at states 1 and 2 of the printed monopole antenna in Figure 2.9 at (a) 2.35 GHz (state 1), (b) 2.62 GHz (state 2), and (c) at state 3 at 2.2 GHz in the circular polarization case. Solid lines: simulation, dashed lines: measurement. *Source:* Liu et al. [41]. © 2018, IEEE.

radiating element. By perturbing the shape of the radiating elements, the antenna can be designed to operate beyond its fundamental mode [44, 46–48]. For an example, simulated model and photograph of the fabricated prototype of an E-shaped planar antenna is shown in Figure 2.12a and b in which by creating the slots in the radiating patch, additional mode is generated. To make the antenna frequency agile, a tuning element such as a varactor can be placed suitably to change the electrical length of the antenna by varying the capacitance of the reverse-biased varactor diode [49]. Frequency can be varied from 1.47 to 1.87 GHz (24% frequency agility) only by altering the reverse bias voltage of varactors from 0 to 20 V (Figure 2.12c). When the bias voltages of varactors are lesser than 10 V, this antenna could not obtain impedance matching below $S_{11} = -10$ dB. The diode can also be used to balance the matching level of the antenna. The varactor diode also alters the surface current distribution on the patch and its effect is shown in Figure 2.13 at 1.42 and 1.83 GHz. From Figure 2.13a, we also can see that, when this antenna is operating at low frequency, the induced currents on the adjustable arms 1 and 2 have contributed to the far-field radiation, which means that, the operating frequency can be shifted to much lower frequency by increasing the length of the adjustable arms 1 and 2. Thus, we can obtain wider bandwidth performance however, with the increase in the length of the adjustable arms 1 and 2, the radiation pattern of this antenna will be asymmetrical too. Similarly, since the value of C_j is only 0.35 pF (20 V reverse bias voltage), the induced currents on main radiation patch is much larger than the induced currents on the adjustable arms 1, 2, 3, and 4; therefore, the antenna operates towards high frequency end, i.e., 1.83 GHz. This capacitance of C_j was 3.2 pF at 0 V reverse bias voltage; therefore, both the central patch and the adjustable arms 1 and 2 contribute to radiation mechanism, and the antenna operates at 1.42 GHz (low frequency end).

In [50], both wide frequency tunability and polarization reconfigurability are demonstrated using four varactor diodes between a circular patch and a concentric ring patch around it. The geometry of the antenna with the biasing of the varactor is shown in Figure 2.14. The capacitance value that can be controlled by varactors is used to control the coupling between the patch and the ring and make the antenna radiate at various frequencies. For a smaller value of the capacitance, the coupling is less and the antenna system operates at higher frequency. As the value of the capacitance is increased, the coupling between the patch and the ring increases making the patch and the ring look like a single radiating element. The electrically larger size of the combined antenna lowers the resonant frequency. Skyworks varactor diode (Model # SMV 1139-011LF) is used for its low parasitic resistance (0.6 Ω) and adequate capacitance variation (1.9–8 pF) with low reverse bias voltage (12–0 V). With the varactor capacitance, C = 1.9 pF, the resonant frequency of the antenna is 1.58 GHz, whereas for C = 8 pF, the antenna operates at 1.17 GHz.

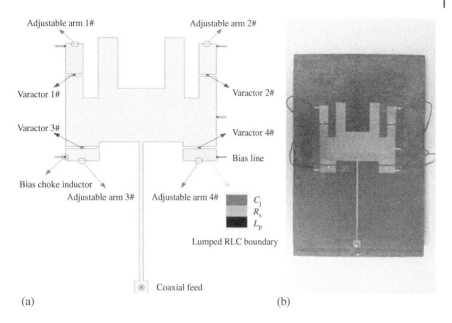

(a)

(b)

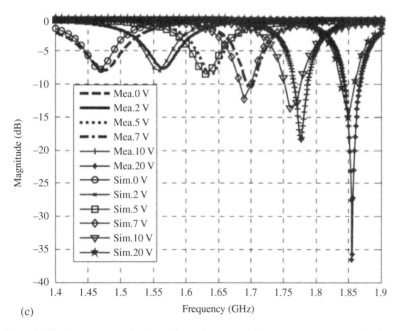

(c)

Figure 2.12 Varactor-tuned microstrip patch antenna: (a) antenna geometry with placement of varactor diodes, (b) photograph of the fabricated antenna, and (c) simulated and measured frequency agility response by varying varactor bias voltage. *Source:* Meng et al. [49].

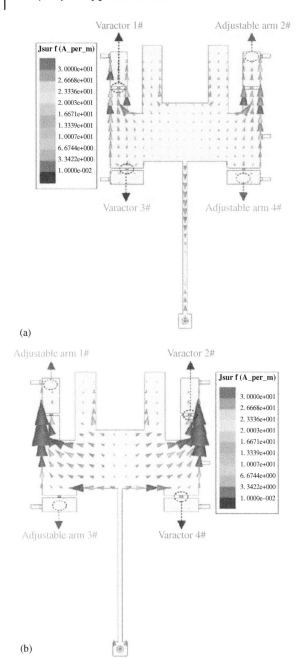

(a)

(b)

Figure 2.13 Vector current distributions on the proposed antenna at (a) 1.42 GHz and (b) 1.83 GHz. *Source:* Meng et al. [49].

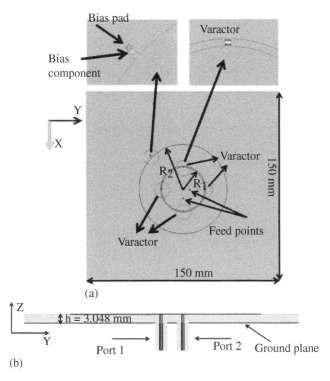

Figure 2.14 The geometry of the concentric circular microstrip patch antenna with varactor diodes and biasing components: (a) top view and (b) side view. *Source:* Babakhani and Sharma [50].

The fabricated prototype of the antenna with varactors and its feed network are shown in Figure 2.15. This feed is capable of providing vertical and horizontal polarizations as well as circular polarization. The frequency tunability with varactor diodes is shown in Figure 2.16, where measured S-parameters (S_{11} and S_{22}) of the antenna are shown, which were obtained when the same reverse bias voltage was applied to the varactors through the feed network for the two linear polarization cases. In the case of horizontal polarization, input port of the feed network is connected to port 1 of the antenna and S_{11} was measured. For the vertical polarization, its input port is connected to port 2 of the antenna, and S_{22} was measured. Varying the biasing voltage, the operating frequency of the antenna can be tuned from 1.17 to 1.58 GHz (around 30% tunability) where each of the tuned bands is around 20–30 MHz range. In order to investigate the effect of varactor resistance on peak realized gain and total antenna efficiency, simulations for different varactor resistance values were performed. Figure 2.17a shows the variation in the peak realized gain and Figure 2.17b shows the variation in total antenna efficiency for

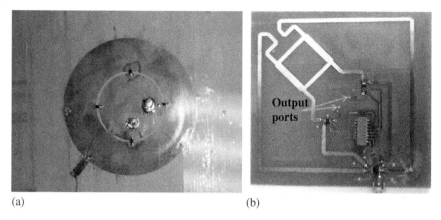

(a)   (b)

Figure 2.15 (a) Fabricated antenna with feed ports, varactor diodes, and bias pads. (b) The feed network, DC control lines, and bias pads. *Source:* Babakhani and Sharma [50].

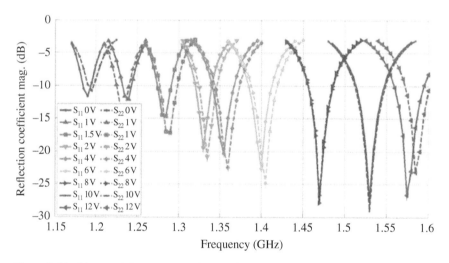

Figure 2.16 Measured S-parameters (S_{11} and S_{22}) controlled through feed network for different applied reverse bias voltages, corresponding to various capacitances. *Source:* Babakhani and Sharma [50].

various varactor diode resistance values. It is evident that increasing the resistance of the varactor decreases the gain and the efficiency of the antenna. This effect is more severe toward the lower frequency end than the higher frequency in the case of both gain and efficiency performance. This radiation pattern polarization reconfiguration performance of this antenna will be discussed in Chapter 3.

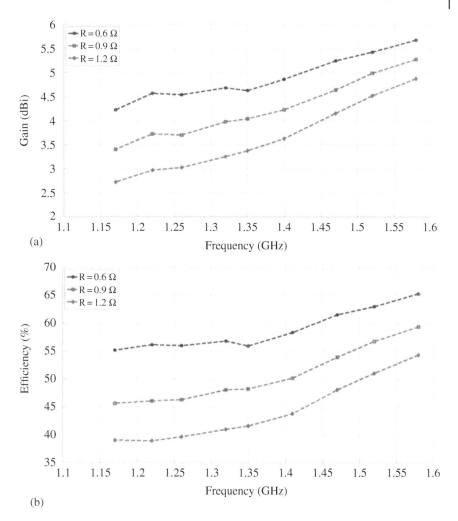

Figure 2.17 Simulation results for the effect of varactor resistance (R) on (a) peak realized gain and (b) total antenna efficiency of the antenna over the tunable range when only port 1 is excited (linear polarization) and port 2 is match terminated. *Source:* Babakhani and Sharma [50].

The following design guidelines can be used to design a similar frequency tunable antenna with polarization reconfiguration [50]:

• Choose an appropriate antenna element that can offer a specific radiation response such as broadside pattern.

• Amount of matching bandwidth in an antenna element can dictate possible individual tuning bandwidths when implemented in a frequency agile antenna.

- Identify how a tuning/frequency agile element can be incorporated into the antenna configuration. For example, at the radiator level or between the radiator and the ground plane. Proper bias network placement should be part of the tuning component so its effect is taken in the antenna response.
- By varying tuning component values (or capacitance), investigate frequency agile response.
- To achieve simultaneous polarization reconfiguration, the radiating element should include appropriate feed points and feed network that can be automatically controlled by applying proper bias voltage at the control lines. If possible, feed network should be placed below the ground plane so feed network radiation can be eliminated from the antenna radiation.

2.3.1.3 Micro-Electro-Mechanical-System (MEMS) Switches

RF MEMS has been integrated with antennas to obtain frequency reconfigurability for decades, which rely on the mechanical movement of these switches to achieve reconfiguration [51, 52]. The isolation of RF MEMS is quite high, and they require minimal power consumption. However, their switching speed is relatively low for some applications compared to PIN or varactor diodes. MEMS-based switches are used for FRAs, arrays, and frequency selective surfaces (FSS). Figure 2.18 shows the model of MEMS-bridge over a slot used to completely short it for or keep the slot open to control the frequency of operation [53]. If the bridge is not suspended over the slot, it operates at the slot resonance frequency. Otherwise, the frequency of operation shifts to a much higher frequency depending on the bridge dimension. The concept can be used in arrays to develop an adaptive surface with transparent state or reflective state.

A smooth tuning of frequency is also possible using the same bridge over a slot in the case of a FSS [54]. The movable metallic MEMS bridge over the slot acts as a capacitive load. By controlling the deflection of the bridge over the slot, the capacitance can be changed, and thus the resonance frequency. The displacement of the bridge toward the ground plane is controlled by a voltage. The fabricated bridge is shown in Figure 2.19. The cell is placed inside a waveguide simulator by sandwiching the substrate between two HP X281C coaxial-to-waveguide adaptors with customized flanges, as shown in Figure 2.20a. Bias lines that are needed to actuate the MEMS bridge are placed in grooves on X-band flanges to ensure that they do not touch the waveguide flange, as shown in Figure 2.20b. Measured and simulated reflection coefficients of the FSS unit cell for different heights of the MEMS bridge are shown in Figure 2.20c. Results show more than 1.7 GHz frequency shift in the X-band from 8.54 to 10.26 GHz, achieved by using only one MEMS bridge.

2.3.1.4 Optical Switches

Optical-switch-controlled reconfigurable antennas provide unique advantages compared to electrical switches, such as easy integration into optical systems and absence of bias lines, which essentially help in eliminating unwanted interference

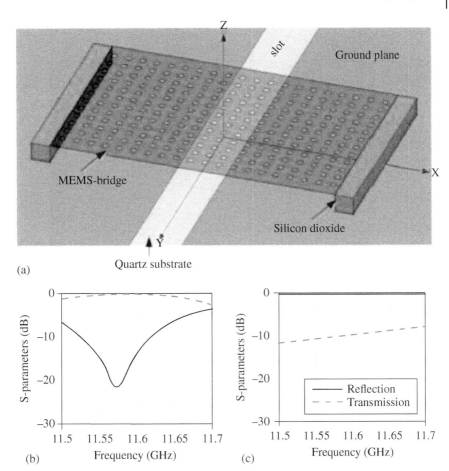

Figure 2.18 (a) Model of a MEMS bridge over a slot and its scattering parameters to make it (b) transparent or (c) reflective, by controlling the frequency of operation. *Source:* Latif et al. [53].

and losses. Reconfigurability by the optical switches is based on the photoconductive principal of semiconductor materials. An optical switch has mainly two states: "ON/OFF" when the semiconductor material is illuminated by a laser beam, thus provides discrete tuning.

An optically controlled reconfigurable antenna for millimeter wave applications is presented in [25]. It is based on a slotted-waveguide antenna array, and uses two photoconductive switches that are used to control the slot electrical length. The slot length, the inter-slot distance, and the space between the last slot and the waveguide end are approximately one-half of the guided wavelength. This

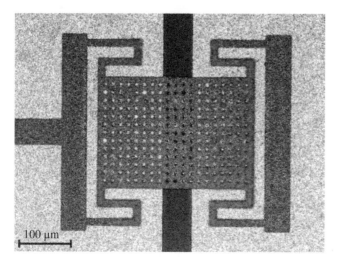

Figure 2.19 Photograph of the fabricated MEMS bridge over a slot for frequency tuning. *Source:* Safari et al. [54].

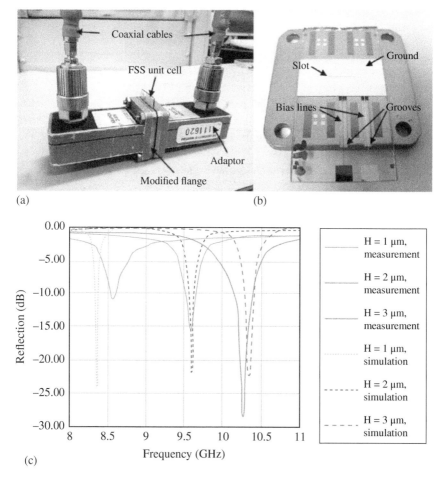

Figure 2.20 Test setup. (a) FSS unit cell under test, (b) FSS unit cell on top of the modified flange, and (c) measured and simulated reflection coefficients of the FSS unit cell vs. height of the MEMS bridge. *Source:* Safari et al. [54]. © 2015, IEEE.

particular design consists of four slots with two different lengths, 5.35 and 3.95 mm, which are slightly shorter than one-half of the guided wavelength (8.14 and 4.75 mm for 28 and 38 GHz, respectively), as shown in Figure 2.21a. Two intrinsic silicon photoconductive switches with 0.3 mm thickness, two 808-nm lasers, one half-power optical splitter, and three standard single-mode optical fibers are used to obtain optical reconfigurability. The lengths of the photoconductive switch 1 (PS_1) and photoconductive switch 2 (PS_2) are 5.5 and 1.4 mm, respectively. The photograph of the optically controlled reconfigurable slotted waveguide antenna array is shown in Figure 2.21b, where the photoconductive switches are labeled as PS_1 and PS_2. The switches are on "ON" state when illuminated by a laser beam, and on "OFF" state if no light comes onto them. When the switch is illuminated by light from the top, silicon changes from an insulator to a near-conducting state due to the creation of electron–hole pairs. Two configurations are considered to obtain reconfigurability so that the arrays work at 28 and 38 GHz, and are depicted in Figure 2.21c. First configuration is when PS_1 is on "ON-state" illuminated by the optical fiber 2 to cover the right part of slot 1 (PS_1) and PS_2 on "OFF-state," enabling the antenna operation in the 28-GHz frequency band. Second configuration is with PS_2 on "ON-state" illuminated by the optical fiber 1 for covering the left part of slot 1 (PS_1) and PS_2 on "ON-state" illuminated by the optical fiber 3 to cover the left part of slot 2 (PS_2). This configuration allows the antenna to operate in the 38-GHz frequency band. The corresponding frequency responses are shown in Figure 2.22.

2.3.1.5 Ground Plane Membrane Deflection

Frequency reconfiguration can also be obtained using ground plane membrane deflection such as discussed in [52, 55] where patch antenna frequency is reconfigured at higher frequency band (16.8–17.82 GHz). An antenna layout and its frequency reconfiguration are shown in Figure 2.23 where the reconfiguration depends on the ground plane deflection and for a ground plane deflection from 0 to 138 μm, the antenna reconfigures its frequency by 6.0% [55]. This technique is dependent upon DC voltage-based micro-machined membrane deflection. This technique can be a good option for millimeter wave frequency bands where a little deflection can be sufficient for providing reasonable frequency tunability. This technique works based on the effective permittivity variation as bias voltage is applied and thereby providing frequency control.

2.3.2 Frequency Reconfigurability Using Special Materials

Frequency reconfigurability has been demonstrated by introducing change in the material used in the antenna fabrication. These so-called "agile" or "smart" materials change their physical size or property by external means, such as voltage and temperature. Liquid crystalline material is widely used as antenna substrate

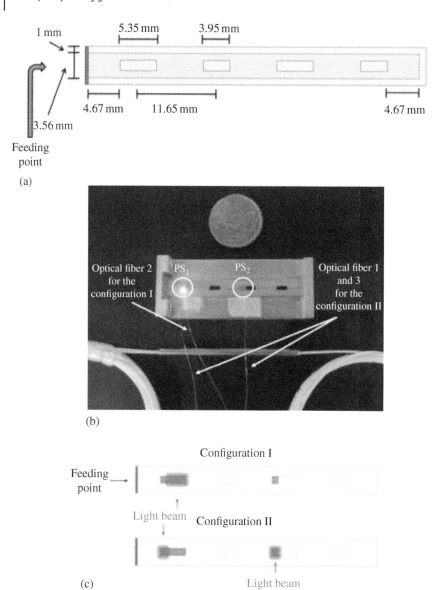

Figure 2.21 (a) Geometry of the multiband optically controlled reconfigurable slotted waveguide antenna array based on two different slot lengths, (b) photograph of the optically controlled reconfigurable slotted waveguide antenna array, and (c) two configurations of the optically controlled reconfigurable slotted waveguide antenna array *Source:* da Costa et al. [25]. © 2017, IEEE.

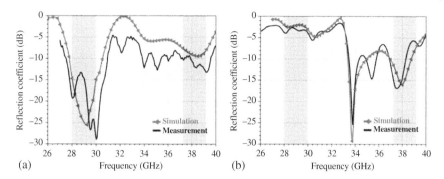

Figure 2.22 Optically controlled reconfigurable slotted waveguide array reflection coefficients for two configurations: (a) first configuration and (b) second configuration. *Source:* da Costa et al. [25]. © 2017, IEEE.

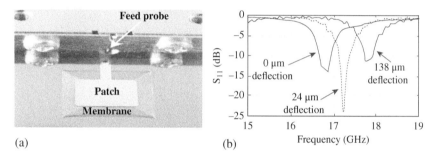

Figure 2.23 MEMS-controlled microstrip patch antenna and its frequency response. *Source:* Al-Dahleh et al. [52] and Shafai et al. [55].

material, which is changed mechanically or thermally to alter its properties so that frequency tunability can be achieved [1]. The radiating element can also be modified with the help of special material such as liquid crystals to obtain frequency and pattern reconfigurability.

2.3.2.1 Liquid Crystals

A FRA based on self-morphing liquid crystalline elastomers (LCEs) is presented in [56]. LCEs are anisotropic polymers that maintain orientational and/or positional order named as smectic or nematic phases. When heat is applied, nematic or nematic LCEs undergo a phase transition to an isotropic state, which is accompanied by a large strain parallel to the nematic director. Adhering metal layers to LCE films yields reconfigurable antennas that can change their electromagnetic performance. In [56], a thin aluminum layer is attached to LCE strips using optical glue in order to form antennas, as shown in Figure 2.24. This combination

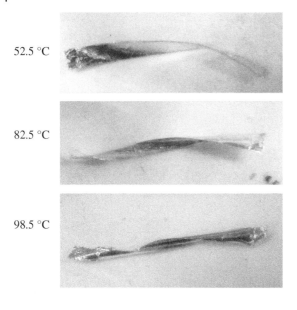

52.5 °C

82.5 °C

98.5 °C

Figure 2.24 Liquid crystalline elastomers (LCEs) with aluminum showing the different stages of the polymer changing its structure because of the temperature rise. *Source:* Gibson et al. [56].

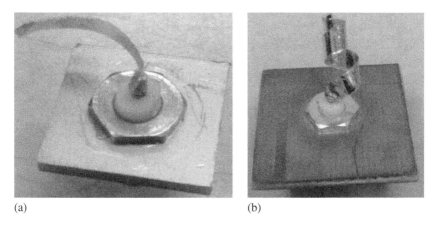

(a)   (b)

Figure 2.25 Prototype of LCE frequency-reconfigurable antenna at different temperatures: (a) 0.5-turn loop at 80 °C and (b) 1.5-turn helix at 104 °C. *Source:* Gibson et al. [56].

changes the number of turns, as shown in Figures 2.24 and 2.25, at varying temperatures to radiate differently and at different frequencies. Measured and simulated reflection coefficients indicating the frequency reconfigurability of this self-morphed antenna are shown in Figure 2.26.

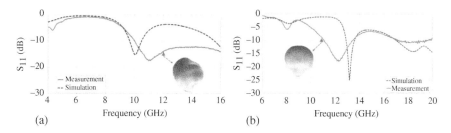

Figure 2.26 S_{11} vs. frequency of the LCE frequency-reconfigurable antenna obtained self-morphed at different temperatures: (a) 0.5-turn loop at 80 °C and (b) 1.5-turn helix at 104 °C. *Source:* Gibson et al. [56]. © 2016, IEEE.

2.3.2.2 Graphene

Communication at THz frequencies has attracted the attention of many antenna researchers especially after the rise of graphene as an antenna design material [1]. The behavior of the surface conductivity of graphene in the THz region enables the creation of plasmonic modes in radiating structures. The design of antennas using graphene provides a good total efficiency, good matching, high miniaturization, and inherent reconfiguration capabilities [57–60]. The antenna structures take advantage of resonant plasmonic modes propagating on graphene sheets to implement actual radiators, able to couple electromagnetic energy from small THz sources to free space. Furthermore, the surface impedance of graphene material can be tuned by applying an external DC bias voltage, which enables obtaining one of two extreme values that emulate the ON and OFF states of common switches, such as PIN diodes or RF switches. This allows modifying the electrical length of the current on the antenna conductor and can be utilized to discretely tune the resonant frequency of the antenna providing frequency reconfigurability.

Plasmonic antennas are optical nano-antenna operating at THz frequencies [61]. The concept of a dual-band reconfigurable terahertz patch antenna using a graphene-stack-defined backing cavity is presented in [62]. The proposed design employs patch resonance based on backing cavity defined by interleaved graphene/Al_2O_3 stacks, which can be dynamically frequency-tuned to dual resonances on large range about 1 THz via electrostatic gating on the graphene stack. The microstrip substrate is stacked on the top of the backing cavity, and the rectangular square patch with side length a_{patch} is printed on the microstrip substrate with dielectric constant ε_1 and height h_1, which is fed through a microstrip line, as shown in Figure 2.27a. The top layer of the cavity substrate with dielectric constant ε_2 and height h_2 is a circular opening with radius R_{inner} underneath the patch, as illustrated in Figure 2.27b. Many graphene stack columns are embedded into the cavity substrate space along the outer circular with radius R_{outer} and connect the top layer and the ground layer of the backing cavity. Each graphene stack

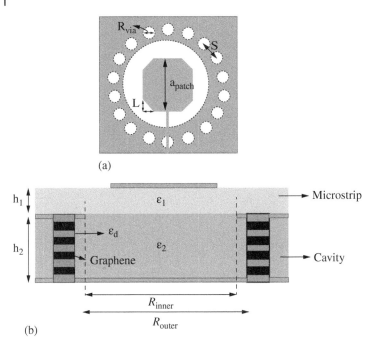

Figure 2.27 Topology of the patch antenna using a graphene-stack-defined backing cavity: (a) top view and (b) cross-sectional view. Some design parameters of the microstrip and the graphene backing cavity are: $h_1 = 0.1\,\mu m$, $h_2 = 0.254\,\mu m$, $a_{patch} = 15.2\,\mu m$, and $\varepsilon_1 = \varepsilon_2 = 3.9$. All electrical conductors including patch, top layer, and ground layer are A_g ($\sigma = 6.3 \times 10^7$ s/m). *Source:* Dong et al. [62]. © 2016, IEEE.

column is comprised of alternating graphene sheet. The effective dielectric constant of the cavity substrate can be varied by changing the dielectric constant, ε_g, of the graphene stack column with various chemical potentials, which essentially implies that the patch resonance characteristics can be altered.

The antenna is modeled and simulated in the full-wave EM software CST Microwave Studio. The variation in the reflection coefficient of the antenna for the considered 0.16–0.26 eV range of chemical potential is shown in Figure 2.28. Evidently, there are two resonant frequency bands over the entire frequency range, from 4 to 5 THz and from 6.5 to 7.5 THz. The first and second resonant frequencies can be controlled within a large range varying the chemical potential. This design is simulation-based and arguably the implementation of graphene-based reconfigurable antennas is difficult. A practical application of graphene for consumer electronics is demonstrated in [63]. This will open up the practical implementation of this type of antenna and the graphene-based reconfiguration technique more versatile.

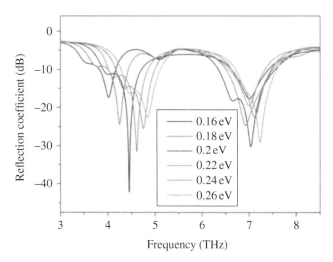

Figure 2.28 Reflection coefficient magnitudes of the patch antenna backed by a graphene-stack-defined cavity with different chemical potential μ_c. *Source:* Dong et al. [62]. © 2016, IEEE.

2.3.3 Frequency Reconfigurability by Mechanical Changes

Mechanically reconfigurable antennas have emerged to overcome limitations of reconfigurability by electrical switches or MEMS-based reconfigurable antennas [64–66]. This type of reconfigurability does not require lumped components which cause huge losses in antennas. Therefore, this reconfiguration technique can offer higher radiation efficiency of the antenna. Moreover, the frequency tuning process is linear unlike with varactors, which exhibit nonlinear behavior. The mechanically reconfigurable antennas can also handle more power than FRAs using lumped components. In some severe weather conditions, for example, very high temperature, the mechanically reconfigurable antennas resist better than the electronically reconfigurable antennas. However, mechanically reconfigurable antennas have the drawback of lower tuning and switching speed, and the use of motors or actuators may increase the overall antenna size. Mechanical means, such as stretching, bending, and pressuring have been applied successfully to obtain frequency reconfigurability.

2.3.3.1 Actuators
An actuator-based mechanically FRA is presented in [67], where a frequency-tunable or frequency reconfigurable half-wavelength dipole antenna is realized using an array of electrically actuated liquid-metal pixels, shown in Figure 2.29. Liquid metal is kept in reservoirs and actuated electrically by manipulating its

Liquid-metal pixelated dipole

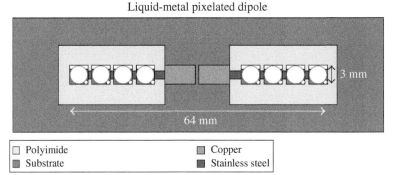

Figure 2.29 Liquid-metal pixelated dipole prototype. *Source:* Sarabia et al. [67]. Licensed under CC-BY-4.0.

surface tension using continuous electro-wetting. When the liquid metal is immersed in sodium hydroxide (NaOH) solution, it forms an electrical double layer (EDL) at the metal–NaOH interface. By applying a voltage on the EDL, a surface tension imbalance is created on the liquid metal, which results in a pressure imbalance, actuating the liquid metal. While a pixel is considered "ON" when the liquid metal is actuated to the top-side reservoir of the antenna, it is considered "OFF" when the liquid metal is actuated to the bottom-side reservoir. The steps for actuating the liquid metal are shown in Figure 2.30. A 4 V square wave with a +1 V DC offset at a frequency of 30 Hz is applied to the electrodes to actuate the liquid metal from a reservoir buried below to add pixels. The liquid metal is then actuated back to the reservoir by swapping the applied voltage polarities on the electrodes to remove pixels. When metal pixels are added, the dipole length is increased on each arm decreasing the resonant frequency of the antenna. With the lengthening of the antenna, the incremental frequency shift decreases as the inverse square of the antenna length. The tuning of the dipole to the following frequencies has been observed, as different pixels are activated: 2.51, 2.12, 1.85, and 1.68 GHz.

2.3.3.2 Motors

In order to achieve frequency reconfigurability by physically rotating the ground plane or the antenna metal, low-power motors are used. Mechanically reconfigurable dual-band slot dipole antennas with wide tuning ranges using motors are presented in [68]. While the dual-band operation is obtained by inserting parallel slots with various lengths, the tuning is implemented by incorporating a dual rack and pinion mechanism that helps slide parasitic patches to vary the slot lengths

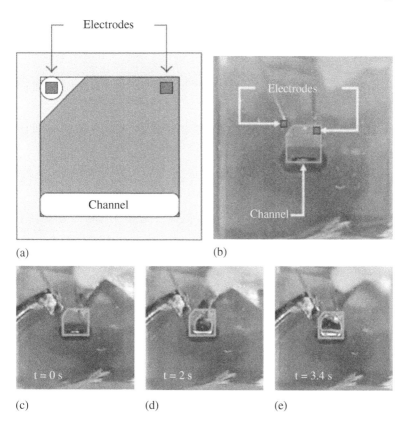

Figure 2.30 (a) Layout of the bottom side of the dipole pixel with an actuation circuit, (b) fabricated prototype pixel with an outlined pixel and electrodes, and steps of actuation, (c) application of voltage to actuate liquid metal from a reservoir buried below, (d) liquid metal actuating, and (e) completion of liquid-metal actuation. *Source:* Sarabia et al. [67]. Licensed under CC-BY-4.0.

enabling the variation of the frequency of operation. The design uses two motors to slide parasitic patches over each slot pair independently to achieve a frequency band ratio between 1 and 2.6.

Figure 2.31 shows the CAD illustration of the prototype antenna including linkage, which is composed of a pinion and dual racks that convert rotational motion into linear motion and is actuated by a small stepper motor from the bottom side. The sliding motion of the racks is guided by two supports that are mounted on the top surface of the antenna substrate. The nature of the dual rack and pinion mechanism allows for the horizontal sliding of the two racks in opposite directions. The antenna is fabricated with 3-D printing technology on acrylonitrile butadiene styrene (ABS) substrate. The linkage is printed of ABS material as well to minimize

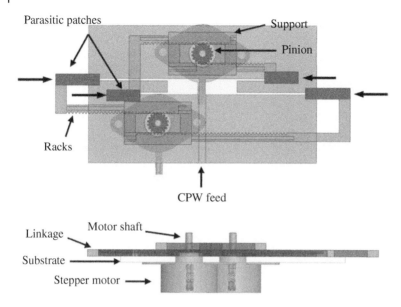

Figure 2.31 CAD illustration of the mechanically reconfigurable dual-band antenna with arbitrary frequency ratio: top view (top) and side view (bottom). *Source:* Nassar et al. [68]. © 2015, IEEE.

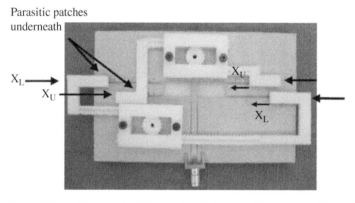

Figure 2.32 Photograph of the mechanically reconfigurable dual-band slot antenna with arbitrary frequency ratio. *Source:* Nassar et al. [68].

the weight. A photograph of the fabricated antenna is shown in Figure 2.32, where X_L represents the inward compression of the longer slot pair and X_U represents the decrease in the shorter slot pair. Figure 2.33 compares the measured and simulated S_{11} for different X_L values with $X_U = 0$. By increasing X_L, the frequency can be tuned from 2 to 4 GHz without affecting the upper band. Again, with $X_L = 0$, by

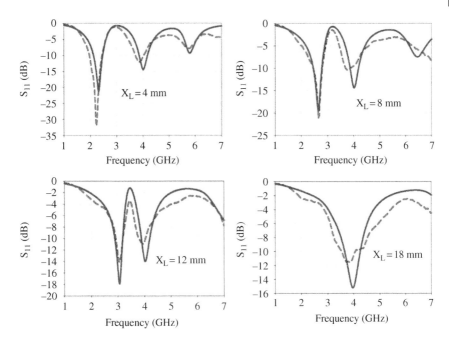

Figure 2.33 Measured (dashed) and simulated (solid) reflection coefficient (S_{11}) of the mechanically reconfigurable slot antenna, for $X_U = 0$ mm and different X_L values. *Source:* Nassar et al. [68]. © 2015, IEEE.

increasing X_U, the frequency can be tuned from 4 to 5.25 GHz without affecting the lower band frequency. Thus, by varying X_L and X_U independently, the frequency band ratio can be varied significantly between 1 and 2.6.

2.3.4 Frequency Reconfigurability Using Special Shapes

FRAs using special shapes, for example, origami or fractal, reconfigurable antennas, have recently been developed. This type of antenna is a physically reconfigurable antenna that can change its performance by changing its geometry.

2.3.4.1 Origami Antennas

An origami frequency and pattern reconfigurable antenna is presented in [69], where the design and development of a conical spiral antenna (CSA) is discussed. This antenna is based on the origami Nojima wrap pattern that enables the antenna to morph from a planar dipole to a conical spiral. The prototype of the developed antenna is shown in Figure 2.34. It is manufactured using 0.1-mm thick copper tape on 0.2-mm thick sketching-paper substrate without any coating.

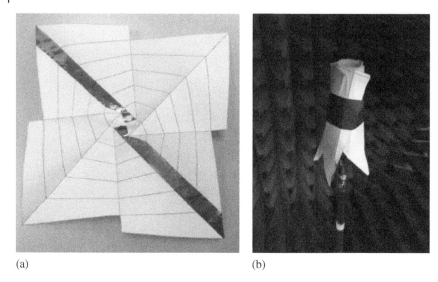

(a)   (b)

Figure 2.34 Prototype of the Nojima square CSA at (a) unfolded state and (b) fully folded state. *Source:* Yao et al. [69].

The copper tape was glued on the paper substrate and creased with it, so that it would stay attached to the paper substrate while the antenna was being folded or unfolded. The permittivity of the paper substrate was considered 3.2 when the antenna was simulated. The folded and unfolded versions of the fabricated antenna are shown in Figure 2.34. A microstrip balun was used to transform the unbalanced coaxial cable to the balanced Nojima square CSA structure. Figure 2.35a shows the comparison of the simulated and measured reflection coefficient of the Nojima square CSA at the unfolded state, where it operates as a dipole antenna. Figure 2.35b compares its simulated and measured reflection coefficients at the fully folded state. $S_{11} = -10\,dB$ bandwidth of this antenna is from 2.1 to 3.5 GHz in both simulation and measurement at this state, which indicates its broadband operation at higher frequencies. When the Nojima antenna is unfolded, its radiation pattern is linear like an ordinary half-wavelength dipole operating at 0.48 GHz. When it is folded, it exhibits a directional radiation pattern over the wide frequency bandwidth.

2.3.4.2 Fractal Shapes

Due to self-similar property and compactness, several fractal shapes, namely Koch, Hilbert, Peano, Minkowski, and Sierpinski, have been applied to antenna and array designs [70]. Fractal antennas demonstrate miniaturization and multiband, and in some occasions, wideband performances, when compared to

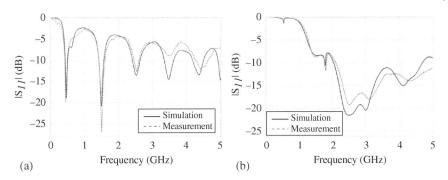

Figure 2.35 (a) Simulated and measured S_{11} of the Nojima square CSA at the unfolded state. (b) Simulated and measured S_{11} of the Nojima square CSA at the folded state. *Source:* Yao et al. [69]. © 2017, IEEE.

same-sized Euclidean geometry-based antenna. Fractal shapes have also found application in FRAs, where reconfigurability is achieved using typical tools, such as PIN diodes or MEMS switches and compactness or multiband performance is obtained using the advantage of their special shapes [35, 71–79].

A frequency reconfigurable dual-polarized fractal-shaped slot-ring antenna/ array for operation in S- (1.8–3.7 GHz) and C- (4.5–8.2 GHz) bands is presented in [71], where the aperture of an S-band antenna is reconfigured to a 2×2 C-band antenna array by turning on eight PIN diodes. In order to achieve wide-band performance, a modified Koch fractal geometry is incorporated into this slot-ring configuration, as shown in Figure 2.36. A portion of the fractals provides additional resonances that are merged with the fundamental resonances of S- and C-bands from the slot rings to significantly increase the antenna band-width. A metal wire is placed at the center of the antenna with minimum electric field, which provides a DC bias voltage for all the PIN diodes. When all the diodes are open, i.e. reverse-biased, the antenna is excited by port 1 providing horizontal polarization or port 2 providing vertical polarization and tuned to the S-band frequency (S-band state). By turning on all the PIN diodes, i.e. forward-biased, a 2×2 slot-ring antenna array is formed tuning the antenna to the C-band frequency (C-band state). Ports 3–6 or ports 7–10 are excited in this state to provide horizontal polarization or vertical polarization, respectively. Microstrip lines are used for feeding this antenna/array and shown in Figure 2.36 along with all ports.

The generation of the fractal shapes for one frequency in the S-band is shown in Figure 2.37, which shows the slot-ring antenna geometries with different iteration orders (IO = 0, 1, and 2) and for iteration factor of 0.3. As the IO increases, the

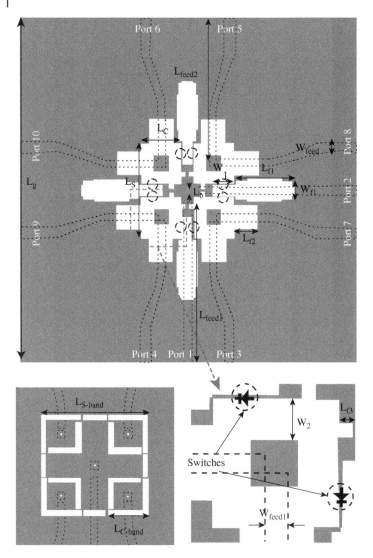

Figure 2.36 (Top) Configuration of the frequency reconfigurable wideband fractal slot-ring antenna/array. (Bottom left) Configuration of the switchable-frequency single polarized slot-ring antenna. (c) (Bottom right) Zoomed-in view of the C-band unit cell. *Source:* Shirazi et al. [71]. © 2018, IEEE.

average electrical length of the slot ring also increases that lowers the resonant frequency of the slot-ring antenna at the S-band state. In addition to that, as more fractal geometries are incorporated, the number of resonant bands increases, which results in the significant increase in the antenna operating bandwidth.

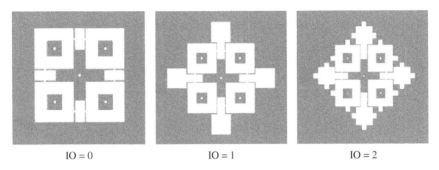

IO = 0   IO = 1   IO = 2

Figure 2.37 S_{11} of the fractal-shaped slot-ring antenna in the S-band operating state with different IO. *Source:* Shirazi et al. [71]. © 2018, IEEE.

(a)   (b)

Figure 2.38 Photos of the fabricated reconfigurable dual-polarized wideband slot-ring antenna/array prototype: (a) top view and (b) bottom view with port numbers shown. *Source:* Shirazi et al. [71].

Figure 2.38 shows the fabricated antenna, where it can be seen that eight PIN diodes are mounted on the radiating aperture on the top side of the antenna/array and the feed lines are on the bottom side. The antenna was fabricated on a Rogers RT/Duroid 5880 substrate ($\varepsilon_r = 2.2$ and tan $\delta = 0.0009$) having a thickness of 0.79 mm. The simulated and measured S_{11} and S_{44} versus frequency for the fractal antenna in the S- and C-band states are shown in Figure 2.39a and b, respectively. In the S-band state, the antenna shows $S_{11} < -10$ dB bandwidth of 69.1%, as shown in Figure 2.39a when the diodes are OFF. When the PIN diode are forward-biased or ON, S_{11} has a higher value over -1.5 dB in the S-band indicating that most of the energy is reflected back in the S-band state and no operation is possible.

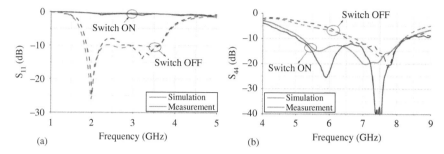

Figure 2.39 Simulated and measured (a) S_{11} and (b) S_{44} of the reconfigurable dual-polarized wideband slot-ring antenna/array when all the PIN diodes are ON (solid lines) or OFF (dashed lines). *Source:* Shirazi et al. [71]. © 2018, IEEE.

On the other hand, in the C-band state, $S_{44} < -10\,dB$ bandwidth of 58.3% is observed for the ON states of the switches. It can be observed in Figure 2.39b that the antenna is mismatched, especially in the lower portion of the C-band when the switches are OFF.

2.4 FRAs in the Future: Applications in Emerging Technologies

As new and challenging applications are emerging, FRAs are becoming more prominent candidates in those systems. For example, in the past one decade, cubesat development has seen a rapid growth, especially in academia. With crowded UHF/VHF and S-band frequency spectrum, it will be necessary to use higher frequency spectrum, e.g. X-band. A common scenario is to use low-frequency spectrum with low data rate for the uplink for telemetry and control, and high-frequency spectrum with high data rate for downlink to download scientific mission data. Instead of using two antennas operating at two different frequencies, a FRA will be best suited for this case to utilize the small room available in a cubesat. Many wire and patch-based cubesat antennas, some with deployable option, have been proposed [80]. Some reconfigurable antennas with multiple functions for cubesats can be found in the literature [81, 82], but FRAs have not seen a widespread use in cubesats due to complexity, cost, and unreliability.

In the upcoming fifth generation of mobile communication (5G) systems, the use of millimeter wave frequencies is proposed. As more spectrum is available for the communications, traditional antenna solutions will not be sufficient to meet the demand of emerging technologies, for example, carrier aggregation (CA), massive MIMO, etc., the device will have to transmit or receive on several channels simultaneously for increased data rate. This makes frequency tuning more difficult, as any antenna has to be tuned to several frequencies at the same time. Conventional FRAs operate at a particular frequency at a given time and will fail in such applications. The concept of frequency reconfigurability obtained through a cluster of mutually coupled antenna elements excited with frequency-specific weights using distributed transceivers will be useful in the future [83–85]. This concept is based on connecting the antenna array to multiple transceivers instead of single transceiver to make coupled waves interfere destructively with the reflected waves by weighting the feed signals optimally. Also, in massive MIMO systems equipped with massive switches in FRA arrays, innovative optimization techniques have to be developed to reduce the number of switches so that they take less space in the chip and are low loss [86]. Fluidically controlled FRAs are another alternative if conventional switches are to be avoided [87–89]. However, their practical use in millimeter wave frequencies has to be implemented.

2.5 Conclusion

This chapter discusses the technique to obtain frequency reconfigurability in modern communication and other emerging applications. Recent developments in this area are presented to update any readers. Frequency reconfigurable, along with polarization and pattern reconfigurable, antennas are the core of any communication system which undergoes changes. The technology will continue to evolve to newer methodologies to meet the demand of emerging technologies. Future millimeter wave systems and space communications, along with many biomedical applications, will require antennas that can adapt and configure to a changing environment and FRAs will be an integral part of these systems.

References

1 Costantine, J., Tawk, Y., and Christodoulou, C.G. (2016). Reconfigurable antennas. In: *Handbook of Antenna Technologies* (eds. Z.N. Chen, D. Liu, H. Nakano, et al.). Singapore: Springer.
2 Guo, Y.J. and Qin, P.Y. (2016). Reconfigurable antennas for wireless communications. In: *Handbook of Antenna Technologies* (eds. Z.N. Chen, D. Liu, H. Nakano, et al.). Singapore: Springer.
3 Huff, G.H. and Bernhard, J.T. (2011). Reconfigurable antennas. In: *Modern Antenna Handbook* (ed. C.A. Balanis). Hoboken, NJ: Wiley.
4 Costantine, J., Tawk, Y., Barbin, S.E., and Christodoulou, C.G. (2015). Reconfigurable antennas: design and applications. *Proc. IEEE* 103 (3): 424–437.

5 Bernhard, J.T. (2007). *Reconfigurable Antennas*. San Rafael, CA: Morgan and Claypool Publishers.

6 Costantine, J., Tawk, Y., and Christodoulou, C.G. (2013). *Design of Reconfigurable Antennas Using Graph Models*. San Rafael, CA: Morgan and Claypool.

7 Latif, S.I. and Shafai, L. (2011). Investigation on the EM-coupled stacked square ring antennas with ultra-thin spacing. *IEEE Trans. Antennas Propag.* 59: 3978–3990.

8 Chu, F.H. and Wong, K.L. (2010). Planar printed strip monopole with a closely-coupled parasitic shorted strip for eight-band LTE/GSM/UMTS mobile phone. *IEEE Trans. Antennas Propag.* 58: 3426–3431.

9 Zhang, T., Li, R., Jin, G. et al. (2011). A novel multiband planar antenna for GSM/ UMTS/LTE/Zigbee/RFID mobile devices. *IEEE Trans. Antennas Propag.* 59: 4209–4214.

10 Dai, X.W., Wang, Z.Y., Liang, C.H. et al. (2013). Multiband and dual-polarized omnidirectional antenna for 2G/3G/LTE Application. *IEEE Antennas Wirel. Propag. Lett.* 12: 1492–1495.

11 Henderson, K.Q., Latif, S.I., Lazarou, G.Y. et al. (2018). Multi-slot antennas excited by novel dual-stub loaded microstrip lines for 4G LTE bands. *Progr. Electromagn. Res. M* 75: 1–12.

12 Latif, S.I., Shafai, L., and Sharma, S.K. (2005). Bandwidth enhancement and size reduction of microstrip slot antennas. *IEEE Trans. Antennas Propag.* 53 (3): 994–1003.

13 Rajgopal, S.K. and Sharma, S.K. (2009). Investigations on ultrawideband pentagon shape microstrip slot antenna for wireless communications. *IEEE Trans. Antennas Propag.* 57 (5): 1353–1359.

14 Sharma, S.K., Shafai, L., and Jacob, N. (2004). Investigations of wide band microstrip slot antenna. *IEEE Trans. Antennas Propag.* 52: 865–872.

15 Michel, A., Nepa, P., Gallo, M. et al. (2017). Printed wideband antenna for LTE-Band automotive applications. *IEEE Antennas Wirel. Propag. Lett.* 16: 1245–1248.

16 Wang, H., Wang, Y., Wu, J. et al. (2016). Small-size reconfigurable loop antenna for mobile phone applications. *IEEE Access* 4: 5179–5186.

17 Borda-Fortuny, C., Tong, K.F., and Chetty, K. (2018). Low-cost mechanism to reconfigure the operating frequency band of a Vivaldi antenna for cognitive radio and spectrum monitoring applications. *IET Microwaves Antennas Propag.* 12 (5): 779–782.

18 Tawk, Y., Albrecht, A.R., Hemmady, S. et al. (2010). Optically pumped frequency reconfigurable antenna design. *IEEE Antennas Wirel. Propag. Lett.* 9: 280–283.

19 Tawk, Y., Costantine, J., Hemmady, S. et al. (2012). Demonstration of a cognitive radio front end using an optically pumped reconfigurable antenna system (OPRAS). *IEEE Trans. Antennas Propag.* 60 (2): 1075–1083.

20 Pendharker, S., Shevgaonkar, R.K., and Chandorkar, A.N. (2014). Optically controlled frequency-reconfigurable microstrip antenna with low photoconductivity. *IEEE Antennas Wirel. Propag. Lett.* 13: 99–102.

21 Zheng, S.H., Liu, X., and Tentzeris, M.M. (2014). Optically controlled reconfigurable band-notched UWB antenna for cognitive radio systems. *Electron. Lett.* 50 (21): 1502–1504.

22 Zheng, S., Liu, X., and Tentzeris, M.M. (2014). A novel optically controlled reconfigurable antenna for cognitive radio systems. *Proceedings of the 2014 IEEE Antennas and Propagation Society International Symposium (APSURSI)*, Memphis, TN (6–12 July 2014) 1246–1247.

23 Andy, A., Alizadeh, P., Rajab, K.Z., Kreouzis, T., and Donnan, R. (2016). An optically-switched frequency reconfigurable antenna for cognitive radio applications. *Proceedings of the 2016 10th European Conference on Antennas and Propagation (EuCAP)*, Davos (10–15 April 2016), 1–4.

24 daCosta, I.F., Sodré, A.C. Jr., Páez, J.S.R. et al. (2017). Photonics-assisted wireless link based on mm-wave reconfigurable antennas. *IET Microwaves Antennas Propag.* 11 (14): 2071–2076.

25 da Costa, I.F., Sodré, A.C., Spadoti, D.H. et al. (2017). Optically controlled reconfigurable antenna array for mm-wave applications. *IEEE Antennas Wirel. Propag. Lett.* 16: 2142–2145.

26 Costa, I.F.D., Spadoti, D.H., Sodré, A.C. et al. (2017). Optically controlled reconfigurable antenna for 5G future broadband cellular communication networks. *J. Microwaves Optoelectron. Electromagn. Appl.* 16: 208–217.

27 Jiang, Z. and Yang, F. (2013). Reconfigurable sensing antennas integrated with thermal switches for wireless temperature monitoring. *IEEE Antennas Wirel. Propag. Lett.* 12: 914–917.

28 Su, W., Nauroze, S.A., Ryan, B., and Tentzeris, M.M. (2017). Novel 3D printed liquid-metal-alloy microfluidics-based zigzag and helical antennas for origami reconfigurable antenna "trees". *Proceedings of the 2017 IEEE MTT-S International Microwave Symposium (IMS)*, Honolulu, HI (4–9 June 2017), 1579–1582.

29 Sessions, D., Fuchi, K., Pallampati, S. et al. (2018). Investigation of fold-dependent behavior in an origami-inspired FSS under normal incidence. *Progr. Electromagn. Res. M* 63: 131–139.

30 Hayes, G.J., Liu, Y., Genzer, J. et al. (2014). Self-folding origami microstrip antennas. *IEEE Trans. Antennas Propag.* 62 (10): 5416–5419.

31 Liu, X., Yao, S., Cook, B.S. et al. (2015). An origami reconfigurable axial-mode bifilar helical antenna. *IEEE Trans. Antennas Propag.* 63 (12): 5897–5903.

32 Yao, S., Bao, K., Liu, X., and Georgakopoulos, S.V. (2017). Tunable UHF origami spring antenna with actuation system. *Proceedings of the 2017 IEEE International Symposium on Antennas and Propagation & USNC/URSI National Radio Science Meeting*, San Diego, CA (9–14 July 2017), 325–326.

33 Liu, X., Yao, S., and Georgakopoulos, S.V. (2017). A frequency tunable origami spherical helical antenna. *Proceedings of the 2017 IEEE International Symposium on Antennas and Propagation & USNC/URSI National Radio Science Meeting*, San Diego, CA (9–14 July 2017), 1361–1362.

34 Hussain Shah, S.I. and Lim, S. (2017). Frequency switchable origami magic cube antenna. *Proceedings of the 2017 IEEE Asia Pacific Microwave Conference (APMC)*, Kuala Lumpar (13–16 November 2017), 105–107.

35 Choukiker, Y.K. and Behera, S.K. (2017). Wideband frequency reconfigurable Koch snowflake fractal antenna. *IET Microwaves Antennas Propag.* 11 (2): 203–208.

36 Rayno, J.T. and Sharma, S.K. (2012). Wideband frequency-reconfigurable spirograph planar monopole antenna (SPMA) operating in the UHF band. *IEEE Antennas Wirel. Propag. Lett.* 11: 1537–1540.

37 Rayno, J.T. (2012). Design and analysis of frequency reconfigurable compact spirograph planar monopole antenna (SPMA) elements for a beam scanning array. M.Sc. Thesis. San Diego State University, San Diego, CA.

38 Christodoulou, C.G., Tawk, Y., Lane, S.A., and Erwin, R.S. (2012). Reconfigurable antennas for wireless and space applications. *Proc. IEEE* 100: 2250–2261.

39 Kulkarni, A.N. and Sharma, S.K. (2013). Frequency reconfigurable microstrip loop antenna covering LTE bands with MIMO implementation and wideband microstrip slot antenna all for portable wireless DTV media player. *IEEE Trans. Antennas Propag.* 61: 964–968.

40 Sharma, S.K., Thyagarajan, M.R., Kulkarni, A.N., and Shanmugam, B. (2013). Frequency reconfigurable compact spiral loaded planar dipole antenna. *Microwave Opt. Technol. Lett.* 55: 313–316.

41 Liu, J., Li, J., and Xu, R. (2018). Design of very simple frequency and polarisation reconfigurable antenna with finite ground structure. *Electron. Lett.* 54 (4): 187–188.

42 Nguyen-Trong, N., Hall, L., and Fumeaux, C. (2016). A frequency- and pattern-reconfigurable center-shorted microstrip antenna. *IEEE Antennas Wirel. Propag. Lett.* 15: 1955–1958.

43 Zainarry, S.N.M., Nguyen-Trong, N., and Fumeaux, C. (2018). A frequency- and pattern-reconfigurable two-element array antenna. *IEEE Antennas Wirel. Propag. Lett.* 17 (4): 617–620.

44 Hum, V. and Xiong, H.Y. (2010). Analysis and design of a differentially-fed frequency agile microstrip patch antenna. *IEEE Trans. Antennas Propag.* 58: 3122–3130.

45 Li, H., Xiong, J., Yu, Y., and He, S. (2010). A simple compact reconfigurable slot antenna with a very wide tuning range. *IEEE Trans. Antennas Propag.* 58: 3725–3728.

46 Korosec, T., Ritosa, P., and Vidmar, M. (2006). Varactor-tuned microstrip-patch antenna with frequency and polarisation agility. *Electron. Lett.* 42 (18): 1015–1016.

47 White, C.R. and Rebeiz, G.M. (2009). Single- and dual-polarized tunable slot-ring antennas. *IEEE Trans. Antennas Propag.* 57 (1): 19–26.

48 Tawk, Y., Costantine, J., and Christodoulou, C.G. (2012). A varactor based reconfigurable filtenna. *IEEE Antennas Wirel. Propag. Lett.* 11: 716–719.

49 Meng, F., Sharma, S.K., and Babakhani, B. (2016). A wideband frequency agile fork-shaped microstrip patch antenna with nearly invariant radiation patterns. *Int. J. RF Microwave Comput. Aid Eng.* 26: 623–632.

50 Babakhani, B. and Sharma, S. (2015). Wideband frequency tunable concentric circular microstrip patch antenna with simultaneous polarization reconfiguration. *IEEE Antennas Propag. Mag.* 57: 203–216.

51 Zhou, L., Sharma, S.K., and Kassegne, S.K. (2008). Reconfigurable microstrip rectangular loop antennas using RF MEMS switches. *Microw. Opt. Technol. Lett.* 50: 252–256.

52 Al-Dahleh, R., Shafai, C., and Shafai, L. (2004). Frequency-agile microstrip patch antenna using a reconfigurable MEMS ground plane. *Microw. Opt. Technol. Lett.* 43: 64–67.

53 Latif, S.I., Abadi, M.S.H., Shafai, C., and Shafai, L. (2014). Development of adaptive structures incorporating MEMS devices to be used as reflectarrays or transmitarrays. *Microw. Opt. Technol. Lett.* 56: 935–938.

54 Safari, M., Shafai, C., and Shafai, L. (2015). X-band tunable frequency selective surface using MEMS capacitive loads. *IEEE Trans. Antennas Propag.* 63: 1014–1021.

55 Shafai, C., Shafai, L., Al-Dahleh, R., Chrusch, D.D., and Sharma, S.K. (2005). Reconfigurable ground plane membranes for analog/digital microstrip phase shifters and frequency agile antenna. *Proceedings of the 2005 International Conference on MEMS, NANO, and Smart Systems (ICMENS)*, Banff, AB(24–27 July 2005), 287–289.

56 Gibson, J.S., Liu, X., Georgakopoulos, S.V. et al. (2016). Reconfigurable antennas based on self-morphing liquid crystalline elastomers. *IEEE Access* 4: 2340–2348.

57 Correas-Serrano, D. and Gomez-Diaz, J.S. (2017). Graphene-based antennas for terahertz systems: a review. *Forum Electromagn. Res. Methods Appl. Tech. (FERMAT)* 20: 1–2.

58 Correas-Serrano, D., Gomez-Diaz, J.S., Alù, A., and Álvarez Melcón, A. (2015). Electrically and magnetically biased graphene-based cylindrical waveguides: analysis and applications as reconfigurable antennas. *IEEE Trans. Terahertz Sci. Technol.* 5 (6): 951–960.

59 Azizi, M.K., Ksiksi, M.A., Ajlani, H., and Gharsallah, A. (2017). Terahertz graphene-based reconfigurable patch antenna. *Progr. Electromagn. Res. Lett.* 71: 69–76.

60 Núñez Álvarez, C., Cheung, R., and Thompson, J.S. (2017). Performance analysis of hybrid metal-graphene frequency reconfigurable antennas in the microwave regime. *IEEE Trans. Antennas Propag.* 65 (4): 1558–1569.

61 Savelev, R.S., Sergaeva, O.N., Baranov, D.G. et al. (2017). Dynamically reconfigurable metal-semiconductor Yagi-Uda nanoantenna. *Phys. Rev. B* 95 (23): 235409.

62 Dong, Y., Liu, P., Yu, D. et al. (2016). Dual-band reconfigurable terahertz patch antenna with graphene-stack-based backing cavity. *IEEE Antennas Wirel. Propag. Lett.* 15: 1541–1544.

63 Scidà, A., Haque, S., Treossi, E. et al. (2018). Application of graphene-based flexible antennas in consumer electronic devices. *Mater. Today* 21: 223–230.

64 Mehdipour, A., Denidni, T.A., Sebak, A.-R. et al. (2013). Mechanically reconfigurable antennas using an anisotropic carbon-fibre composite ground. *IET Microwaves Antennas Propag.* 7: 1055–1063.

65 Boukarkar, A., Lin, X.Q., Jiang, Y. et al. (2018). Compact mechanically frequency and pattern reconfigurable patch antenna. *IET Microwaves Antennas Propag.* 12: 1864–1869.

66 Jang, T., Zhang, C., Youn, H. et al. (2017). Semitransparent and flexible mechanically reconfigurable electrically small antennas based on tortuous metallic micromesh. *IEEE Trans. Antennas Propag.* 65 (1): 150–158.

67 Sarabia, K.J., Yamada, S.S., Moorefield, M.R. et al. (2018). Frequency-reconfigurable dipole antenna using liquid-metal pixels. *Int. J. Antennas Propag.* 2018: Article ID 1248459, 6 pages.

68 Nassar, I.T., Tsang, H., Bardroff, D. et al. (2015). Mechanically reconfigurable, dual-band slot dipole antennas. *IEEE Trans. Antennas Propag.* 63 (7): 3267–3271.

69 Yao, S., Liu, X., and Georgakopoulos, S.V. (2017). Morphing origami conical spiral antenna based on the Nojima wrap. *IEEE Trans. Antennas Propag.* 65: 2222–2232.

70 Werner, D. and Ganguly, S. (2003). An overview of fractal antenna engineering research. *IEEE Antennas Propag. Mag.* 45 (1): 38–57.

71 Shirazi, M., Li, T., Huang, J., and Gong, X. (2018). A reconfigurable dual-polarization slot-ring antenna element with wide bandwidth for array applications. *IEEE Trans. Antennas Propag.* 66 (11): 5943–5954.

72 Varamini, G., Keshtkar, A., and Naser-Moghadasi, M. (2018). Compact and miniaturized microstrip antenna based on fractal and metamaterial loads with reconfigurable qualification. *AEU Int. J. Electron. Commun.* 83: 213–221.

73 Ali, T., Fatima, N., and Biradar, R.C. (2018). A miniaturized multiband reconfigurable fractal slot antenna for GPS/GNSS/Bluetooth/WiMAX/X-band applications. *AEU Int. J. Electron. Commun.* 94: 234–243.

74 Tripathi, S., Mohan, A., and Yadav, S. (2017). A compact frequency-reconfigurable fractal UWB antenna using reconfigurable ground plane. *Microw. Opt. Technol. Lett.* 59: 1800–1808.

75 Tripathi, S., Mohan, A., and Yadav, S. (2016). A compact fractal UWB antenna with reconfigurable band notch functions. *Microw. Opt. Technol. Lett.* 58: 509–514.

76 Reddy, V.V. and Sarma, N.V.S.N. (2015). Circularly polarized frequency reconfigurable Koch antenna for GSM/Wi-Fi applications. *Microw. Opt. Technol. Lett.* 57: 2895–2898.

77 Kang, S. and Jung, C.W. (2015). Dual band and beam-steering antennas using reconfigurable feed on Sierpinski structure, *Int. J. Antennas Propag.* vol. 2015, Article ID 492710, 1–8.

78 Imen, B.T., Rmili, H., Floch, J.M. et al. (2015). Planar square multiband frequency reconfigurable micro-strip fed antenna with quadratic Koch-Island fractal slot for wireless devices. *Microw. Opt. Technol. Lett.* 57 (1): 207–212.

79 Li, L., Wu, Z., Li, K. et al. (2014). Frequency-reconfigurable quasi-Sierpinski antenna integrating with dual-band high-impedance surface. *IEEE Trans. Antennas Propag.* 62 (9): 4459–4467.

80 Gao, S., Rahmat-Samii, Y., Hodges, R.E., and Yang, X. (2018). Advanced antennas for small satellites. *Proc. IEEE* 106 (3): 391–403.

81 Zhang, M.-T., Gao, S., Jiao, Y.-C. et al. (2016). Design of novel reconfigurable reflectarrays with single-bit phase resolution for Ku-band satellite antenna applications. *IEEE Trans. Antennas Propag.* 64 (5): 1634–1641.

82 Padilla, J., Rosati, G., Ivanov, A. et al. (2011). Multi-functional miniaturized slot antenna system for small satellites. *Proceedings of the EuCAP*, Rome, 2170–2174.

83 Hannula, J.-M., Holopainen, J., and Viikari, V. (2017). Concept for frequency reconfigurable antenna based on distributed transceivers. *IEEE Antennas Wirel. Propag. Lett.* 16: 764–767.

84 Hannula, J.-M., Saarinen, T., Holopainen, J., and Viikari, V. (2017). Frequency reconfigurable multiband handset antenna based on a multichannel transceiver. *IEEE Trans. Antennas Propag.* 65 (9): 4452–4460.

85 Hannula, J.M., Kosunen, M., Lehtovuori, A. et al. (2018). Performance analysis of frequency-reconfigurable antenna cluster with integrated radio transceivers. *IEEE Antennas Wirel. Propag. Lett.* 17 (5): 756–759.

86 Wu, C., Yang, Z., Li, Y. et al. (2018). Methodology to reduce the number of switches in frequency reconfigurable antennas with massive switches. *IEEE Access* 6: 12187–12196.

87 Pourghorban Saghati, A., Singh Batra, J., Kameoka, J., and Entesari, K. (2016). A microfluidically reconfigurable dual-band slot antenna with a frequency coverage ratio of 3:1. *IEEE Antennas Wirel. Propag. Lett.* 15: 122–125.

88 Bhattacharjee, T., Jiang, H., and Behdad, N. (2016). A fluidically tunable, dual-band patch antenna with closely spaced bands of operation. *IEEE Antennas Wirel. Propag. Lett.* 15: 118–121.

89 Ghosh, S. and Lim, S. (2018). Fluidically reconfigurable multifunctional frequency-selective surface with miniaturization characteristic. *IEEE Trans. Microwave Theory Tech.* 66 (8): 3857–3865.

3

Radiation Pattern Reconfigurable Antennas

Sima Noghanian and Satish K. Sharma

3.1 Introduction

The pattern reconfigurable antennas are a major group under reconfigurable antennas. By pattern change we mean either radiation pattern main beam movement or scanning. In this process, the pattern polarization can also be changed. For example, one state of the antenna's function might be in linear polarization and one in circular polarization (CP). The change in pattern might happen by switching or scanning the beam.

Like other types of reconfigurable antennas, the switching between states can be done in multiple ways. Some of these include the use of electronically controlled switches such as PIN diodes or metal–semiconductor field-effect transistor (MESFET) transistors, micro-electro-mechanical system (MEMS) switches, photonic switches, band gap structures, and the mechanical changes in the antenna structure. The changes might be applied to different sections of a single element antenna, parasitic parts of the antenna, or to an array of elements. Under array antenna category, we might also consider changes in the reflect array elements. In this chapter, some examples of various designs and methods of implementation of pattern reconfigurability will be discussed.

3.2 Pattern Reconfigurable by Electronically Changing Antenna Elements

In [1], a meta-surface is proposed to provide radiation pattern and polarization reconfigurability. The antenna and meta-surface consist of three conducting layers and two dielectric layers. The first layer is the meta-surface layer as shown in Figure 3.1a. The second and third layers form a slot antenna. The first dielectric

Multifunctional Antennas and Arrays for Wireless Communication Systems, First Edition.
Edited by Satish K. Sharma and Jia-Chi S. Chieh.

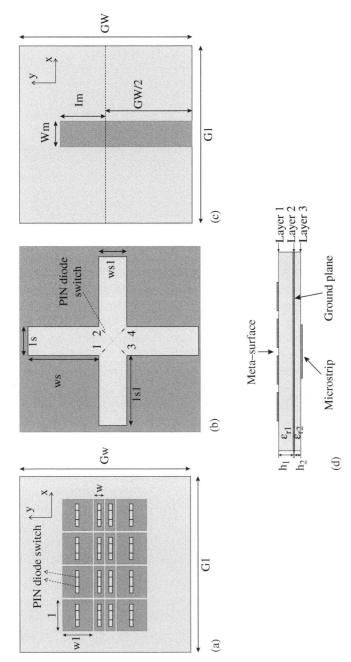

Figure 3.1 Different layers of meta-surface reconfigurable slot antenna [1]: (a) Layer 1, (b) layer 2, (c) layer 3, and (d) layer 4; dimensions in mm: L = 14, W = 4, W_1 = 13.8, g = 0.9, G_m = G_l = 78, W_f = 1.8, L_{f1} = 10, L_{f2} = 4, L_s = 3, W_s = 7.03, L_{s1} = 7.13, W_{s1} = 3, L_m = 20.6, W_m = 10.8. *Source:* Chen et al. [1]. © 2018, IEEE.

layer between conducting layers 1 and 2 is Arlon DiClad 880, with a dielectric constant of 2.2. The second dielectric layer between conducting layers 2 and 3 is Arlon 25N with a dielectric constant of 3.38. The meta-surface is a matrix of 4×4 conducting patches with slots. PIN diodes are used as switches to control the slot lengths. By changing the slot lengths on the meta-surface layer, the main beam can be moved to different angles.

The middle layer acts as a ground with slots. PIN diodes are located along the metallic strips that connect the slots. By turning ON/OFF the PIN diode, it is possible to change the polarization between right- and left-handed circular polarizations (RHCP/LHCP). Each PIN diode was simulated as a 3Ω resistor for the ON state and $0.025 \, pF$ capacitor for the OFF state.

By switching the slot lengths on the meta-surface, the main beam direction can move from $-20°$ to $20°$ around the z-axis. The left- and right-hand circular polarization (LHCP/RHCP) was also possible. This antenna was designed to operate at $5 \, GHz$. The main lobe gain was between 7 and 8 dBi.

A similar design is suggested in [2], where a dual slot antenna is also controlled with PIN diodes for changing the radiation pattern and polarization. However, the antenna is placed on a cavity that is formed by a substrate-integrated-waveguide (SIW) cavity, as shown in Figure 3.2. The slots are fed on the same plane using a grounded coplanar waveguide (CPW). The microstrip line is added at the end of the feed-line to provide a means of connecting the antenna to the test equipment. The cavity at the back is made by an array of

Figure 3.2 Cavity-backed slot antenna controlled by PIN diodes. *Source:* Ge et al. [2]. © 2017, IEEE.

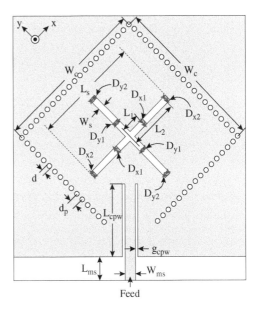

Table 3.1 Dimensions of polarization reconfigurable cavity-backed slot antenna, all in mm.

Parameter	L_{ms}	W_{ms}	L_{cpw}	g_{cpw}	D	d_p
Value	10	4.7	30	1	2	3.3
Parameter	W_c	L_s	W_s	L_1	L_2	
Value	66	46	3	8	22	

Source: Ge et al. [2]. © 2017, IEEE.

Table 3.2 Different polarization states.

Polarization	Frequency	D_{x1}	D_{x2}	D_{x3}	D_{y1}	D_{y2}	D_{y3}
xoz-LP	Low	ON	ON	ON	OFF	OFF	OFF
	Middle	ON	ON	ON	OFF	OFF	ON
	High	ON	ON	ON	OFF	ON	ON
yoz-LP	Low	OFF	OFF	OFF	ON	ON	ON
	Middle	OFF	OFF	ON	ON	ON	ON
	High	OFF	ON	ON	ON	ON	ON
LHCP	Low	OFF	OFF	ON	OFF	OFF	OFF
	High	OFF	ON	ON	OFF	OFF	ON
RHCP	Low	OFF	OFF	OFF	OFF	OFF	ON
	High	OFF	OFF	ON	OFF	ON	ON

Source: Ge et al. [2]. © 2017, IEEE.

metallized vias on a Rogers 5870 substrate with the thickness of 1.575 mm, and a dielectric constant of $\varepsilon_r = 2.33$. The cavity is added to reduce the back radiation and have a directional pattern. The crossed slot is bridged by four pairs of switches. The antenna is capable of radiating in two different linear polarized (xz and yz, slots are in x- and y-directions), right-handed and left-handed circular polarized fields. Table 3.1 provides the dimensions of the antenna and Table 3.2 summarizes the states of polarization.

In the antenna design proposed in [3], a dipole antenna is surrounded by unit parasitic cells in a hexagonal arrangement. The unit cell is shown in Figure 3.3a.

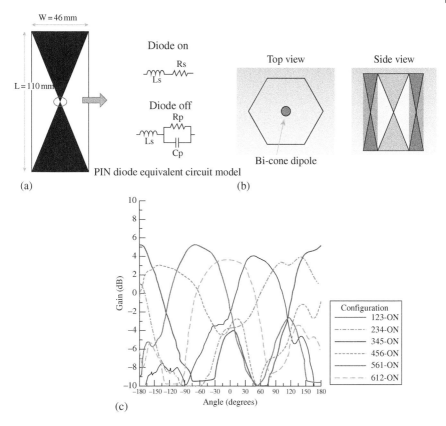

Figure 3.3 (a) Unit cells used on each wall and the PIN diode equivalent circuit, ON state: R_s = 2.1 Ω, L_s = 0.6 nH; OFF state: C_p = 0.17 pF and R_p = 3000 Ω; (b) top and side view of the antenna structure; (c) beam switching on the azimuth plane.
Source: Tsai et al. [3]. © 2012, IEEE.

The bow-tie is acting as a frequency selective surface (FSS). The arrangement is shown in Figure 3.3b. When the PIN diode is ON, the FSS surface is transparent. When the diode is OFF, the surface is opaque. The sample of operation when three switches are ON and three are OFF is shown in Figure 3.3c. The system operates at 2.5 GHz.

A similar design in [4] is presented where a dipole antenna is surrounded by six parasitic elements. The main element in this design is a dipole antenna working at 4.8–5.2 GHz. The parasitic elements are placed on a 3D structure built using 3D printing of acrylonitrile butadiene styrene (ABS) with relative permittivity of 2.1

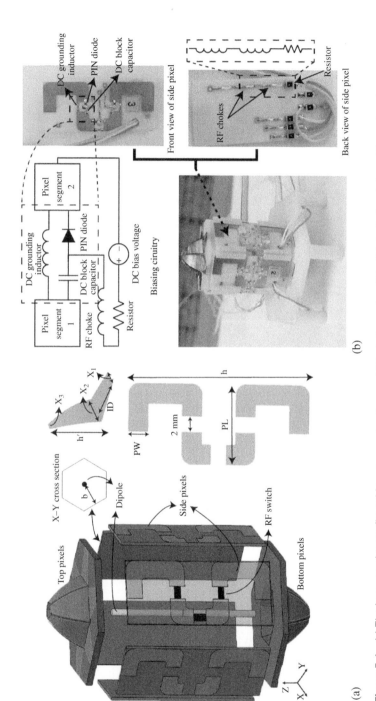

Figure 3.4 (a) The hexagonal reconfigurable antenna structure and (b) photographs of the fabricated prototype. Dimensions in mm: $b = 10$, $P_L = 10$, $P_W = 2.5$, $h = 23$, $h' = 5.5$, $ID = 5$, $X_1 = 2.5$, $X_2 = 5$, and $X_3 = 1$. *Source:* Hossain et al. [4]. © 2017, IEEE.

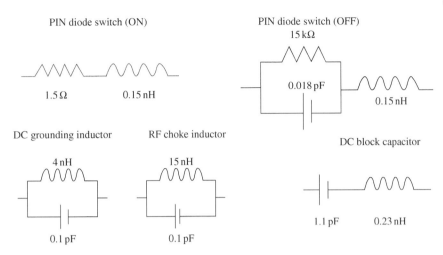

Figure 3.5 Equivalent circuit model of the lumped components and PIN diode. *Source:* Hossain et al. [4]. © 2017, IEEE.

and loss tangent of 0.05. The parasitic elements were printed on rectangular layers of microwave laminates with relative permittivity of 3.35 and loss tangent of 0.0025 and thickness of 0.5 mm (Figure 3.4). The top and bottom are covered by hexagonal domes. The reconfigurable pattern over the horizon ($0° < \phi < 360°$) is controlled by the parasitic elements on the rectangular sides, while the pattern can scan the elevation angle of $-18° < \theta < 18°$, using the top and bottom pixels. All pixels were connected by DC grounding inductors. PIN diodes were used to connect the parts of parasitic elements (pixels). DC-blocking capacitors were added between the pixel and PIN diodes. The diodes were MA4AG910 form MACOM. Figure 3.4b shows the biasing circuit and Figure 3.5 shows the equivalent circuits.

Figure 3.6 shows four modes of operation and the simulated radiation patterns of each. When all the switches are OFF, the parasitic elements are transparent; therefore, pattern is omnidirectional (Mode 1). To direct the beam to different horizontal angles, the pixels that are diametrically along the direction of the main beam are all in the ON state to create reflectors for the dipole (Mode 2). To create an elevation tilted beam, the reflector is connected to the adjacent top or bottom pixel of the diametrically opposite side (Mode 3). A diversity mode (Mode 4) is created by turning on combining the switches of three types of patterns mentioned before. Figure 3.7 shows patterns at 5 GHz with various beam tilting and beam steering.

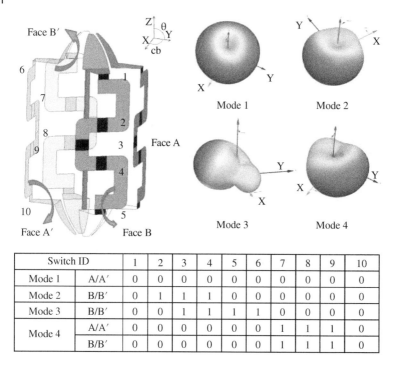

Switch ID		1	2	3	4	5	6	7	8	9	10
Mode 1	A/A′	0	0	0	0	0	0	0	0	0	0
Mode 2	B/B′	0	1	1	1	0	0	0	0	0	0
Mode 3	B/B′	0	0	1	1	1	1	0	0	0	0
Mode 4	A/A′	0	0	0	0	0	0	1	1	1	0
	B/B′	0	0	0	0	0	0	1	1	1	0

Figure 3.6 Switch configuration for four different modes of operation, "0" means the OFF state of the switch and "1" indicates ON state. *Source:* Hossain et al. [4]. © 2017, IEEE.

A beam steering Yagi antenna with a similar concept of parasitic elements is reported in [5]. In this design, shown in Figure 3.8, a planar dipole antenna is fed through a CPW. Directors (parasitic elements) are placed on two sides of the dipole to direct the beam to either −20° or +20°. Therefore, the antenna can work in three states. The diode was modeled in the simulation program (CST Microwave Studio) as a 4 Ω resistor for the ON state and a parallel circuit of a capacitor of 0.04 pF and a resistor of 20 Ω for the OFF state. Simulated and measured E-plane radiation patterns for the three states at 5 GHz are shown in Figure 3.9. The H-plane pattern remains the same as the states are changed. The main beam pointing angle for 5 GHz were measured 20° and −17°, while the simulation was showing 17° and −15°. The antenna showed small gain variation between the three states. The measured realized gains for the three states were 7.6–9.03, 7.5–9.38, and 8–10.03 dBi, respectively, for the frequency range of 5–5.2 GHz.

A similar concept was used to switch the beam using a slot configuration [6]. The radiating slot was located at the center and two non-radiating parasitic slots are placed at the two sides of the slot, which can be loaded with open or short-circuited stubs. The radiating slot was fed through a microstrip line. The slots and feed line were fabricated on a substrate with the dielectric constant of 2.55, loss tangent of

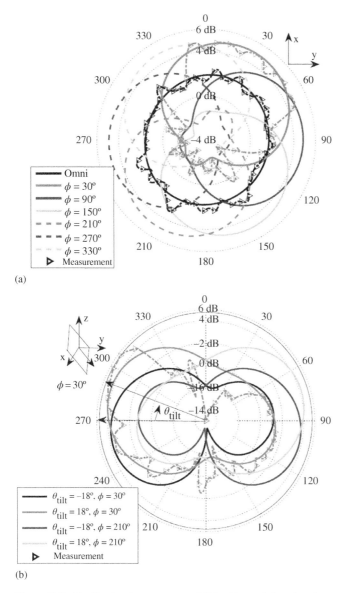

Figure 3.7 Realized gain patterns at 5 GHz: (a) omnidirectional patterns at different azimuth steering beams and (b) patterns at different elevation tilts. *Source:* Hossain et al. [4]. © 2017, IEEE.

0.0022, and thickness of 0.76 mm (Figure 3.10). To find the optimized reactive load for the parasitic elements, [6] presents a modeling method. The substrate and ground plane was 50×50 mm^2. The radiating and parasitic slot sizes were 2.2×17 mm^2 and 1.6×14.7 mm^2, respectively. By using the developed model, two

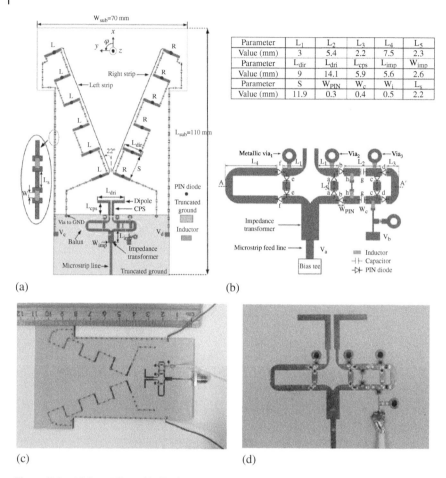

Figure 3.8 (a) Reconfigurable Yagi antenna, (b) reconfigurable balun, (c) fabricated Yagi antenna, and (d) fabricated dipole and balun. *Source:* Qin et al. [5]. © 2013, IEEE.

states of reactive load for switching the beams, from +25° and +155° to −25° and −155°, were found to be 176 and −209 Ω. This reactive load was realized by using a stub and MEMS switches. The overall lengths of stubs were 6.6 and 3 mm. The MEMS switches were modeled as a 2 Ω resistor in the ON state and 5 fF capacitor in the OFF state. Figure 3.11 shows the simulated and measured radiation patterns at 5.6 GHz. The gain at +25° was measured at 5.7 dBi and for −25° it was 5.4 dBi.

A simple design of a dipole or loop antenna on a planar configuration that incorporates a parasitic element in a reflector or director mode is presented in [7]. Figure 3.12 illustrates the antenna arrangement. Figure 3.12a shows a planar monopole and as the switches are turned ON, the parasitic elements will act as directors and change the direction of the main lobe. In Figure 3.12b, a similar concept is used to make a planar loop with reflectors. Figure 3.12c and d shows how the pattern gets affected by turning ON/OFF the switches.

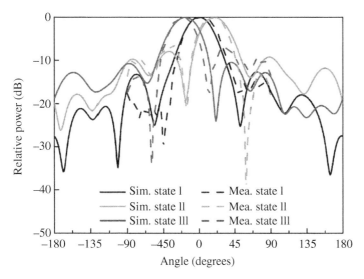

Figure 3.9 Simulated and measured E-plane normalized radiation patterns at 5.0 GHz. *Source:* Qin et al. [5]. © 2013, IEEE.

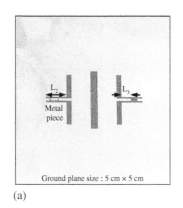

(a)

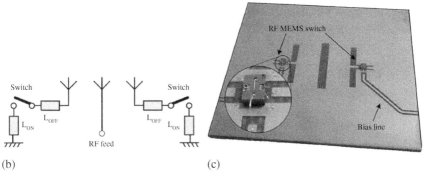

(b)   (c)

Figure 3.10 MEMS-based parasitic antenna: (a) photograph of the antenna without switches, (b) switched-load concept for reconfigurability, and (c) photograph of the reconfigurable antenna with MEMS switches. *Source:* Petit et al. [6]. © 2006, IEEE.

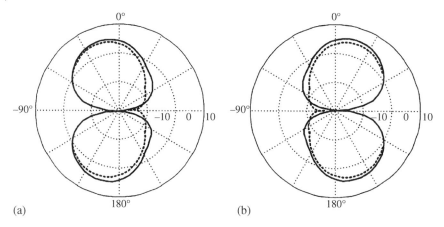

Figure 3.11 Simulated (solid) and measured (dashed) radiation patterns at 5.6 GHz for the two states of the MEMS switches. *Source:* Petit et al. [6]. © 2006, IEEE.

Authors of [7] suggest using the PIN and FET switches. PIN switch in forward bias model has a resistor of 3 Ω. In the reverse bias, a parallel combination of 25 kΩ resistor and 0.12 pF capacitor was used to model the PIN diode switch. The FET switch in the ON state was modeled as an inductor of 2.6 nH and a resistor of 4.3 Ω. In the OFF state, the resistor was replaced by a parallel circuit of a capacitor of 0.49 pF and a resistor of 5.9 kΩ. Figure 3.13 shows the simulated and measured patterns for both antennas by using either PIN or FET switches. The antennas were fabricated on 0.8 mm-thick FR4 with relative permittivity of 4.6 and loss tangent of 0.02. Switches were Microsemi MPP4203 or Agilent ATF-34143. In general, switches will reduce the efficiency of the antenna. Table 3.3 compares the maximum gain and front-to-back ratio (FBR) of these antennas.

Another planar Yagi-Uda antenna geometry along with design parameters is shown in Figure 3.14, where PIN diodes were used to reconfigure radiation patterns in [8]. The feed element is a planar dipole with dipole arms positioned on top and bottom of the substrate. It is excited by a 50 Ω SMA connector. A 130 mm × 130 mm size Arlon AD 250 ($\varepsilon_r = 2.5$) substrate of h = 1.6 mm is used. The antenna operates around 1.5 GHz. The dimensions selected are as follows: $L_0 = 14.5$ mm, $L_1 = 89.17$ mm, $L_2 = 81$ mm, $L_3 = 50.97$ mm, $L_4 = 48.44$ mm, $L_5 = 27.7$ mm and, for all elements, W = 4.51 mm. The PIN diodes used were the SMP1340-040LF. When the diode is in the ON state, it is modeled as an inductor (L) and resistor (R_s) in series. When the diode is modeled in the OFF state, the model uses an inductor (L) in series with a parallel high-value resistor (R_p) and capacitor (CT). The component values are L = 0.45 nH, $R_S = 1.2$ Ω, $R_P = 5$ MΩ, and CT = 0.14 pF. Bias lines are needed to activate PIN diodes. A biasing circuit with a blocking capacitor and choke inductor was placed between the gaps to avoid DC currents and interferences to the RF circuit. The component values for a RF biasing network are: L = 2.2 nH,

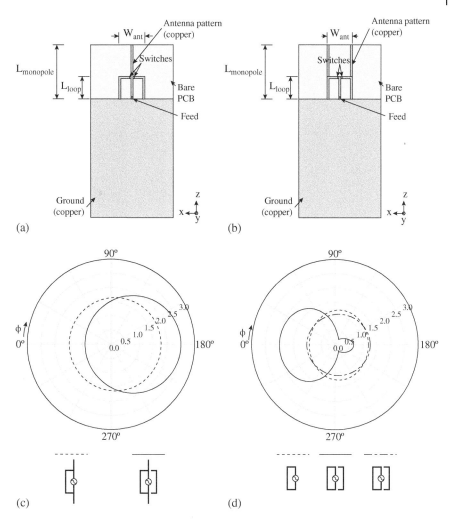

Figure 3.12 (a) Planar monopole antenna with directors, (b) planar loop with reflectors, (c) dipole-loaded loop and open wire, and (d) a loop and open wire. *Source:* Kang et al. [7]. © 2012, EMW Publishing.

CB = 15 pF, and R = 47 Ω, which were determined after careful simulations so that it does not affect antenna performance adversely. Figure 3.14b shows the photograph of the fabricated antenna. PIN diodes and biasing networks were realized within the antenna using surface-mount component soldering.

Yagi-Uda array antenna works as an end-fire pattern antenna that uses mutual coupling between the elements (director, reflector, and fed dipole) to produce a unidirectional pattern. Reconfigurable pattern characteristics achieved in [8] consist of changing the beam peak direction to opposite sides in the end-fire mode of radiation

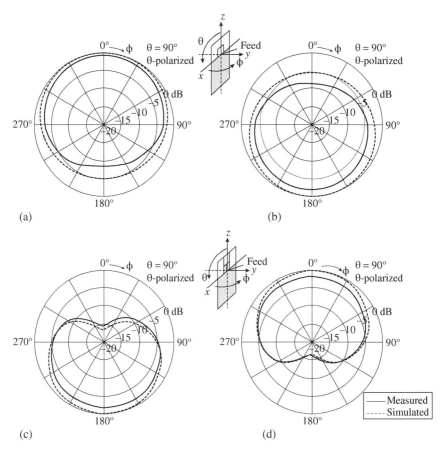

Figure 3.13 Simulated and measured gain patterns of the antennas (this page: PIN diode switch, next page: FET switches), (a) director-type, forward bias; (b) director-type, reverse bias; (c) reflector-type, forward bias; (d) reflector-type, reverse bias. *Source:* Kang et al. [7]. © 2012, EMW Publishing.

while maintaining the matching bandwidth around 10%. This was achieved by changing the electrical lengths of the parasitic elements (reflector and director) and spacing between the fed element and reflector and director so that the reflector and director are correctly positioned which generate end-fire mode radiation patterns. PIN diodes were inserted between the gaps in the reflector and director elements to control the radiated beam direction when they are switched ON or OFF states. When diodes 1 and 2 are switched ON and diodes 3 and 4 are in OFF states, it is defined as "side 1" and the beam is focused onto –x axis direction (Figure 3.15a). On the other hand, when diodes 3 and 4 are activated in ON states and diode 1 and 2 are in OFF states, it is defined as "side 2" and the beam is directed to +x direction (Figure 3.15b). In either case, antenna geometry also includes additional parasitic elements on

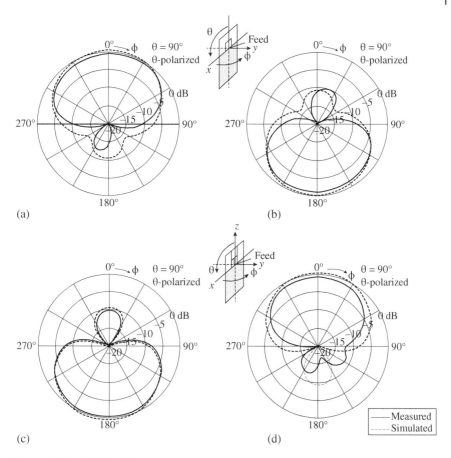

Figure 3.13 *(Continued)*

Table 3.3 Measured maximum gain and FBR of antennas with PIN diodes and FET switches.

Antenna type	Switch type	Max gain (dBi)		FBR (dB)	
		Bias 1	Bias 2	Bias 1	Bias 2
Director	PIN	1.18	0.18	7.78	6.55
	FET	0.81	0.97	13.00	10.63
Reflector	PIN	0.22	−0.03	14.48	17.39
	FET	0.11	−0.29	9.63	13.42

Source: Kang et al. [7]. © 2012, EMW Publishing.
Bias 1, forward bias (PIN diodes), FET1 ON (FET's); Bias 2, reverse bias (PIN diodes), FET2 ON (FET's).

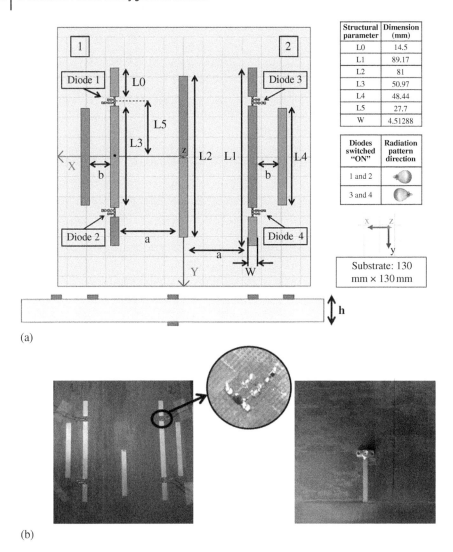

(a)

(b)

Figure 3.14 (a) Geometry of the reconfigurable planar Yagi-Uda antenna, including coordinate system, on a 1.6 mm-thick substrate material: top view and side view. (b) Photograph of the fabricated antenna: front view, surface-mounted PIN diode and biasing network and back view showing one dipole arm fed-in using a 50 Ω SMA. *Source:* Sharma et al. [8].

either sides of the antenna which assists in meeting the beam peak direction to end-fire mode. When all diodes are in ON or in OFF states, patterns are like in a dipole antenna (Figure 3.15c, d). The simulated and measured results for the impedance matching are shown in Figure 3.15e. The simulated and measured matching bandwidths ($S_{11} \leq -10$ dB) are from around 1.44–1.58 GHz which is in good agreement

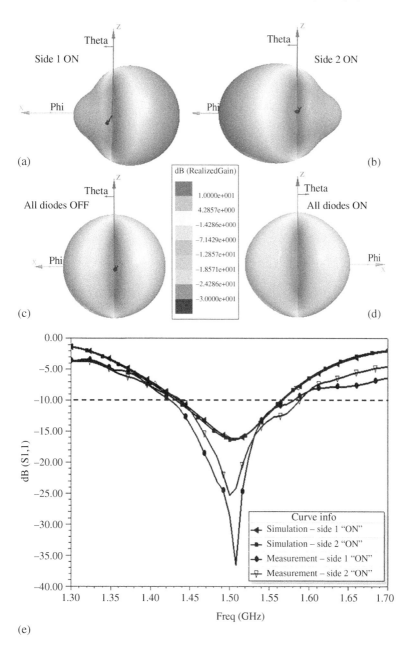

Figure 3.15 Modes of operation of the proposed Yagi-Uda planar antenna when (a) side 1 is ON, (b) side 2 is ON, (c) all diodes are in ON state, (d) all diodes are in OFF state, and (e) comparison of the simulated and measured reflection coefficient magnitudes when sides 1 and 2 are in the ON state. *Source:* Sharma et al. [8].

(Figure 3.15e). When side 1 is in the ON state, the end-fire beam focuses in −xz direction and when side 2 is in the ON state, the beam focuses on +xz direction.

Figure 3.16a–d presents the simulated normalized radiation patterns for the both sides 1 and 2 for the both elevation planes (xz-plane, Figure 3.16a, b) and azimuthal plane (xy-plane, Figure 3.16c, d) all for 1.45, 1.5, and 1.56 GHz within the matching bandwidth. Measured normalized radiation patterns are shown in Figure 3.17a–d for the elevation plane (+xz and −xz planes) and azimuthal plane (xy plane with beam peaks in $\phi = 0°$ and 180° directions) for 1.45, 1.50, and 1.56 GHz for both cases: side 1 and side 2. From Figure 3.17a and b, it can be observed that as expected end-fire beam peaks are located toward $\theta = -90°$ and 90° and FBRs are around 20, 16, and 12 dB at 1.45, 1.50, and 1.56 GHz, respectively. Similarly, Figure 3.17c and d shows the azimuthal plane patterns with acceptable FBR as found earlier. Measured patterns (Figure 3.17a, b) show that, beam peak is located toward around $\theta = -90°$ and 90° with some asymmetry in the pattern quality. Beam peaks are shifted by 5°–10° away from the expected angles hence backlobes are also shifted by the same angles. Similarly, Figure 3.17c and d shows azimuthal or xy-plane measured normalized patterns with beam peak asymmetry. Beam peaks are shifted by 20° away from the expected angles hence backlobes are also shifted by the same angles. FBR is also not the same as expected from the simulations. These are attributed to possible misalignment errors present during testing of the antenna and scattering effects from RF cables and RF bias network wires including battery placements for turning ON and OFF the PIN diodes. While in simulations, bias wires were assumed straight, in measurement, this could not be maintained. Thus, there were more than one unknown than the simulated results. Measured FBR is changing between 9 and 15 dB which is inferior to the simulated data. The measured gain is varying between 6 and 7.2 dBi. Overall, the measured patterns verify the end-fire beam peak reconfiguration from $-90° \leq \theta \leq 90°$.

In addition to PIN diodes or MEMS switches, other type of switches may be used to alter the antenna configuration. An example of photoconducting switch to reconfigure frequency or pattern of a dipole antenna is presented in [9]. In this design, two silicon photo switches are connecting or disconnecting parts of a dipole fed at the center (Figure 3.18). The dipole is fed through a CPW that is then transformed to a coplanar stripline (CPS) via a circular balun. An infrared laser diode was used to turn ON and OFF the switches. The light was transmitted to the switches through fiber-optic cables. The dipole was printed on a substrate with 1.17 mm thickness and permittivity of 2.2 (TLY-5), and it was fed through an SMA connector. The length of the dipole should be half of effective wavelength (λ_{eff}), that is free-space wavelength divided by square root of effective permittivity ($\lambda/\sqrt{\varepsilon_{eff}}$). The length of dipole includes the parts added via switches was 62.3 mm. ε_{eff} was around 1.2. Two slots located at 14.4 mm from each end were added. The silicon dice were $1 \times 1 \times 0.3$ mm^3. If switches are simultaneously turned ON or OFF, the pattern remains the same as dipole pattern, but the frequency changes

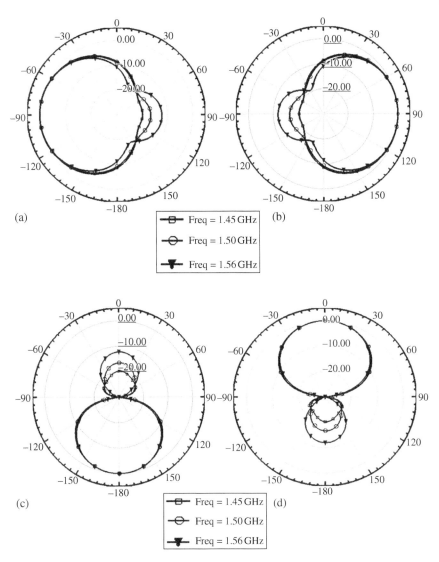

Figure 3.16 Simulated normalized radiation patterns at 1.45, 1.50, and 1.56 GHz for (a) xz-plane for side 2 ON case, (b) xz-plane for side 1 ON case, (c) xy-plane for side 1, and (d) xy-plane for side 2 ON case. *Source:* Sharma et al. [8].

according to the length of the dipole. For the ON state, the resonance frequency was 2.26 GHz and if both switches are OFF, the frequency shifts to 3.15 GHz. The patterns for these states are shown in Figure 3.19. The antenna gain for the ON state was about 2.9 dBi and for the OFF state it was 4 dBi. The pattern

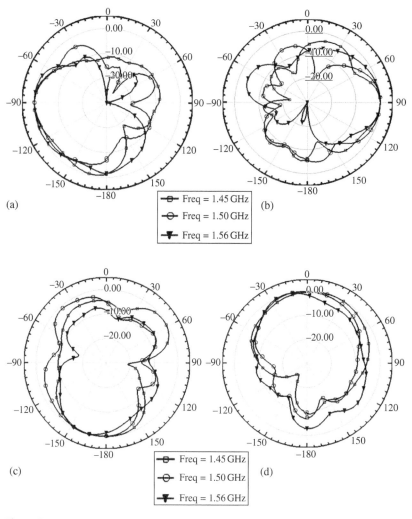

Figure 3.17 Measured normalized radiation patterns at 1.45, 1.50, and 1.56 GHz for (a) xz-plane for side 2 ON case, (b) xz-plane for side 1 ON case, (c) xy-plane for side 1, and (d) xy-plane for side 2 ON case. *Source:* Sharma et al. [8].

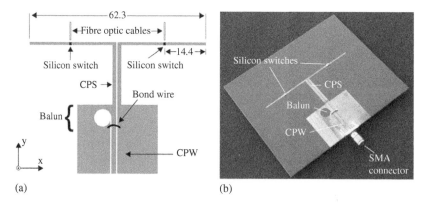

Figure 3.18 The proposed reconfigurable antenna with two photo-conductive switches on each arm: (a) top view of the structure and (b) fabricated antenna. *Source:* Panagamuwa et al. [9]. © 2006, IEEE.

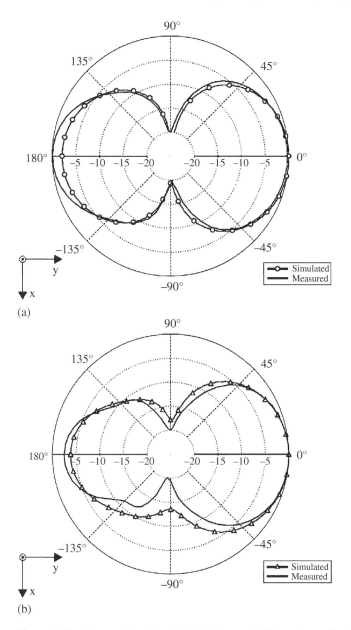

Figure 3.19 Measured and simulated E-plane radiation patterns for the photo-activated switches with both switches: (a) ON and both switches and (b) OFF. *Source:* Panagamuwa et al. [9]. © 2006, IEEE.

reconfiguration happens when only one switch is ON and the other one is OFF. The maximum angle was shifted to 12° off boresight when the right switch was ON, and 7° off boresight when the left switch was ON. Patterns are shown in Figure 3.20. In this case, the resonance frequency was around 2.6 GHz. The boresight gain was about 3.1 dBi.

3.3 Pattern Reconfigurable by Electronically Changing Feeding Network

One method of reconfiguring an antenna is through modification of the feeding of the antenna system. In this method, the antenna shape is unchanged; however, different elements or different parts of one element may be fed in various combinations to provide polarization or pattern change. For example, in [10], a three-element combination is proposed that can be switched between omnidirectional and sectorial directional patterns. The basic element of this design is a wideband planar dipole. As shown in Figure 3.21, two elements are arranged in a cross shape arrangement on the top and bottom layers of a substrate. The dipole elements are printed on the opposite sides of a FR4 substrate with relative permittivity $\varepsilon_r = 4.3$, loss tangent of 0.025, and thickness 1.6 mm. The optimized dimensions are: $U_{in} = 19$ mm, $U_{out} = 30$ mm, $V_{in} = 7$ mm, $V_{out} = 21$ mm, $S = 8$ mm, $W = 3$ mm, $L_d = 52$ mm, and $P = 40$ mm. Each sectorial antenna generates a directive pattern that can be working at a different frequency (Figure 3.22). Since the antennas are wideband, they can work at different frequencies. If all the antennas are fed with the same magnitude and phase, they provide an omnidirectional pattern (Figure 3.23).

Another example of reconfiguring the feed is presented in [11] where a reconfigurable feed network is used in an array of 2×2 patch antennas to switch between left-hand and right-hand CP (Figure 3.24). By changing the DC bias of PIN diodes in the feed network, the state of modes is changed. The design includes four circular patch antennas and three substrates (Figure 3.25). The patch antennas are located on the top layer of dielectric with the dielectric constant of 3.55, loss tangent of 0.0027, and thickness of 32 mil (Roger 4003). The top and middle dielectric layers are the same and an air layer is added to improve the bandwidth. The bottom layer that is used for the feed network has the dielectric constant of 10.2, loss tangent of 0.0035, and thickness of 50 mils (Rogers 3010). The stacked layers are connected through copper vias. The overall size of the antenna array is $154 \times 154 \times 8.8$ mm^3 ($1.28 \times 1.28 \times 0.07$ λ_0^3). The feed network consists of seven Wilkinson power dividers, four phase shifters, and 16 RF PIN diodes. Each antenna element has three feed points. The patches are capacitively coupled to small patches located under the top dielectric layer. The patches have rotational symmetry to generate CP. By generating 90° phase shift, as shown in Figure 3.26,

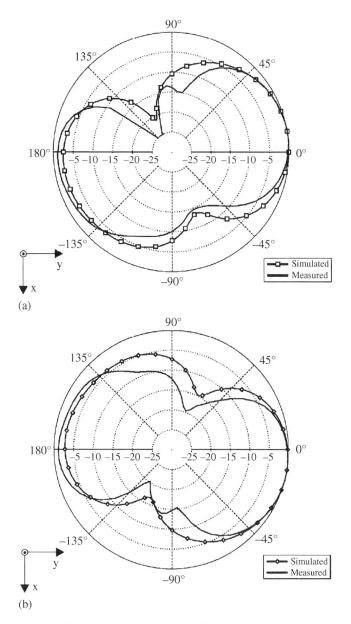

Figure 3.20 Measured and simulated E-plane radiation patterns the photo-activated switches with (a) only left switch ON and (b) only right switch ON. *Source:* Panagamuwa et al. [9]. © 2006, IEEE.

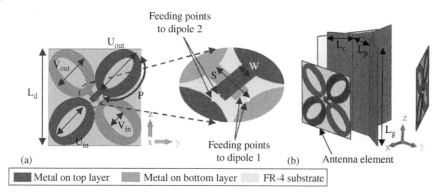

Figure 3.21 (a) Geometry of two dipole antenna and (b) sectorial design of the antenna. *Source:* Alieldin et al. [10]. © 2018, EMW Publishing.

two states of polarizations are possible. Radiation patterns for two polarizations are similar and are shown in Figure 3.27.

Feed switching is also proposed for a dielectric resonator antenna (DRA) [12]. The antenna that is proposed consists of a hollow cylindrical dielectric resonator with relative permittivity of 6, fed by four probes. The inner radius had a diameter of 4.6 mm and the outer radius was 6.6 mm. The inner height of the cylinder was 3.9 mm and the outer one was 6.8 mm. Four probes were made out of gold with the radius of 0.653 mm and height of 3.88 mm. The ground plane size was $5.5 \times 5.5 \text{ cm}^2$. The gap of 1.6 mm was between the centers of the electrodes to the inner surface of the cylinder. The structure is shown in Figure 3.28. The center frequency was 11 GHz and over 30% bandwidth was achieved. The method of switching the beam is shown in Figure 3.28a. Two methods are proposed. In the first method, some probes are activated while the other ones are left open. In the second method, all probes are excited, but a phase difference is applied to them. The paper [12] presents the simulation results of the radiation patterns. The first method is demonstrated in Figure 3.29 for two cases and at two frequencies. For the second method, 11 cases are listed in Table 3.4. For example, H-plane patterns of four cases that were simulated at 10 GHz are shown in Figure 3.30.

3.4 Mechanically Controlled Pattern Reconfigurable Antennas

Mechanically controlled antennas make another group of reconfigurable antennas. In these antennas, the structural features of the antennas are mechanically changed to change antenna radiation properties. One group of these antennas is

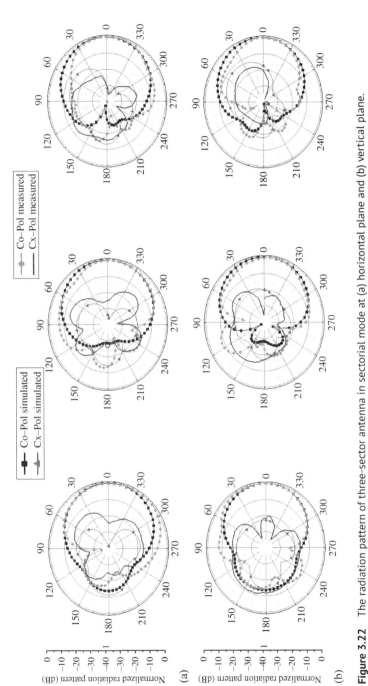

Figure 3.22 The radiation pattern of three-sector antenna in sectorial mode at (a) horizontal plane and (b) vertical plane.

Source: Alieldin et al. [10]. © 2018, EMW Publishing.

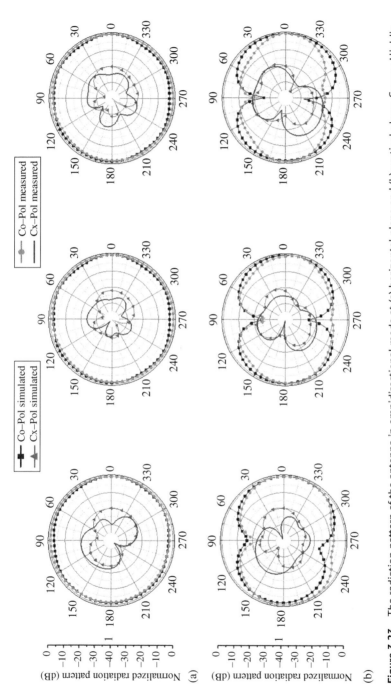

Figure 3.23 The radiation pattern of the antenna in omnidirectional mode at (a) horizontal plane and (b) vertical plane. *Source:* Alieldin et al. [10]. © 2018, EMW Publishing.

① Input port   ⑤ Coupled patch
② Feed network   ⑥ Radiation patch
③ Ground plane   ⑦ Probe
④ DC bias circuit ⑧ Substrate

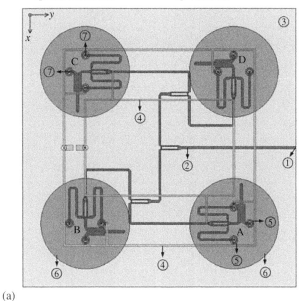

(a)

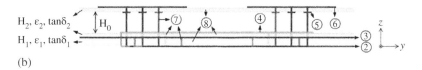

(b)

Figure 3.24 Configuration of array antenna: (a) top view and (b) side view. *Source:* Liu et al. [11]. © 2018, IEEE.

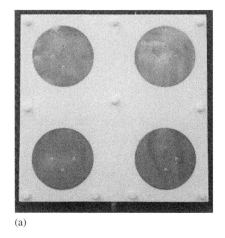

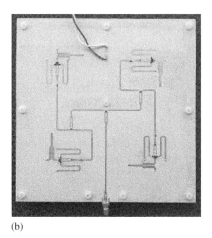

(a)   (b)

Figure 3.25 Fabricated antenna: (a) patch and (b) feed network. *Source:* Liu et al. [11].

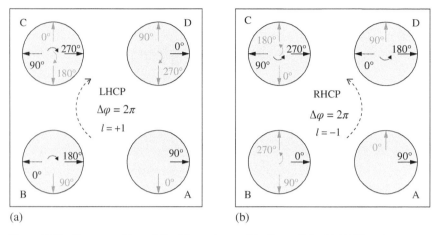

Figure 3.26 Decomposition of the CPs and recombination of LPs: (a) LHCP and (b) RHCP. *Source:* Liu et al. [11]. © 2018, IEEE.

called "origami antennas." "Origami" is the art of paper folding. In origami-based antennas, often a planar shape structure is folded to form a three-dimensional (3D) structure with different radiation characteristics. The control mechanism might be through a mechanical movement, or materials that are activated using heat, light, fields, or other means of control.

In [13], an accordion-shaped antenna is proposed and tested (Figure 3.31). The accordion is placed on a ground plane and the height of the antenna is changed by folding and unfolding the structure. This causes the resonance frequency to be changed. To fabricate this antenna, a flat paper is folded to create creases. Then the paper is folded and the two ends are attached (Figure 3.32).

In [14], a different approach is taken toward origami antenna. The proposed design in this paper is based on a cube design. The so-called "magic cube" can be folded and unfolded, providing two different resonance frequencies. Figure 3.33 shows the antenna in the folded and unfolded state. To build the antenna, two pieces of paper with $140 \times 140\,\text{mm}^2$ size and 0.2 mm thickness were folded in several steps and the monopole was built using metal strips on the top (Figure 3.33). The antenna has two resonance frequencies in each state. In the flat state, it resonates at 1.5 and 2.5 GHz, and in the cube shape, it works at 900 MHz and 2.3 GHz.

Another application of origami antenna is to switch between the single element and antenna array. For example in [15], a push–pull mechanism is used to switch between a single-element patch antenna and three-element patch array. Patch antennas were printed on paper and installed on a 3D printed structure that can be folded using the push–pull method. Figure 3.34 illustrates the fabricated antennas.

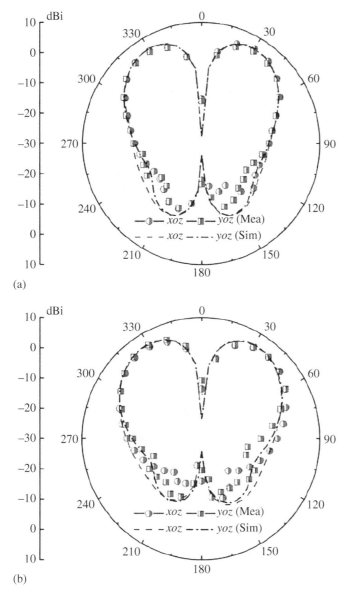

(a)

(b)

Figure 3.27 Radiation patterns: (a) LHCP and (b) RHCP. *Source:* Liu et al. [11]. © 2018, IEEE.

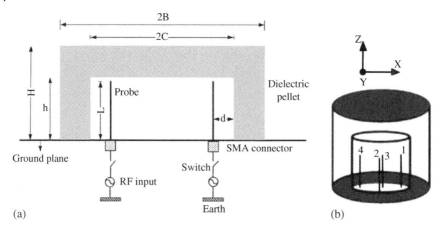

Figure 3.28 (a) Side view of reconfigurable dielectric resonator antenna and (b) 3D view. *Source:* Fayad and Record. [12]. © 2007, EMW Publishing.

Figure 3.35 shows the patterns of a single antenna and the array.

The combination of 3D printed structure and ink-jet printed antennas is also presented in [16]. In this paper, the origami antenna is used for power harvesting. The unit includes two antennas on the sides of a cube that is built using 3D printing technology. The cubic shape package is printed on a planar structure with "smart" shape-memory hinges, shown in Figure 3.36. The cube was made using 3D printing with the size of $5.7 \times 5.7\,\text{cm}^2$ sides and 3 mm thick. Two similar size patches were printed on top of 3D printed surfaces of two sides. The patch size was W = 2.79 cm and L = 3.59 cm. The feeding was done through feed point at 0.63 cm from the center along L. The patches were printed using a combination of diammine silver acetate (DSA) and silver nano particle (SNP). SNP was first applied in the shape of small drops to avoid overspreading of DSA. Then DSA was printed on top of SNP drops. 2 SNP layers and 20 DSA layers on a 3D-printed slab formed the patch antennas. Figure 3.37 shows the simulated and measured realized gain.

The folding was realized by shape memory polymer (SMP). SMPs are polymeric smart materials that can be shaped (deformed shape) by applying temperature and load, and can be returned to the original shape by an external temperature and removing the load. In this design, SMP was used at the edges (Figure 3.38).

The origami antenna may be printed entirely on paper substrate. An example is given in [17]. In this example, the unfolded paper acts as a dipole antenna, as shown in Figure 3.39a, as solid line box. The other parts are inactive in this mode. When the paper is folded into a cube, it will operate as a loop antenna. The loop antenna's resonance frequency is at 1.85 GHz. For the dipole mode, the dipole is one wavelength at around the same frequency. The ink-jet printing on a Kodak

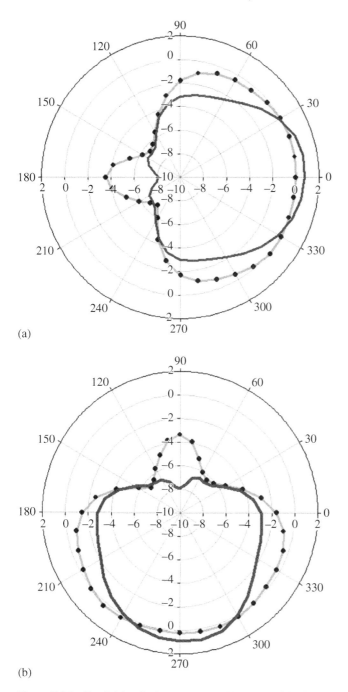

(a)

(b)

Figure 3.29 Far-field radiation patterns in H-plane at 9.53 GHz (dotted line) and 11.1 GHz (solid dark grey line): (a) probe 1 is activated and (b) probe 2 is activated. *Source:* Fayad and Record. [12]. © 2007, EMW Publishing.

Table 3.4 Different cases of beam switching using different phases on each probe at 10 GHz.

Case	Phase				Beam direction (°)
	Probe 1	Probe 2	Probe 3	Probe 4	
1	0	0	0	0	Omnidirectional
2	0	180	0	0	270
3	0	0	180	0	90
4	0	0	0	180	0
5	180	0	0	0	180
6	45	45	0	0	225
7	0	0	45	45	45
8	0	45	0	45	315
9	0	180	45	0	270
10	180	0	0	45	180
11	0	220	0	0	90 and 270

Source: Fayad and Record [12]. © 2007, EMW Publishing.

photo paper of 0.254 mm was used to fabricate the antenna. The conductive ink contained silver nanoparticles Novacentrix JS-B25P printed by a commercial inkjet printer (EPSON WF-7011). The printed antenna is shown in Figure 3.39b. Figure 3.40a shows the printed antenna's reflection coefficient in dipole and loop modes. Figure 3.40b shows the measured radiation patterns for each mode.

3.5 Arrays and Optimizations

Reconfigurable arrays and reflect-arrays are of particular interest since they usually occupy a large volume and having the same physical setup for work for multiple purposes can save space and cost dramatically. Additionally, the mechanical scanning is usually not desired due to the speed and mechanical vibration. Similar methods of reconfiguring the pattern of antenna elements (PIN switches, MEMS, etc.) might be used for the arrays and reflect-arrays.

In addition, the configuration of switches may be optimized for various pattern directions and gain. Genetic algorithm and other optimization methods have been used in optimization of the switch arrangements in several papers. For example in [18], an arrangement of dipoles is considered as shown in Figure 3.41.

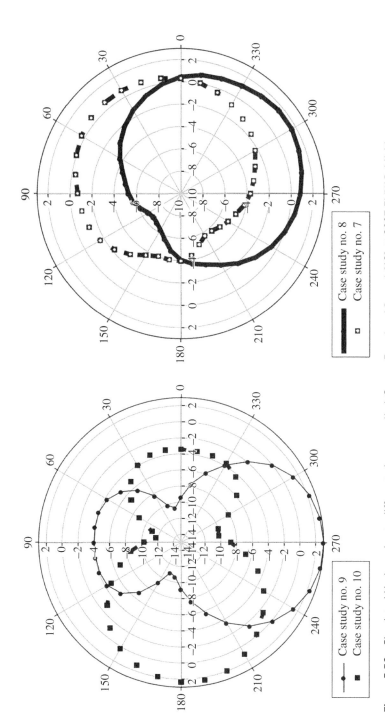

Figure 3.30 Simulated H-plane due to different phases applied. *Source:* Fayad and Record. [12]. © 2007, EMW Publishing.

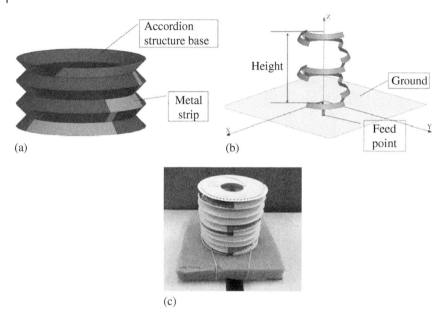

(a)

(b)

(c)

Figure 3.31 (a) Accordion shape antenna, the height is changed by folding the structure; (b) simulation setup; (c) fabricated antenna. *Source:* Yao et al. [13]. © 2014, IEEE.

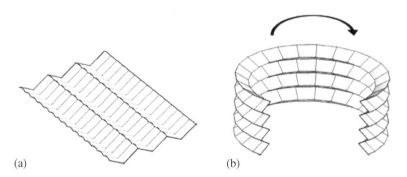

(a)

(b)

Figure 3.32 (a) Creased paper and (b) folding the creased paper. *Source:* Yao et al. [13]. © 2014, IEEE.

The arrangement is a matrix of dipoles that are either active and are considered to be driven elements, or they are not active and are considered as parasitic elements. If there were total of N half-wavelength dipoles, they were modeled as $N - P$ active dipoles and 2P quarter wavelength parasitic dipoles. A model for the radiation pattern as the summation of the array element patterns was developed.

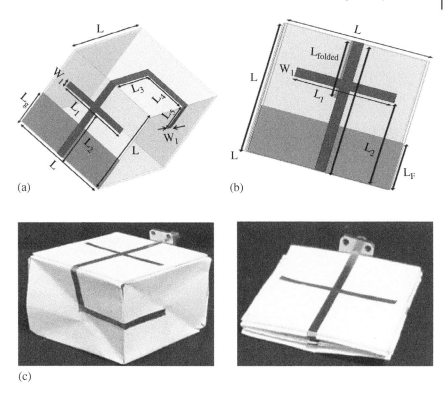

Figure 3.33 Magic cube antenna: (a) folded state, (b) unfolded state, (c) fabricated antenna; dimensions in mm: L = 50, L_1 = 24, L_g = 20, L_2 = 30, L_3 = L_4 = 25, L_5 = 15, W_1 = 2, and L_{folded} = 25. *Source:* Shah and Lim [14]. Licensed under CC-BY-4.0.

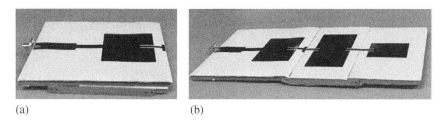

Figure 3.34 Push–pull origami antenna: (a) folded state, single antenna; (b) array. *Source:* Shah and Lim [15].

In this model, N + P ports needed to be determined. Method of moments (MoM) was used to calculate the pattern. If a ground plane was used, the image theory was used. Using a binary genetic algorithm (BGA), the proposed solution was coded by binary genes. Each possible solution, that is each chromosome, was

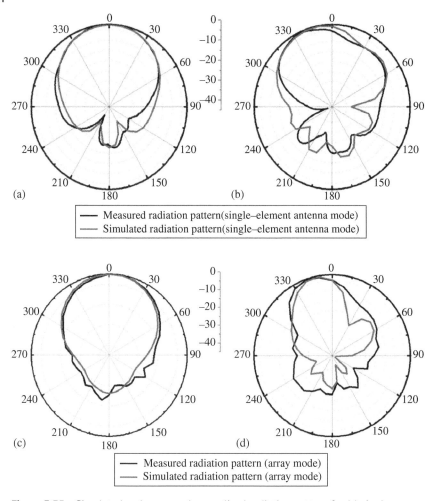

Figure 3.35 Simulated and measured normalized radiation pattern for (a) single-element antenna mode on the xz plane, (b) single-element antenna mode on the yz plane, (c) array mode on the xz plane, and (d) array mode on the yz plane. *Source:* Shah and Lim [15]. Licensed under CC-BY-4.0

made up of $2N+2$ genes. The first $2N$ bits were used for two genes. These genes were used to code the switch configuration. The next $2N$ genes, each with 6 bits, coded the elements amplitude, and the next $2N$ genes, with 6 bits each, were used for coding the phase of elements. A sample chromosome is shown in Figure 3.42. The BGA was then used to optimize for a fitness function that included the goal of minimizing the side lobe levels (SLL) for a pencil beam or a flat beam, and

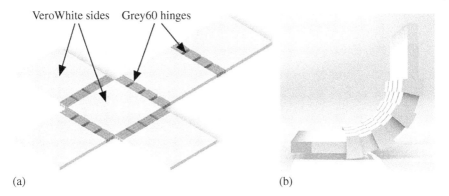

VeroWhite sides   Grey60 hinges

(a)   (b)

Figure 3.36 (a) Flat configuration of the cube, shown where the hinges are placed; (b) rendering of the hinge in a 90° fold. *Source:* Kimionis et al. [16]. © 2015, IEEE.

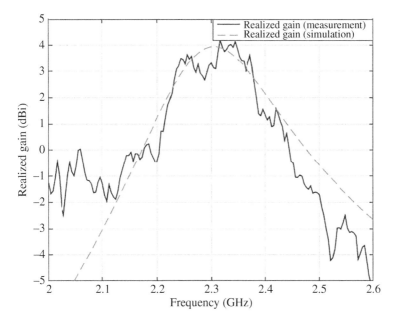

Figure 3.37 Measured and simulated realized gain of ink-jet printed single-patch antenna. *Source:* Kimionis et al. [16]. © 2015, IEEE.

minimizing ripples for a flat beam. The algorithm can work for a linear array as well as two-dimensional array. For example, Figure 3.43 shows an array of 30 elements with half wavelength dipoles was optimized for both flat beam and pencil beam patterns. The flat beam has 8 active elements and the penciled beam has 28.

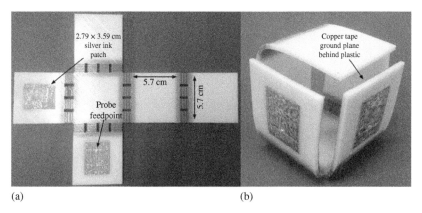

(a)   (b)

Figure 3.38   Inkjet-printed patch antenna on unfolded 3D-printed cube and (b) origami-folded cube after heating, folding, and cooling down. *Source:* Kimionis et al. [16]. © 2015, IEEE.

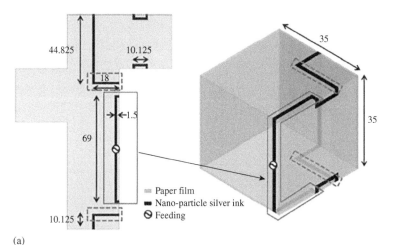

(a)

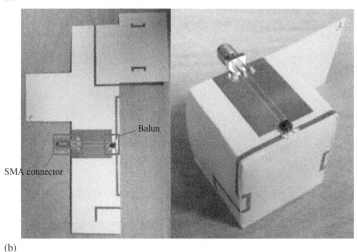

(b)

Figure 3.39   (a) Origami cube antenna in the unfolded and folded states and (b) ink-jet printed antenna. *Source:* Lee et al. [17]. © 2016, John Wiley & Sons.

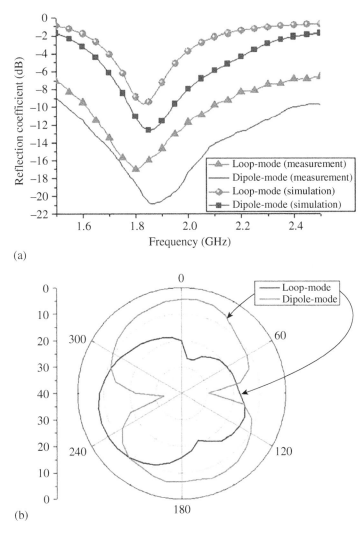

Figure 3.40 (a) Measured and simulated reflection coefficients and (b) measured radiation patterns of the cubic origami antenna. *Source:* Lee et al. [17]. © 2016, John Wiley & Sons.

A reconfigurable aperture (RECAP) was developed in the Georgia Institute of Technology that has multiple metallic small patches and an array of electronically controlled switches [19]. The antenna can be reconfigured by changing the switches (Figure 3.44). The metallic patches are much smaller than a wavelength. The frequency of operation is between 0.85 and 1.45 GHz. Each patch has the size

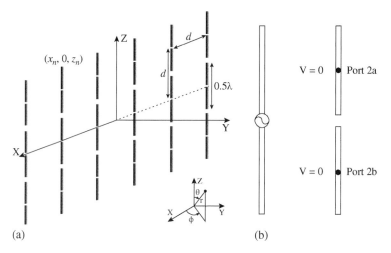

(a)   (b)

Figure 3.41 (a) Geometry of dipole array considered for optimization and (b) equivalent model of driven element (left) and parasitic element (right). *Source:* Ares et al. [18]. © 2008, EMW Publishing.

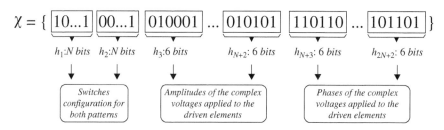

Figure 3.42 A sample chromosome for binary GA to optimize switch configuration. *Source:* Ares et al. [18]. © 2008, EMW Publishing.

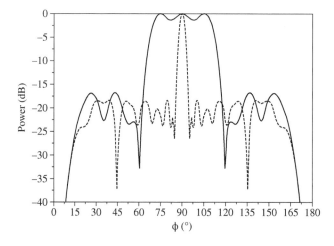

Figure 3.43 Example of two beams generated for a 30-element array of dipole. *Source:* Ares et al. [18]. © 2008, EMW Publishing.

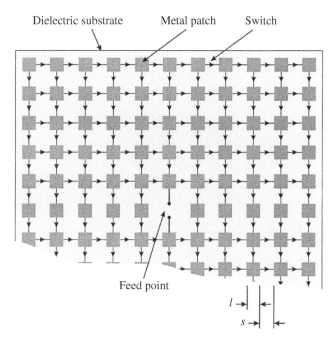

Dielectric substrate   Metal patch   Switch

Feed point

$l \rightarrow$

$s \rightarrow$

Figure 3.44 RECAP configuration with metallic patches and switches. *Source:* Pringle et al. [19]. © 2004, IEEE.

of $1 \times 1 \, \text{cm}^2$. The dielectric substrate used had the thickness of 1.7 mm, relative permittivity of 4.27, and loss tangent of 0.07. A Finite Difference Time Domain (FDTD) code was used to simulate and optimize the arrangements of on/off switches. An example of a bidirectional design with the bandwidth of 38% (0.85–1.25 GHz) and a unidirectional design with end-fire pattern (with 9.5% bandwidth for 1–1.1 GHz) are shown in Figure 3.45. The bidirectional broadside design has left–right symmetry. Figure 3.46 shows their corresponding simulated and measured radiation patterns.

A genetic algorithm optimization was utilized to optimize the arrangements of the FET switches for the given goal. The algorithm shows which switches should be open and which ones should be closed. The theoretical model used in the optimization did not include circuit elements and only included the metallic plates, dielectric layer, and an equivalent circuit model for the FET. The circuit model for the closed state (ON) is a series of $4.3 \, \Omega$ resistor with a 2.6 nH inductor. For the open state (OFF) the resistor is replaced by a parallel combination of a $5.9 \, \text{K}\Omega$ resistor and a 0.49 pF capacitor. For a narrow band design of a directive antenna at 1.25 GHz, for example, the goal was that the realized gain

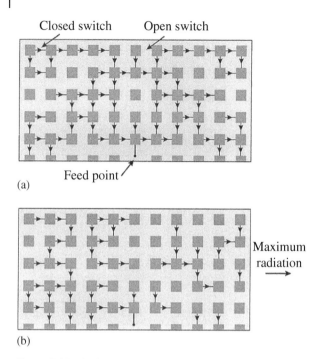

(a)

(b)

Figure 3.45 RECAP antenna in a monopole arrangement: (a) a broadband bidirectional design and (b) a narrowband unidirectional end-fire design. *Source:* Pringle et al. [19]. © 2004, IEEE.

of RECAP antenna should be equal or exceed the gain of a uniformly distributed current on a sheet of the same size and 6 cm away from a ground reflector of infinite size. By changing the goal to have the same gain, but at 45° angle, the switch locations are determined. A broadband antenna was also designed by assigning similar goal for three frequencies. The switch arrangements are shown in Figure 3.47. Corresponding simulated and measured patterns are shown in Figure 3.48.

The switches were controlled electronically and with infrared. The patches were fabricated on FR4 with the permittivity of 4.62 and loss tangent of 0.05. A honeycomb spacer separated this layer from a metallic reflector layer that acted as ground plane. The spacer was 83% air and it was fabricated using Vantico SL7510 with the permittivity of 3.0 and loss tangent of 0.10. Light-emitting diodes (LEDs) were used to send lights through holes that hits photodetectors located on the backside of the dielectric. When the light was on, it activated the switch and turned it on to connect two adjacent patches. Figure 3.49 shows the arrangement of the switch and Figure 3.50 shows its equivalent circuit.

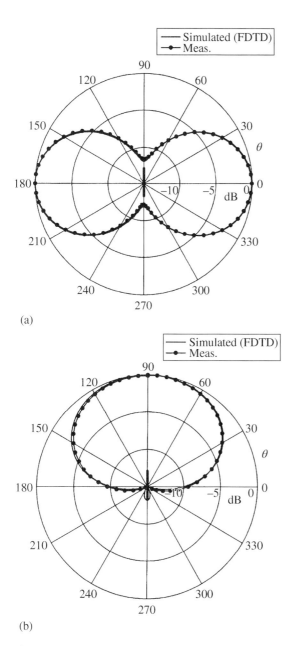

(a)

(b)

Figure 3.46 Simulated and measured radiation patterns of (a) broadband bidirectional antenna and (b) narrow band unidirectional end-fire antenna. *Source:* Pringle et al. [19]. © 2004, IEEE.

Closed switch   Open switch

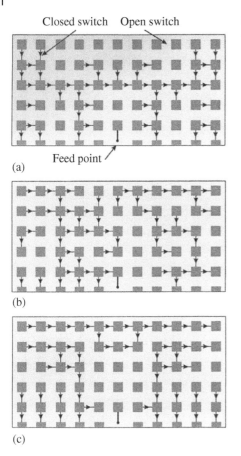

(a)   Feed point

(b)

(c)

Figure 3.47 RECAP switch arrangements of (a) narrow-band broadside design, (b) narrow-band design steered to 45°, and (c) broadband, broadside design. *Source:* Pringle et al. [19]. © 2004, IEEE.

3.6 Reconfigurable Wearable and Implanted Antennas

Wearable and implanted wireless devices are other applications of pattern reconfigurable antennas. Due to the nature of this application, there is little control over the orientation and location of the antenna. Usually omnidirectional patterns are required for communicating on-body devices, and between on-body and off-body devices, directional patterns are required. However, due to the complex surrounding environment, it is often difficult to keep the patterns as the antenna is bent or changes orientations. In this section, a few examples of reconfigurable antennas proposed for medical wearable and implantable wireless devices are presented.

A reconfigurable monopole antenna is presented in [20] for medical implants. The antenna works at Medical Device Radio communications Service band (MedRadio band: 401–406 MHz). Since in medical devices the available volume is

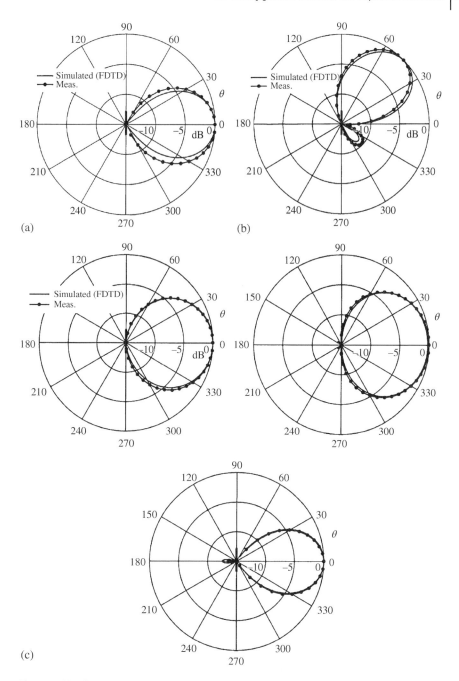

Figure 3.48 RECAP radiation patterns for (a) narrow-band broadside design, (b) narrow-band design steered to 45°, and (c) broadband, broadside design (top right: 1.1 GHz, top left: f = 1.27 GHz, bottom center: f = 1.45 GHz). *Source:* Pringle et al. [19]. © 2004, IEEE.

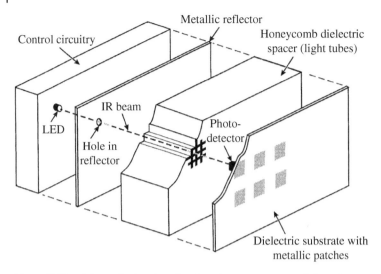

Figure 3.49 Schematic drawing showing the arrangement of photodetector, patches, reflector, and the hole. *Source:* Pringle et al. [19]. © 2004, IEEE.

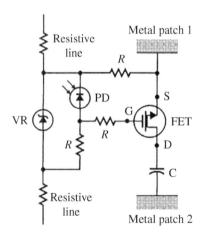

Figure 3.50 Circuit diagram for the FET-based electronic switch: PD, photodetector; VR, voltage reference; R = 8.2 KΩ; C = capacitor (47 pF). *Source:* Pringle et al. [19]. © 2004, IEEE.

very limited, it is advantageous if the pattern can be controlled electronically. The proposed antenna was fabricated on FR4 and placed in an air-filled tube, then tested in a skin phantom. The dimensions of the antenna were $28 \times 11.5 \times 0.6$ mm^3 (193.2 mm^3). Skin-mimicking phantom, made of sugar 54.63%, NaCl 2.77%, and H$_2$O 42.60%, had the dielectric properties of $\varepsilon_r = 46.74$, $\sigma = 0.69$ S/m. Figure 3.51 depicts the antenna structure with different views. Antenna dimensions are given in the figure caption. Figure 3.52 shows the fabricated antenna and its placement

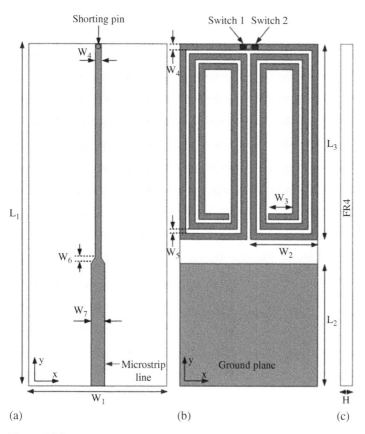

Figure 3.51 Antenna structure: (a) top view, (b) bottom view, and (c) side view. Dimensions are $L_1 \times W_1 \times H = 28 \times 11.5 \times 0.6$, $L_2 = 10$, $L_3 = 16$, $W_2 = 5.6$, $W_3 = 2$, $W_4 = 0.3$, $W_5 = 0.1$, $W_6 = 0.5$, and $W_7 = 1.13$, all in mm. *Source:* Nguyen and Jung [20]. © 2016, IEEE.

in the skin phantom. The top layer consists of a $50\,\Omega$ transmission line that is terminated with a shorting pin. The bottom layer consists of a partial ground plane and two spirals and two switches. Three states were examined. State 1 (S1) is when both switches are ON. State 2 (S2) is when switch 1 is ON and switch 2 is OFF. State 3 (S3) is when switch 1 is OFF and switch 2 is ON. The antenna was simulated while in the skin tissue background using Ansys HFSS simulation program. The 3D simulated patterns and a 2D measured pattern are shown in Figure 3.53. The main beam peak gain was at $0°$, $+46°$, and $-33°$, with the measured gain of -18.24, 18.04, and $-17.94\,$dBi, for S0, S1, and S2, states, respectively.

The next design is for a wearable device, Fitbit Flex wristband. The concept is very similar to the previous design in [20]. The design presented in [21] (shown in Figure 3.54) consists of a monopole and a loop antenna. The loop antenna is

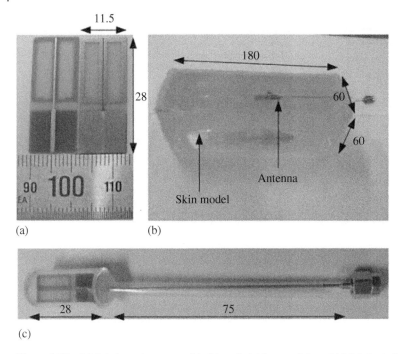

(a) (b)

(c)

Figure 3.52 (a) Fabricated antenna, (b) skin-mimicking model, and (c) fabricated antenna with the plastic pipe (unit: mm). *Source:* Nguyen and Jung [20].

connected to a PIN diode that can be controlled to close (ON) or open (OFF) the loop. The substrate is FR4. Two switches are shown as SW1 and SW2. CASE1 means SW1 is ON and SW2 is OFF, while CASE2 indicates SW1 is in OFF and SW2 is ON. Three angles for the termination of the line are suggested as 30°, 90°, and 150°. For 90° and 150° terminations, operation frequency is 2.4 GHz. For 30°, the resonance frequency shifts to 3.3 GHz. The fabricated antennas are shown in Figure 3.55. An inductor of 0.27 μH is added to provide DC block. The simulated and measured patterns for 90° for two cases of the switch are shown in Figure 3.56. The measured maximum gain was 1.43 dBi.

The last design presented here is a wearable reconfigurable antenna using textile material and based on metamaterial [22]. The antenna was designed for 2.45 GHz and fabricated on a felt fabric with relative permittivity of 1.3, loss tangent of 0.044, and thickness of 3 mm. The conductive part was made of conductive textile with approximate conductivity of 1.18×10^5 S/m and thickness of 0.17 mm. Figure 3.57 shows the geometry of the antenna. The six vias that are located around the patch could be connected or disconnected through a switch. The paper suggests the use of PIN diodes; however, the paper gives the results of

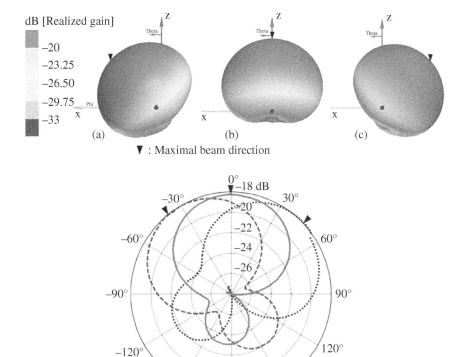

Figure 3.53 3D simulated radiation patterns at states: (a) S1, (b) S0, (c) S2, and (d) measured radiation pattern at 403.5 MHz. *Source:* Nguyen and Jung [20]. © 2016, IEEE

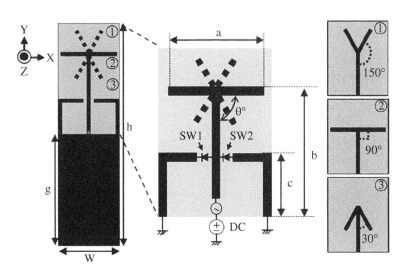

Figure 3.54 Antenna geometry. Dimensions in mm are: a = 12.8, b = 18.5, c = 8, h = 50, W = 14, and thickness = 1. *Source:* Lee and Jung [21]. © 2015, IEEE.

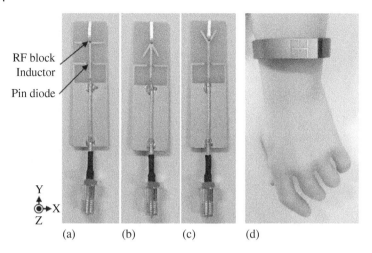

Figure 3.55 (a)–(c) Fabricated antennas and (d) testing on a hand phantom of ε_r = 25.7, σ = 1.32 S/m. *Source:* Lee and Jung [21]. © 2015, IEEE.

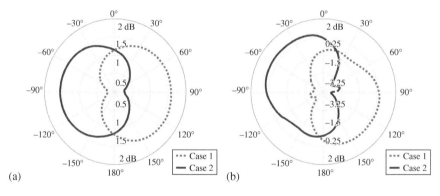

Figure 3.56 (a) Simulated and (b) measured radiation patterns for the two cases for 90° termination antenna at 2.4 GHz. *Source:* Lee and Jung [21]. © 2015, IEEE.

measurements when the vias were connected through a conductive material. The fabricated samples of each case (connected and not connected) are shown in Figure 3.58. When the switches are OFF and the vias are not connected, the antenna works as a regular patch antenna with a broadside pattern. When the switches are ON and vias are connected, the vias operate as a shunt inductor and generate a zeroth-order-resonance (ZOR) patch antenna that creates an omnidirectional pattern similar to a monopole. Figure 3.59 shows these patterns for two different planes.

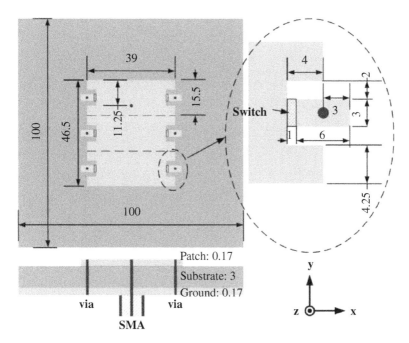

Figure 3.57 Wearable textile reconfigurable antenna. Dimensions are in mm, vias diameters = 1.25 mm. *Source:* Yan and Vandenbosch [22]. © 2016, IEEE.

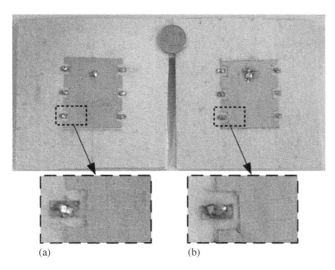

(a)   (b)

Figure 3.58 Prototypes of (a) antenna with connected vias (patch mode) and (b) antenna with disconnected vias (monopole mode). *Source:* Yan and Vandenbosch [22]

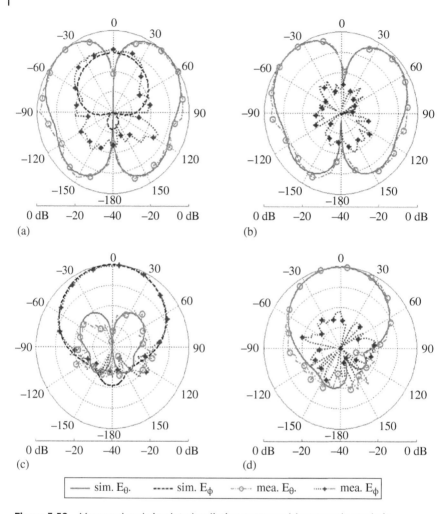

Figure 3.59 Measured and simulated radiation patterns: (a) monopole mode in xz-plane, (b) monopole in yz-plane, (c) patch mode in xz-plane, and (d) patch mode in yz-plane. *Source:* Yan and Vandenbosch [22]. © 2016, IEEE.

For any antenna in the vicinity of human body, it is important to characterize the maximum power that is absorbed by the tissues. The specific absorption rate (SAR) is defined as the power absorbed per unit mass of the tissue. There are various standards to limit this value. IEEE C95.3 standard requires an average SAR of 10 g of biological tissue to be less than 1.6 W/kg. The antenna's SAR on a cubical tissue model is 3 mm skin, 7 mm fat, 60 mm muscle with an area of 300 mm × 300 mm. The antenna was located 5 mm above the skin surface. The calculated SAR (for 0.5 W of input power) and other antenna performance parameters are summarized in Table 3.5.

Table 3.5 Antenna performance and SAR simulations.

		Center frequency (GHz)	Bandwidth (MHz)	Realized gain (dBi)	Radiation efficiency (%)	SAR (W/kg)
Patch mode	Simulation	2.448	207	2.9	49.5	0.05
	Measurement	2.442	157	2.0	45.0	—
Dipole mode	Simulation	2.450	134	4.5	40.6	0.01
	Measurement	2.445	119	3.9	38.0	—

Source: Yan and Vandenbosch [22]. © 2016, IEEE.

3.7 Conclusion

The versatility of pattern reconfigurable antennas has made them practical and useful in many applications. The presented designs in this chapter are only handful examples of how reconfigurability can be incorporated into various antennas. Although there has been enormous research in this field, there is still a lot of room for advancing it. With the inclusion of machine learning and artificial intelligence, it is possible to create multifunctional antennas that can interact with the environment and shape the beam based on the feedback from the environment without any human-in-the-loop.

References

1 Chen, Z.Z.A., Ning, X., and Wang, L. (November 2017). A design of radiation pattern and polarization reconfigurable antenna using metasurface. In: *2017 IEEE Asia Pacific Microwave Conference (APMC)*, 108–111. IEEE.

2 Ge, L., Li, Y., Wang, J., and Sim, C.Y.D. (2017). A low-profile reconfigurable cavity-backed slot antenna with frequency, polarization, and radiation pattern agility. *IEEE Trans. Antennas Propag.* 65 (5): 2182–2189.

3 Tsai, Y.L., Hwang, R.B., and De Lin, Y. (2012). A reconfigurable beam-switching antenna base on active FSS. In: *2012 15th International Symposium on Antenna Technology and Applied Electromagnetics, ANTEM*, 1–4. IEEE.

4 Hossain, M.A., Bahceci, I., and Cetiner, B.A. (2017). Parasitic layer based radiation pattern reconfigurable antenna for 5g communications. *IEEE Trans. Antennas Propag.* 65 (12): 6444–6452.

5 Qin, P.Y., Guo, Y.J., and Ding, C. (2013). A beam switching quasi-Yagi dipole antenna. *IEEE Trans. Antennas Propag.* 61 (10): 4891–4899.

6 Petit, L., Dussopt, L., and Laheurte, J.M. (2006). MEMS-switched parasitic-antenna array for radiation pattern diversity. *IEEE Trans. Antennas Propag.* 54 (9): 2624–2631.

7 Kang, W., Ko, K.H., and Kim, K. (2012). A compact beam reconfigurable antenna for symmetrical beam switching. *Prog. Electromagn. Res.* 129: 1–16, 2012.

8 Sharma, S.K., Fideles, F., Kalikond, A. et al. (2013). Planar Yagi-Uda antenna with reconfigurable radiation. *Microwave Opt. Technol. Lett.* 55 (12): 2946–2952.

9 Panagamuwa, C.J., Chauraya, A., and Vardaxoglou, J.C. (2006). Frequency and beam reconfigurable antenna using photoconducting switches. *IEEE Trans. Antennas Propag.* 54 (2): 449–454.

10 Alieldin, A., Huang, Y., Boyes, S.J., and Stanley, M. (2018). A reconfigurable broadband dual-mode dual-polarized antenna for sectorial/omnidirectional mobile base stations. *Prog. Electromagn. Res.* 163: 1–13.

11 Liu, Q., Chen, Z.N., Liu, Y. et al. (2018). Circular polarization and mode reconfigurable wideband orbital angular momentum patch array antenna. *IEEE Trans. Antennas Propag.* 66 (c): 1796–1804.

12 Fayad, H. and Record, P. (2007). Multi-feed dielectric resonator antenna with reconfigurable radiation pattern. *Prog. Electromagn. Res. PIER* 76: 341–356.

13 Yao, S., Liu, X., Georgakopoulos, S.V., and Tentzeris, M.M. (2014). A novel reconfigurable origami accordion antenna. In: *2014 IEEE MTT-S International Microwave Symposium (IMS2014)*, 1–4. IEEE.

14 Shah, S. and Lim, S. (2017). A dual band frequency reconfigurable origami magic cube antenna for wireless sensor network applications. *Sensors* 17 (11) https://www.mdpi.com/1424-8220/17/11/2675/html.

15 Shah, S.I.H. and Lim, S. (2017). Transformation from a single antenna to a series array using push/pull origami. *Sensors (Switzerland)* 9: 17. https://www.mdpi.com/1424-8220/17/9/1968.

16 Kimionis, J., Isakov, M., Koh, B.S. et al. (2015). 3D-printed origami packaging with inkjet-printed antennas for RF harvesting sensors. *IEEE Trans. Microwave Theory Tech.* 63 (12): 4521–4532.

17 Lee, D., Seo, Y., and Lim, S. (2016). Dipole- and loop-mode switchable origami paper antenna. *Microwave Opt. Technol. Lett.* 58 (3): 668–672.

18 Ares, F., Franceschetti, G., and Rodriguez, J.A. (2008). A simple alternative for beam reconfiguration of array antennas. *Prog. Electromagn. Res. PIER* 88: 227–240.

19 Pringle, L.N., Harms, P.H., Blalock, S.P. et al. (2004). A reconfigurable aperture antenna based on switched links between electrically small metallic patches. *IEEE Trans. Antennas Propag.* 52 (6): 1434–1445.

20 Nguyen, V.T. and Jung, C.W. (2016). Radiation-pattern reconfigurable antenna for medical implants in MedRadio band. *IEEE Antennas Wirel. Propag. Lett.* 15: 106–109.

21 Lee, C.M. and Jung, C.W. (2015). Radiation-pattern-reconfigurable antenna using monopole-loop for fitbit flex wristband. *IEEE Antennas Wirel. Propag. Lett.* 14: 269–272.

22 Yan, S. and Vandenbosch, G.A.E. (2016). Radiation pattern-reconfigurable wearable antenna based on metamaterial structure. *IEEE Antennas Wirel. Propag. Lett.* 15: 1715–1718.

4

Polarization Reconfigurable Antennas

Behrouz Babakhani and Satish K. Sharma

4.1 Introduction

Polarization reconfigurable antennas have been used to create polarization diversity in wireless communication systems and hence create a more reliable link with higher data throughput. Nowadays, wireless communication systems have a significant role in our daily lifestyle. In the recent years, the demand for more reliable communication systems with higher data throughput has increased. One way to increase the data transfer rate (data rate or throughput) in wireless communication systems is to create diversity. One form of diversity which has been considered widely since the third generation of wireless communication system (3G) is the polarization diversity. Antenna systems that utilize polarization diversity are gaining popularity. In broadband wireless communication systems, such as wireless local area networks (WLANs) [1, 2], polarization diversity antennas have been used to mitigate the fading caused by multipath. In active read/write microwave tagging systems, polarization diversity antennas provide a reliable modulation scheme [3]. They have also been used to realize frequency reuse for doubling the system capability in satellite communication systems [4]. Such antenna system has been placed in the recent Mars rover, which is a patch antenna with dual-frequency and dual-polarization capabilities [5]. Generally, to add polarization diversity to a wireless communication system, two or more antennas with different radiation polarization have been used. Usually adding antennas with different polarization is the standard method used in base station and larger portable devices. However, for fairly small and cheap devices (such as garage door key fob), this would become impossible if not complicated.

Multifunctional Antennas and Arrays for Wireless Communication Systems, First Edition.
Edited by Satish K. Sharma and Jia-Chi S. Chieh.
© 2021 John Wiley & Sons, Inc. Published 2021 by John Wiley & Sons, Inc.

As discussed earlier, there is a need to install two or more antennas that can possess different radiation characteristics, such as different radiation polarization. In reconfigurable antennas, rather than using multiple antennas, a single antenna with reconfigurable radiation behavior has been used. Therefore, the communication system can be downsized without having to install many antennas.

Several methods have been used to create polarization reconfigurable antennas. The most widely used method is introducing a switching mechanism on the radiator element (antenna) or in its feed network. In the switching mechanism, an RF switch has been used which can connect or disconnect a specific part of the antenna/circuitry. The RF structure of an RF switch can be sorted into three main categories based on the mechanism used in them:

- Mechanical RF switches (RF relays)
- Solid-state switches (PIN diodes, MESFET, etc.)
- Micro-electro-mechanical systems (MEMS)

Besides the switching, other methods have also been investigated. One method is using a variable load. By changing the load, the surface current direction and density change which result in a different polarization radiation. This variable load is usually a varactor diode (also called a varicap diode or variable capacitance diode). Due to its low cost and other superior properties compared to the other variable RF loads, varactors have been widely used to create polarization reconfigurable antennas.

There are also efforts of using metasurfaces (MSs) to create a polarization reconfigurable antenna. MSs are two-dimensional periodic structures, usually consisting of scatterers or apertures, used to create an engineered structure which behaves differently compared to that are found in nature. The transformation of polarization can be achieved by using the engineered transmission polarization converters, such as those with dielectric, grid-plate, meander-lines, and anisotropy MSs.

Mechanical reconfiguration is the least popular way of creating polarization reconfigurable antennas. One of the main reasons is that mechanically reconfigurable antennas require movable parts. If not designed well, the actuator used to produce the mechanical movements will be very complicated and occupy much space, which will lead to a bulky and expensive structure. Moreover, the change of size and/or shape during tuning is the common problem for most mechanically reconfigurable antennas.

At first, we will discuss the different RF switches such as RF relays, PIN diodes, MESFET, MEMS and also variable loads such as varactor diodes including their benefits and drawbacks and compare them based on their electrical and mechanical properties. Then we will focus on the polarization reconfigurable antennas that used these switches and variable loads in their designs. Detailed analysis of the antenna geometry will be given and antenna performance would be compared. Also, we will discuss the antenna designs that use RF switches in their feed network instead of the antenna itself to create a polarization reconfigurable antenna. Using the RF switch in the feed network rather than the antenna structure has the benefit of ease of fabrication; however, it would usually introduce more losses. This will be discussed in more detail in the following. The use of MSs and other artificial magnetic components would also be reviewed and discussed as a method of creating polarization reconfigurable antennas. Mechanical method will also be reviewed and discussed as a method of creating reconfigurable antennas. As mentioned earlier, this is the least favorable method to create reconfigurable antennas; however, successful and interesting designs have been suggested which would be presented and discussed next.

4.2 Polarization Reconfiguration Mechanism Using RF Switches

In this section, we will focus on the polarization (vertical or horizontal linear polarization [LP], right-hand circular polarization [RHCP], and left-hand circular polarization [LHCP]) reconfigurable antenna which uses RF switches. In microwave circuits, RF switches are either used to select a desired signal among multiple sources, or to route a signal to a desired path. The basic block diagram of an RF switch with one input and two outputs (single-pole double-throw [SPDT]) has been shown in Figure 4.1. As shown in Figure 4.1, an RF switch circuitry is a combination of RF lines, switch and a logic, and supply circuitry. RF switches

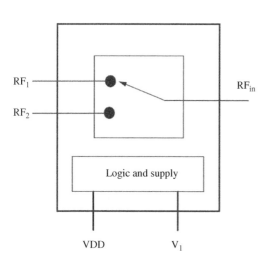

Figure 4.1 A single-pole double-throw (SPDT) RF switch is conceptually simple, with a control signal directing the switch to route the input to either one of two possible outputs.

have been widely used in applications such as diversity antenna systems, radars, and measurement setups. As mentioned before, RF switches come in different forms such as RF relays (mechanical switch), solid-state switches (PIN diodes and MESFET), MEMS, and vanadium dioxide (VO_2) switches (phase-change material).

The RF mechanical switches (also called relays) are usually electromechanical (EM) structures built similar to a non-RF switch. RF relays' mechanism is based on a simple theory of electromagnetic induction and mechanical contacts. Nowadays, The RF relays have been replaced by IC switches (solid state or MEMS), unless there is a need for higher-power, high linearity, or many poles (contacts) with extremely low loss.

A solid-state RF switch is an electronic switching device based on semiconductor technology (e.g. MOSFET, PIN diode). It operates similar to an EM switch except that there are no moving parts. Most RF switches are designed based on GaAs or CMOS technology; however, modern semiconductor technologies, such as UltraCMOS process, have been introduced to increase linearity, power handling, and reduce the power loss in solid-state RF switches. The solid-state RF switch ICs also come in a variety of packaging based on the application and technology of fabrication.

There are also RF switches based on MEMS technology which replicate the EM design but use IC fabrication techniques. RF MEMS switch is a micro-electromechanical system with sub-millimeter-sized parts that provide radio frequency switching functionality. RF MEMS switches offer high isolation, linearity, and power handling with low insertion loss and low Q factor. Ideally, RF MEMS switches do not consume power, but require a high actuation voltage.

In the following sections, we will discuss the antenna and antenna arrays which use one of the above-mentioned RF switches to change their radiation polarization.

4.3 Solid-State RF Switch-Based Polarization Reconfigurable Antenna

In this section, we will focus on the polarization reconfigurable antenna using solid-state RF switches. These switches suffer from the high loss, nonlinearity, and cannot handle high power, but due to its low cost and ease of use, these are the most popular type of RF switch.

To create a polarization reconfigurable antenna, one needs to perturb or redirect the surface current so that magnetic field's polarization can change. In most of the research, single pole single throw (SPST) RF switches have been used to connect or disconnect parts of the antenna or antenna feed network. A PIN diode is the most common form of SPST RF switch. The geometry and the circuit symbol of a PIN diode are given in Figure 4.2. In forward-biased mode, in diode

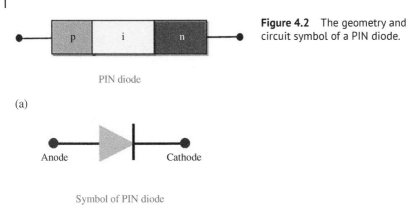

Figure 4.2 The geometry and circuit symbol of a PIN diode.

PIN diode

(a)

Anode   Cathode

Symbol of PIN diode

(b)

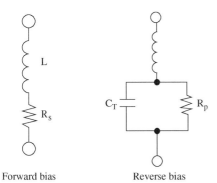

Figure 4.3 Schematic circuit model of a PIN diode in forward and reverse bias modes.

L

R_s

C_T R_p

Forward bias   Reverse bias

structure, the injected carrier concentration is typically much higher than the intrinsic carrier concentration. Due to this high level of injection, a depletion would be created where the electric fields extend deeply into the both region. This electric field helps in speeding up of the transport of charge carriers from the P to the N region, which results in faster operation of the diode, making it a suitable device for high-frequency operations. Under zero- or reverse-bias (the "OFF" state), a PIN diode has a low capacitance. The low capacitance will not pass much of an RF signal. Under a forward bias of 1 mA (the "ON" state), a typical PIN diode will have an RF resistance of less than 1 ohm, making it a good RF conductor. The schematic circuit model of a PIN diode in forward and reverse bias modes is shown in Figure 4.3. The parasitic elements of L and R_p are the result of the packaging and the values are usually given in the diode data sheet. As mentioned above, the PIN diode should be biased to work as a RF switch; therefore, a bias circuitry is needed. This circuitry usually consists of RF choke inductor and DC

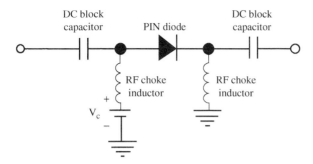

Figure 4.4 A typical bias circuitry used with PIN diodes.

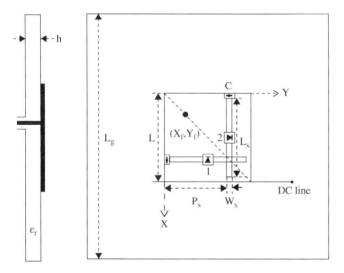

Figure 4.5 Geometry of a patch antenna with RHCP/LHCP polarization reconfiguration. The PIN diodes are biased using DC blocking capacitors and a quarter-wavelength grounded DC line. *Source*: Yang and Rahmat-Samii [6]. © 2002, IEEE.

blocking capacitors to isolate the RF and bias (DC) circuitry from each other. A typical bias circuitry for a PIN diode is illustrated in Figure 4.4. In IC RF switches, the bias network is integrated with the PIN diodes; therefore, the IC comes with RF ports plus the control pins which ease the use of the switch.

Figure 4.5 shows the geometry of a polarization reconfigurable antenna realized by PIN diodes [6]. This design consists of a square patch mounted on a square substrate. The patch is fed using a probe located on the diagonal line of the patch. Two orthogonal and identical slots are incorporated into the patch with two PIN diodes inserted into their center. The radiation mechanism of the antenna can be described using the cavity model. For a normal square patch antenna, when the feeding point is located on its diagonal line, both TM_{01} and TM_{10} modes are excited at the same frequency. Considering the antenna geometry in Figure 4.5, cutting the horizontal slot is only affecting the TM_{10} mode and has negligible effect on the TM_{01} mode, while the vertical slot affects the TM_{01} mode. If any of the PIN diodes in the slots are set to the "ON" mode, that slot is split into two short slots and the electric current of the mode can flow through the diode. Therefore, the slot almost becomes invisible and would not affect the resonant frequency of the mode. If the diode is OFF, the electric current of the mode cannot flow through the diode, thus forcing the current to travel around the slot. Therefore, the effect of the slot becomes significant and the resonant frequency of the mode would greatly decrease. Thus, considering Figure 4.5 antenna geometry, when diode 1 is ON and diode 2 is OFF, the resonant frequencies of the TM_{10} mode and TM_{01} mode are different. This difference in the resonant frequency can be controlled by the slots' length and location on the patch. If frequency difference is properly designed, the radiation fields of the TM_{10} mode and the TM_{01} mode have the same magnitude and are 90° out of phase result, in a RHCP radiation mode. With the same concept, LHCP radiation can be generated by setting diode 1 OFF and diode 2 ON. To control the status of the diodes, a bias circuit is required. In the suggested bias network for this design, a pair of DC blocking capacitors is soldered onto the edges of the slots while a shorted quarter wavelength strip is connected to the right-bottom corner of the patch as a ground at DC but does not affect the behavior of the RF. The control voltage is supplied to the coax probe using a bias tee. As a result, the diodes have different orientations. If a positive voltage is applied to the bias tee, diode 1 is turned OFF while diode 2 is turned ON. Applying a negative voltage would turn ON diode 1 while diode 2 is OFF.

Another geometry to create a polarization reconfigurable patch antenna with capability to switch between LP, RHCP, and LHCP is shown in Figure 4.6 [7]. In this design, a single-feed square microstrip antenna with truncated corners has been considered. The truncated corners provide perturbation to obtain two orthogonal 90° out-of-phase resonant modes which create circular polarization. The technique of truncating the corners of a square patch to obtain single-feed circular polarization operation is well known and has been widely used in practical designs [8]. To create a polarization reconfigurable antenna, four small triangular-shaped parasitic conductors have been added to each corner. These triangular parasitic patches are separated from the main patch with a tiny gap where four PIN diodes are inserted into it. A DC bias circuit is required to control the state of each individual diode. Table 4.1 shows the polarization of the radiation based on the state of the switches. As the table shows, the desired polarization can be switched from LP to RHCP or LHCP based on the state of the PIN

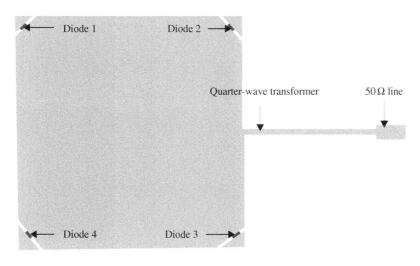

Figure 4.6 Geometry of the corner-truncated square microstrip patch antenna with switchable polarization. *Source*: Sung et al. [7]. © 2004, IEEE.

Table 4.1 Polarization sense of the proposed antenna for four combinations of diode states.

	D_1	D_2	D_3	D_4	Polarization
ANT1	ON	ON	ON	ON	LP
ANT2	OFF	OFF	OFF	OFF	LP
ANT3	ON	OFF	ON	OFF	RHCP
ANT4	OFF	ON	OFF	ON	LHCP

Source: Sung et al. [7]. © 2004, IEEE.

diodes. Figure 4.7 shows the effect of the switches and measured and simulated corresponding reflection coefficient magnitude for each antenna case.

A polarization reconfigurable E-shaped patch antenna using varactor diodes has been presented in [9]. This design can switch its polarization between RHCP and LHCP. This simple structure consists of a single-layer single-feed E-shaped patch and two RF switches placed at appropriate locations in its slots. This antenna was designed targeting the WLAN IEEE 802.11 b/g frequency band (2.4–2.5 GHz) which has been used in various wireless communication systems. The antenna exhibits a 7% effective bandwidth from 2.4 to 2.57 GHz with VSWR below 2 and with maximum gain of 8.7 dBic. The antenna radiation symmetry is maintained upon switching between the two circular polarization modes. Figure 4.8 shows an E-shaped

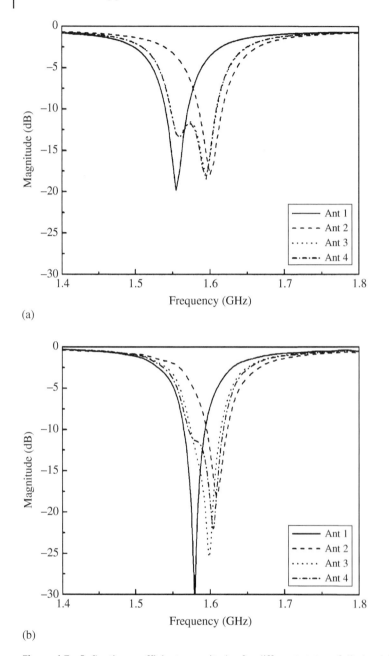

(a)

(b)

Figure 4.7 Reflection coefficient magnitudes for different states of diodes. (a) Simulation result and (b) measurement result. *Source*: Sung et al. [7]. © 2004, IEEE.

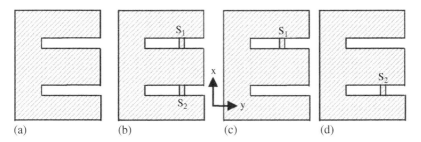

Figure 4.8 Switching states of the reconfigurable E-shaped patch antenna: (a) LP state 1, (b) LP state 2, (c) LHCP state 3, and (d) RHCP state 4. *Source*: Khidre et al. [9]. © 2013, IEEE.

Table 4.2 Polarization sense of the E-shaped proposed antenna in zero for four combinations of diode states [7, 9].

State	Switch 1 (S_1)	Switch 2 (S_2)	Frequency	Polarization
1	OFF	OFF	f_L	LP
2	ON	ON	f_H	LP
3	ON	OFF	f	LHCP
4	OFF	ON	f	RHCP

Source: Khidre et al. [9]. © 2013, IEEE.

patch antenna when it is loaded with two switches in the slots. This results in four possible polarization states listed in Table 4.2. State 1 provides the resonant frequency of the original E-shaped patch (f_L). Such antenna radiates linearly polarized (LP) fields in the y-direction. At state 2, the two switches are ON which allows the surface currents to pass through them, making the current path around the slots shorter. Thus, the electrical length is smaller, and the resonant frequency would be higher than that in state 1 (f_H). Due to the symmetry of the antenna structure, at state 2, the polarization is also linear. Therefore, going from state 1 to state 2 would provide a frequency reconfigurable E-shape patch antenna with LP [10].

By creating an offset in the slot length, asymmetry is introduced which perturbs the field beneath the patch, generating CP mode. At state 3, the E-shaped patch antenna with a switch in the upper slot is effectively having a shorter upper slot than the lower one. Therefore, LHCP could be generated at the resonant frequency of $f_L < f < f_H$. Similarly, state 4 gives RHCP. Thus, with appropriate E-shape patch and switch combination selection, one can create a polarization reconfigurable antenna with RHCP to LHCP switching capability.

Figure 4.9 shows another design based on the above-mentioned concept. In this antenna, a thin substrate is placed above a conductive ground plane (10 mm)

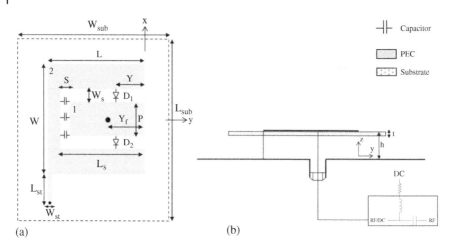

(a)   (b)

Figure 4.9 Geometry of a single-feed reconfigurable E-shaped patch antenna with integrated DC biasing circuit: (a) top view and (b) side view *Source*: Khidre et al. [9]. © 2013, IEEE.

while the E-shape patch is fed using a coaxial cable. Two RF PIN diodes are inserted in the slots of the E-shaped patch antenna. A narrow gap separates the feed from the rest of the patch. This gap is shorted to the RF signal using three DC blocking capacitors. The outer part of the E-shaped patch is DC grounded through a narrow quarter wavelength ($\lambda/4$) transmission line shorted to the ground by a tiny metallic via. This narrow $\lambda/4$ transmission line works as a RF choke which provides high impedance at the E-shaped patch edge and keeps the RF current on the patch surface unperturbed. A bias tee was added to the coaxial cable to provide the required biasing signal. Thus, when the inner part of the E-shaped patch is positively charged, the D2 diode is ON and the D1 diode is OFF, which results in RHCP. For LHCP, the terminals of the DC source are reversed, with the D1 diode ON and the D2 diode OFF. A series resistance with the DC line is used to control the driven current from the DC source as shown in Figure 4.9. Figure 4.10 shows a photograph of the fabricated polarization reconfigurable E-shaped patch antenna prototype along with the associated switching and biasing assemblies. The measured result shows this design has a 7% effective bandwidth with maximum realized gain of 8.7 dBic at 2.45 GHz.

In [11], a single-fed polarization reconfigurable microstrip patch antenna is discussed using PIN diode switches on its feed network. The antenna is excited by the aperture coupling method. The PIN diodes are used to reconfigure the coupling slot and the open stub of the feed line which in turn changes the polarization of the microstrip antenna between vertical and horizontal polarizations. To change the antenna polarization from linear to CP, a perturbation segment needs

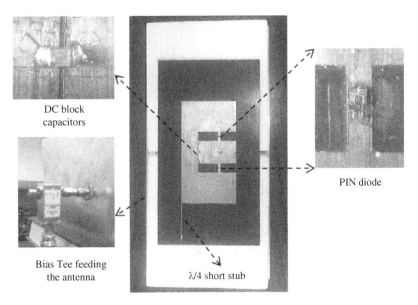

DC block
capacitors

PIN diode

Bias Tee feeding
the antenna

$\lambda/4$ short stub

Figure 4.10 A photo of a fabricated polarization reconfigurable E-shaped patch antenna prototype along with the associated switching and biasing assemblies. *Source*: Khidre et al. [9].

to be introduced. By adding such perturbation using another PIN diode, an antenna with quadri-polarization diversity, including dual orthogonal LPs and two circular polarizations, can be developed from the structure of the switchable circular polarization microstrip antenna, and only three diodes are required. The proposed polarization reconfigurable microstrip antenna is shown in Figure 4.11. This design consists of a circular patch and a rectangular perturbation stub which has been printed on the same layer of a FR4 substrate, and they are connected through a PIN diode (D_1). The antenna is excited through an aperture-coupled feed mechanism in which the microstrip feed line and the coupling slot are fabricated on the opposite sides of another FR4 substrate. A foam material of thickness 5 mm is inserted between them to reduce the antenna quality factor. The open stub of the feed line is divided into two segments and a diode (D_2) is employed to connect them, so that the open-stub length can be reconfigure between l_{s1} and l_{s2}.

In addition, the coupling slot can be reconfigured by another diode that is placed at the angle of α away from the y-axis. Because the three diodes are located at different layers, the layout of their DC-bias circuits is relatively simple.

When all the diodes are OFF, the circular radiating patch is not perturbed. Also, the coupling aperture is a ring slot, and the antenna can produce a y-directional LP radiation by appropriately selecting the l_{s1}. By turning ON the D_2 and D_3 diodes while D_1 is OFF, the coupling aperture becomes an open-ring slot instead of the

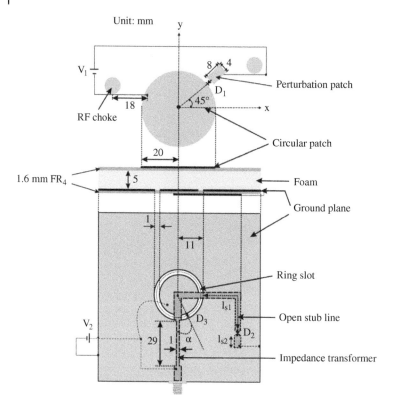

Figure 4.11 Geometry of the reconfigurable quadri-polarization diversity microstrip antenna. *Source*: Chen and Row [11]. © 2008, IEEE.

ring slot, and an x-directional LP radiation can be obtained by tuning l_{s2} and α. Simulation and experimental results from [11] show that the LP modes can be designed at the same operating frequencies, and their impedance matching also can be simultaneously achieved through the impedance transformer. By the above discussions, the polarization of the proposed reconfigurable antenna can be switched between the y- and x-directional LP modes by controlling the states of D_2 and D_3 while D_1 is OFF.

It is well known that the CP radiation can be generated by an inherently LP microstrip antenna when it is perturbed properly. In [11], a perturbation segment is introduced into the LP reconfigurable antenna, and therefore, the dual orthogonal LP modes have been converted to the RHCP and LHCP modes, respectively. One well-known method of realizing a switchable CP microstrip antenna is by embedding an unequal cross into the circular patch. However, the CP modes obtained by this method would have an operating frequency different from those of the original

LP modes due to the embedded slot. Here, an alternative method is proposed using a protruded stub connected to the main patch via a varactor diode D_1, as shown in Figure 4.11. When D_1 is ON, the circular radiating patch is perturbed, causing CP antenna rather than a LP one. Figures 4.12 and 4.13 show the radiation pattern measured for the one linear case and one circular case for the antenna proposed in [11]. Figure 4.14 shows the peak gain measured for each polarization over the frequency. Table 4.3 shows the state of the ports, diodes, and their result on the radiation polarization for the proposed geometry shown in Figure 4.11.

Another polarization reconfigurable antenna was proposed in [12] using a wheel-shaped antenna. The proposed antenna provides wide bandwidth and conical-beam radiation pattern. This wideband circular-monopolar patch was surrounded by eight coupling loop stubs which allow to have a conical pattern with different polarizations using PIN diodes on these coupling loop stubs. The center-fed circular patch operates with the vertically polarized conical-beam radiation while the coupling loop stubs generate a horizontally polarized radiation pattern. By adjusting the amplitude and phase ratio between these two orthogonal radiation polarizations, a circularly polarized conical-beam radiation can be obtained. In addition, a back-plane reflector was suggested to widen the axial ratio bandwidth and enhances the front-to-back ratio of the radiation pattern. This proposed antenna can generate three types of polarizations with the conical-beam radiation pattern including vertical polarization, LHCP, and RHCP by controlling the PIN diodes. Measured impedance and axial ratio bandwidths are 28.6% (3.45–4.6 GHz) and 15.4% (3.6–4.2 GHz), respectively, for the two CP modes. The maximum CP gain is 4.4 dBic. Furthermore, a dual-band operation (3.35–3.44 and 4.5–4.75 GHz) can be observed for the LP mode.

Figure 4.15 shows the proposed antenna geometry. The top and bottom layers are separated by some plastic posts with the distance of 15 mm. The wheel-shaped radiator and its ground plane are fabricated on the top layer of substrate #1 with 3.17 mm thickness. A metallic plane was placed on substrate #2 which serves as a reflector for the top radiator. Around the center feed circular patch on the top, eight shorted vias have been placed in order to merge the TM_{01} and TM_{02} operating modes around the design center frequency. This results in a wide impedance bandwidth. Eight loop stubs with coupling excitation from the monopolar patch are added for the CP conical radiation pattern. The coupling loop stubs are divided into small patches connected by PIN diodes. The other eight plate vias located at the end of the coupling loop stubs are designed for achieving wide axial ratio bandwidth. The DC biasing lines with RF chokes are connected to the eight big patches of the coupling loop stubs.

Figure 4.16a shows how the proposed design antenna operates as a polarization reconfigurable antenna. The monopolar patch radiates a vertically polarized wave same as an electric dipole while the eight coupling loop stubs contribute the

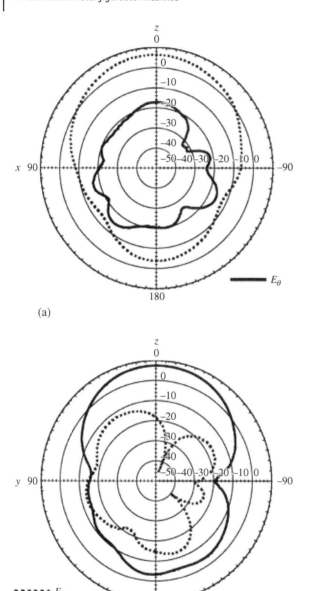

(a)

(b)

Figure 4.12 Radiation patterns measured at 2450 MHz for the example antenna operated at the y-directional LP mode. (a) x–z plane and (b) y–z plane. *Source*: Chen and Row [11]. © 2008, IEEE.

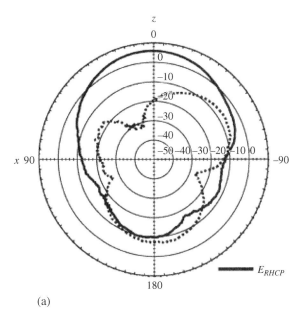

(a)

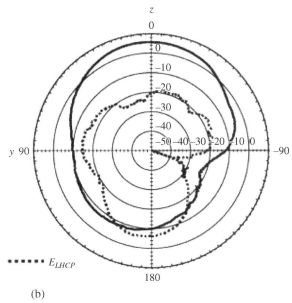

(b)

Figure 4.13 Radiation patterns measured at 2450 MHz for the example antenna operated at the RHCP mode. (a) x–z plane and (b) y–z plane. *Source*: Chen and Row [11]. © 2008, IEEE.

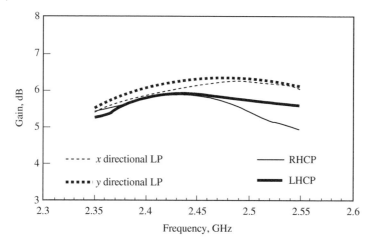

Figure 4.14 Measured gain variations against frequency for the example antenna operated at four different polarization modes. *Source*: Chen and Row [11]. © 2008, IEEE.

Table 4.3 State of the ports, diodes, and their result on the radiation polarization.

Port 1	Port 2	D_1	D_1	D_1	Polarization
OFF	ON	OFF	ON	ON	E_x
OFF	OFF	OFF	OFF	OFF	E_y
ON	ON	ON	ON	ON	E_{RHCP}
ON	OFF	ON	OFF	OFF	E_{LHCP}

Source: Chen and Row [11]. © 2008, IEEE.

horizontally polarized radiation as a magnetic dipole. The combined radiation from both the monopolar patch and the eight coupling loop stubs generates a conical CP radiation when the phase difference between their currents is 90°. This phase difference can be adjusted by the radius of the parasitic loop stubs. Furthermore, the current direction of the coupling loop stubs determines the CP orientation (RHCP or LHCP) of the antenna. PIN diodes are used to reconfigure the orientation of the loop stubs. Each loop stub element is divided into four patches as seen in Figure 4.16b. By appropriately turning ON and OFF the PIN diodes switches, the antenna radiation polarization can be switched between LHCP, RHCP, and vertically LP.

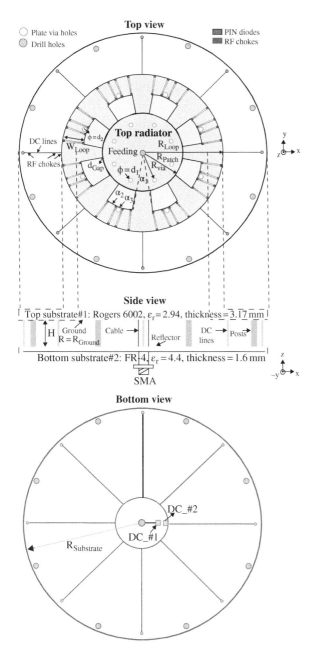

Figure 4.15 Geometry of the proposed design in [12] with conical pattern and three switchable polarizations. *Source*: Lin and Wong [12]. © 2015, IEEE.

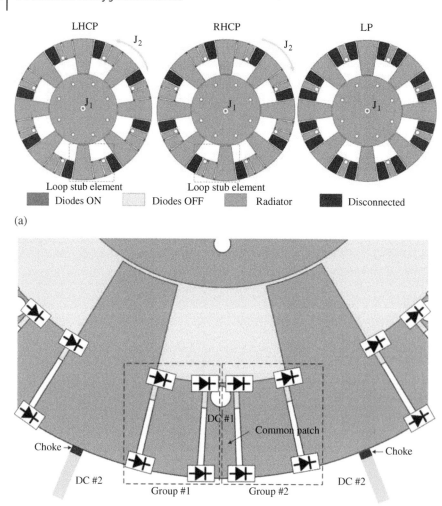

(a)

(b)

Figure 4.16 (a) Demonstration of reconfigurable polarizations realization by controlling PIN diodes. (b) Arrangement of PIN diodes and DC lines. *Source*: Lin and Wong [12]. © 2015, IEEE.

4.4 Mechanical and Micro-electro-mechanical (MEMS) RF Switch-Based Antennas

Another type of switch that has been considered widely in RF is the mechanical and micromechanical (MEMS) RF switches. These types of switches usually operate by applying DC voltage (current) which activate (deactivate) coil (coils) and a

Figure 4.17 A "Hinged" relay most common structure. *Source*: OMRON Corporation [13].

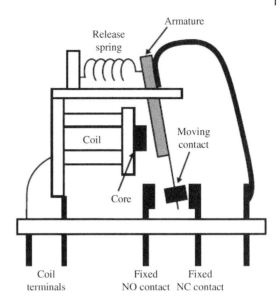

latching system. Therefore, general-purpose RF relays transfer signals through a mechanical action. Using RF relays and MEMS technology, RF switches with extremely low loss, small size, low power consumption, high quality factors, and high linearity compared with conventional semiconductor-based passive has been achieved. The RF relays come in different varieties, but the most basic type is called "Hinged" relays [13]. A Hinged relay switch contacts by the rotating movement of an armature around a fulcrum. A hinged relay structure is shown in Figure 4.17.

In hinged relays, an armature rotates around a fulcrum based on the applied DC power to the coil or the spring force which would open or close a contact. For the "turned ON" state, a DC current flows to the coil which would magnetize the core. As a result, the armature would be attracted to the core. When the armature is attracted to the core, the moving contact touches the fixed contact. One should notice that the release spring is stretched out in this state. Disconnecting the DC power, the input device goes to the "turned OFF" mode. In this state, the current to the magnetic (coil) is cut off and therefore, the force of attraction is lost. The force of the release spring returns the armature to its original position which makes the contacts separated. In the "ON" state, the basic hinged relay needs constant power which increases power consumption and generate heats. To solve this issue, latching relays have been introduced. Latching relays are usually considered for low power consumption or low heat applications. In the latching relays, applying constant DC power is not required which reduces the power

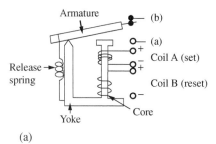

(a)

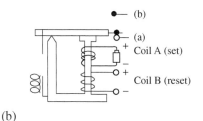

(b)

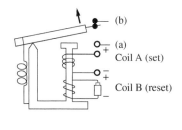

(c)

Figure 4.18 A mechanical latching relay (bistable relay) structure and mechanism of operation using two coils and a spring [13]. (a) Relaxed state (after reset). (b) Operating state (set). (c) Release state (reset) → relaxed state. *Source*: OMRON Corporation [13].

consumption as well as the self-heating of the coil. In latching relays, instead of applying a continuous voltage to the coil, they are operated with short-voltage pulses instead. Latching relays change contact position when a coil voltage is applied and remain in that position even if the voltage is disconnected. To reset a latching relay, another voltage pulse needs to be applied. Initially, mechanical latching relays were introduced first. Figure 4.18 shows a mechanical latching relay (bistable relay) structure and its mechanism of operation using two coils and a spring. The first diagram in Figure 4.18a shows the relay in the relaxed state. The latching relay looks similar to the three hinged relays described previously except that the core, yoke, and armature are made from semi-hard magnetic material and there are at least two coils in the relay. When the relay is in the relaxed state, applying current flow through coil A, which is made of semi-hard material, attracts the armature to the core due to its magnetization. As a result, the moving contact moves away from the normally closed (NC) contact (turns OFF) and makes contact with the normally open (NO) contact (turns ON). This has been shown in Figure 4.18b. In the set state, the residual magnetic flux in the semi-hard magnetic material (material that has properties similar to a permanent magnet) will keep the armature attracted to the core even if a current is no longer applied to coil A. To reset the relay, a voltage is needed to be applied to the coil B which is wound in the opposite direction to coil A as shown in Figure 4.18c. If a current is applied to coil B, the residual magnetic flux in the semi-hard magnetic material will be reduced and the magnetic attraction will weaken. The power of the release spring will become stronger than the magnetic attraction, so the armature will release, and the relay will be in the relaxed state. When the armature has released, there will be almost no residual magnetic flux in the semi-hard magnetic material. In recent communication systems, mechanical latching relays are not very common due to their reliability issue and have been replaced by magnetic latching relays.

In magnetic latching relays, the relays are polarized and use an internal magnet instead of mechanical springs to apply mechanical forces to the relay armature in the reset state. In non-latching relays, one magnet has greater flux compared to the other one causing the drop-out force being much stronger than the flux on the pull-in side. To operate the non-latching relay, a voltage is needed to be applied on the coil and the current through the coil in generating an additional flux on the pull-in side and overpowers the magnetic force on the drop-out side of the relay. For latching relays, the magnet needs to be charged in a different way. Both sides, pull-in and dropout, are magnetized equally. To move the armature from one position to the other, the magnet's force needs to be overpowered by the coil. Once the armature changed positions (SET), a flux in the opposite direction is necessary to switch the relay back (RESET). This can be done by changing polarity on the coil (single coil latching relays) or operating separate SET and RESET coils (dual coil latching relays). Magnetic latching relays do not have the same armature forces than the non-latching counterparts during operation. Therefore, most of them are more sensitive to shock and vibration and probably have lower contact ratings.

Using MEMS technology, micro-size mechanical switches are realized/embedded in electronics devices. RF MEMS switches offer both mechanical and semiconductor advantages and properties. MEMS switches offer low insertion loss and return loss, high isolation, wide bandwidth, low DC power consumption, high linearity, and can be used up to terahertz (THz) frequencies. Nevertheless, one of major disadvantages of RF relays and RF MEMS is low switching speed (5–50 μs) compared to the semiconductor switches (<1 μs). Another important disadvantage of the RF relays and MEMS is their high actuation voltage (5–30 V). Also "hot switching" is another important factor that needs to be addressed for the RF relay and MEMS switches' reliability [14].

Same as the RF relays, the RF MEMS switch can have two mechanisms for changing states: mechanical actuation and the electrical actuation. Methods used to realize mechanical movements in MEMS switches include electrostatic, electromagnetic, thermal, or piezoelectric mechanisms. Among these methods, electrostatic actuation mechanism is more favorable due to its low power consumption, thin layers of material, small electrode size, low switching time, and achievable range of contact forces at 50–200 μN [15]. Despite these advantages, electrostatic actuation MEMS requires a high actuation voltage, typically 5–100 V. Electromagnetic actuation requires lower voltage but with significantly higher current consumption [16]. It uses coil and ferromagnetic armature as an electromagnetic actuator. Permanent magnets or semi-hard magnetic materials allow addition of a self-latching mechanism in electromagnetic actuators same as the above-described method for the RF relays. This type of MEMS is also called electromagnetically actuated micromechanical relay. Instead of magnetic actuator, thermal actuator has been also used [17]. This type of MEMS switches consist of permanent magnets, armature of a soft magnetic material, and thermosensitive magnetic material stators. Another method of actuation is the piezoelectric mechanism where elastic deformation is induced using electrical field stimulation. In these switches, a downforce is introduced to the membrane of the switch using a piezoelectric actuator attached to it. This downforce mimics the ON and OFF conditions. Using this method, a low voltage actuation can be achieved without reducing switching times [18]. Thermal actuators are another mechanism which has been studied. In this method, thermal micro-actuators have been connected to a thin metal which serves as signal lines of coplanar transmission line. In these switches, using a short voltage or current pulse which generates enough heat, the switch can be set ON or OFF despite some advantages; this method is not widely utilized due to its higher DC power consumption compared to the other methods [19].

Many research works have been done to utilize the RF MEMS for realizing reconfigurable antennas. In [20], an antenna system with polarization reconfiguration and frequency switching has been realized using RF MEMS switches. The antenna is a pixelated patch design as shown in Figure 4.19. This reconfigurable pixel-patch antenna is built on a number of printed rectangular-shaped metallic pixels interconnected using RF MEM actuators on a microwave-laminated substrate. The RF MEMS actuator is made of a metallic movable film, suspended on top of a metal stub protruding from adjacent strips, fixed to both ends through metallic posts. An actuation voltage of 50 V can cause an electrostatic force that pulls the suspended membrane down and connect it to the stub (actuator ON) which connects the pixels together. Removing the voltage would make the pixels to disconnect from each other. This architecture allows to deactivate some actuators by keeping at zero biasing while activating the rest of them by applying DC bias voltage. Therefore, the geometry can be modified which results in a

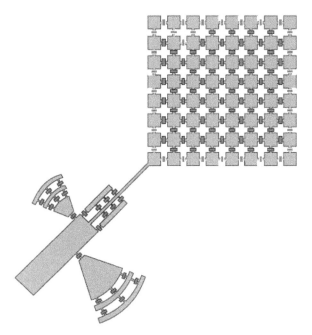

Figure 4.19 Schematic of the proposed reconfigurable pixel-patch antenna architecture using RF MEMS switches. *Source*: Cetiner et al. [20]. © 2004, IEEE.

multi-polarization and dual-frequency operation. As shown in Figure 4.19, the antenna is fed by a microstrip along its diagonal axis. The feed is also made reconfigurable as the input impedance of radiator changes based on the operation mode. The electrical lengths of the stubs implemented in the feed circuitry can be adjusted using the same method (activation/deactivation of MEM actuators) to achieve appropriate impedance matching for each mode of operation.

As the feed line is placed at the diagonal line of the pixel antenna, the antenna radiates in dual LP mode due to the pixels connected in both X- and Y-directions. Linear X- or linear Y-polarizations are obtained by connecting the pixels either in only the X- or only the Y-direction, respectively, as shown in Figure 4.20. To obtain circular polarization, an internal slot with proper dimensions and location was implemented in the antenna geometry. The slots were generated by properly deactivating some of the MEM actuators into the pixel-patch antenna geometry. The slots are calculated to excite X- and Y-polarized modes with equal amplitude and 90° phase difference. Therefore, right- and left-hand circularly polarized (RHCP and LHCP) radiation are achieved based on the implemented slots. The schematics of such reconfiguration are shown in Figure 4.21.

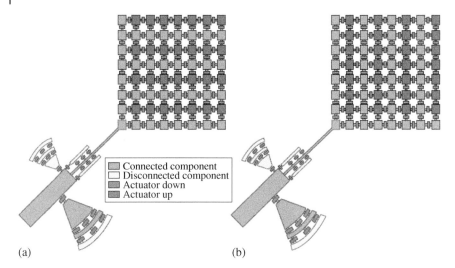

Figure 4.20 Reconfigurable pixel-patch antenna schematics for (a) linear X-polarization and (b) linear Y-polarization. *Source*: Cetiner et al. [20]. © 2004, IEEE.

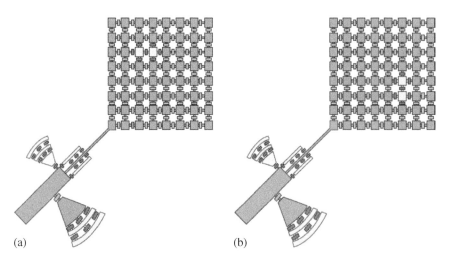

Figure 4.21 Reconfigurable pixel-patch antenna schematics for (a) RHCP and (b) LHCP. *Source*: Cetiner et al. [20]. © 2004, IEEE.

In another research, a K-band linear-polarization (LP)/circular-polarization (CP) switchable reconfigurable antenna using RF MEMS switches based on a novel package platform is introduced in [21]. In this design, the substrate of the package platform is used as the substrate for the antenna structure which would simplify the fabrication process and prevent the radiation performance from

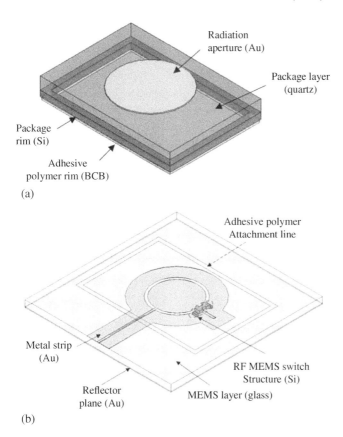

Figure 4.22 Geometry of the K-band polarization reconfigurable antenna integrated with a MEMS switch. (a) Top layer and (b) bottom layer. *Source:* Jung et al. [21]. © 2012, IEEE.

degradation. As shown in Figure 4.22, in the proposed geometry, the radiation aperture is placed on the top side of the package substrate; therefore, an aperture-coupling feed structure is employed. To implement the LP-CP polarization reconfiguration, a stub, acting as the other feed, is added. An RF-MEMS switch was used to connect/disconnect this extra feed line.

Figure 4.22 shows the packaged antenna structure. The proposed structure consists of two layers on top and bottom. The top layer is the substrate on which the radiation aperture is placed. This layer also protects the RF MEMS switch from the external environment. RF MEMS switch is mounted on the bottom layer which also has a signal line and a slot ring structure for aperture coupling. Quartz material has been used for the top layer which has relative permittivity and loss tangent as 3.78 and 0.04, respectively. An air cavity is required to be mounted onto the

RF-MEMS switch as the top and bottom layers get combined. To introduce such cavity, a silicon rim is built on the lower face of the top layer. This silicon ring can be considered as a good cavity due to the low relative permittivity and a tangential loss that is close to zero. The RF-MEMS switch is placed on the bottom layer along with the signal line, ground plate, and ring slot. Glass material was used for the bottom layer which has relative permittivity and loss tangent of 4.8 and 0.05, respectively. This geometry would prevent any undesirable and unmeasurable inductance caused by using wire bonding or soldering to connect the MEMS switch to the antenna. Also, the MEMS switch would not affect the antenna radiation behavior as it is mounted below the radiation aperture. The radiation polarization of the proposed geometry can be controlled by the ring slot, specifically a stub. To generate CP radiating fields, the circular aperture needs to have orthogonal excitation with 90° phase difference between them. Therefore, a stub is added orthogonal to the feed line and the ring slot. This stub operates as a second feed line generating electric fields orthogonal to the main feed line. By carefully adjusting the stub length, we can ensure that the two modes will have equal magnitudes and a 90° phase difference. To switch between the CP and LP radiation modes, a MEMS switch is placed on the stub. When the stub is shorted (switch OFF), no energy coupled through the stub, causing the antenna to radiate in LP mode. By turning ON the MEMS switch, the energy can couple to the patch through the stub which makes the antenna radiate in a LHCP radiation mode. LHCP is induced as the stub is positioned at $\phi = 0°$. RHCP can be generated by placing the stub at $\phi = 180°$. An impedance matching transformer is considered to properly match both the CP and LP modes due to the difference in their frequency response. A reflector plane is developed at the bottom of the glass substrate to compensate for the back radiation and creates a boresight directional pattern.

4.5 Switchable Feed Network-Based Polarization Reconfiguration

In some polarization reconfigurable antenna systems, the reconfiguration happens by changing the feed network structure using RF switches. By putting the RF switches and their control lines into the feed network PCB, their destructive effect on the antenna radiation behavior would be removed. It would also provide more freedom to the designer as there are usually more space available on the feed network to layout lines and components compared to the antenna layer. Besides, the design and integration process would be easier if one can design a reconfigurable feed network rather than putting the switches on the antenna. The drawback of using a reconfigurable feed network is the size of the PCB which would increase as one needs different lines to generate different excitation for each polarization.

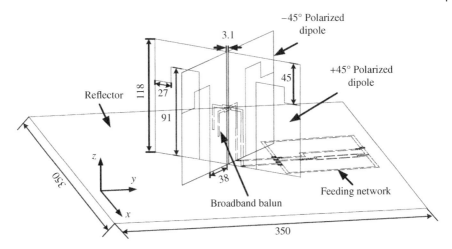

Figure 4.23 Configuration of broadband quad-polarization reconfigurable antenna (unit: mm). *Source*: Cui et al. [22]. © 2018.

A broadband quad-polarization reconfigurable antenna using a reconfigurable feed network was designed and reported in [22]. This reconfigurable antenna consists of a pair of crossed ±45° polarized dipoles excited using a broadband balun. Using a switchable feed network, if the ±45° polarized dipoles are fed separately, a ±45° LP radiation can be achieved. If both dipoles are simultaneously fed with equal amplitude and a +90° or −90° phase difference, an LHCP or RHCP radiation can be achieved. Crossed dipoles are notoriously wideband and could provide a good axial ratio if excited properly. The configuration of a broadband crossed dipole antenna is shown in Figure 4.23. Each dipole is printed on a dielectric substrate and is excited through a coupling mechanism by a broadband balun etched on the other side of the substrate. The crossed dipoles are placed above a reflector ground plane to obtain a directional pattern. The broadband antennas are connected to a reconfigurable feed network shown in Figure 4.24. This feed network consists of nine PIN diode switches to realize the quad-polarization reconfiguration. When switches 1 and 4 (S_1 and S_4) are turned ON while all other switches are OFF, the +45° polarized dipole is excited. Turning switches 1 and 3 (S_1 and S_3) ON while the rest of the switches are OFF would excite the −45° polarized dipole. To create CP with either RH or LH sense of rotation, a Wilkinson power divider is employed. When switches 2, 5, 7, and 9 (S_2, S_5, S_7, and S_9) are turned ON, the feeding path for −45° LP delays by 90° compared with the path for +45° LP; thus, LHCP is obtained. As switches 2, 5, 6, and 8 (S_2, S_5, S_6, and S_8) are turned ON, the feeding path for +45° LP delays by 90° regarding the path for −45° LP; thus, RHCP is achieved. The operating states for the quad-polarization reconfigurable antenna

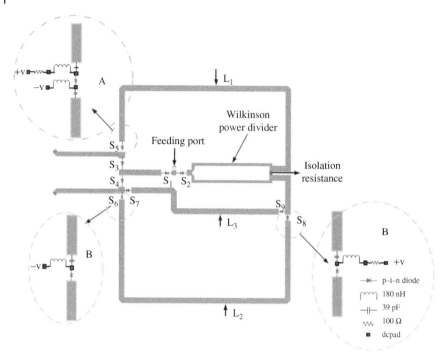

Figure 4.24 Feeding network with biasing circuit for quad-polarization reconfigurable antenna. *Source*: Cui et al. [22]. © 2018.

Table 4.4 Operating states of reconfigurable antenna.

	S_1	S_2	S_3	S_4	S_5	S_6	S_7	S_8	S_9
+45° LP	ON	OFF	OFF	ON	OFF	OFF	OFF	OFF	OFF
−45° LP	ON	OFF	ON	OFF	OFF	OFF	OFF	OFF	OFF
LHCP	OFF	ON	OFF	OFF	ON	OFF	ON	OFF	ON
RHCP	OFF	ON	OFF	OFF	ON	ON	OFF	ON	OFF

are listed in Table 4.4. Simulation and measurement show that the proposed quad-polarization reconfigurable crossed dipole antenna achieves a wideband impedance bandwidth of 37% with return loss >15 dB and an overlapped bandwidth of 36% for axial ratio <3 dB.

A polarization feed network (PFN) which was used in a phased array antenna implementation is shown in Figure 4.25 [23]. Here, by controlling the port excitations, we can generate radiation patterns with different polarization properties. If

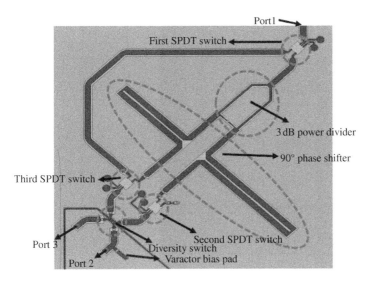

Figure 4.25 A polarization feed network (PFN) to control the ports' excitation of the antenna for achieving polarization reconfiguration. *Source*: Babakhani et al. [23].

only one port is excited at a time, the antenna can generate LP (along the x-axis (LPX) or along the y-axis (LPY) based on which port is excited). If both ports are excited with equal magnitude and ±90° phase difference, the antenna can generate RHCP or LHCP. This PFN includes the following: three SPDT Skyworks AS186-302LF RF switches (less than 1 dB insertion loss and isolation better than 45 dB), one diversity Skyworks 13355-374LF RF switch (less than 0.6 dB insertion loss and isolation better than 22 dB), equal-split Wilkinson power divider, and a modified coupled line Schiffman 90° phase shifter [24]. This feed network was designed in coplanar waveguide (CPW) structure on a 1.58 mm-thick FR4 substrate ($\varepsilon_r = 4.4$, tan δ = 0.02).

To explain the PFN working, the magnitude of surface current density for the feed network is shown in Figure 4.26 for the different polarizations. The state of the switches is defined with arrows for clear understanding. Comparing surface current for the LPs, LPX and LPY, it is seen that the signal is traveling through a simple transmission line up to the diversity switch. State of diversity switch determines which port receives the power and which port is matched terminated. For the RHCP and LHCP cases, the signal is going through the power divider and then the 90° coupled line phase shifter. Thus, there are two equally split signals where one has −90° phase difference compared to the other one at the inputs of the diversity switch. When the diversity switch directs the 0° signal to antenna port 1 and −90° signal to antenna port 2, we get RHCP. The reverse of this action generates LHCP. This feed network is a general design and can be used with any two

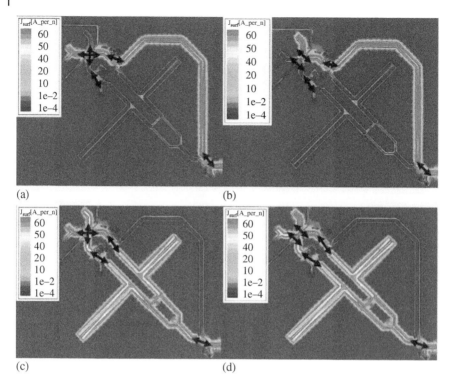

Figure 4.26 Magnitude of the surface current distribution for the different polarizations of (a) LPY, (b) LPX, (c) RHCP, and (d) LHCP at 2.0 GHz. The state of the switches has been defined with arrows. *Source*: Babakhani et al. [23].

Table 4.5 Measured loss for the PFN for the linear and circular polarizations.

Frequency (GHz)	1.5	1.7	1.9	2.0	2.2	2.4
PFN loss for LP (dB)	−2.23	−2.34	−2.95	−2.96	−2.87	−3.68
PFN loss for CP (dB)	−3.13	−3.25	−3.4	−3.27	−3.34	−4.1

Source: Babakhani et al. [23].

port antenna configuration supporting two orthogonal LPs. This PFN suffers loss due to the lossy FR-4 substrate material (tan δ = 0.02), RF switches loss (0.7 dB, nominal value for each switch), biasing components (DC blocking capacitor and RF choke inductors), and quality of the soldering. Loss values are noted for selected frequencies in Table 4.5. One should notice that for LP, the signal would travel through three RF switches while for CP, the signal should go through four RF, which causes more losses.

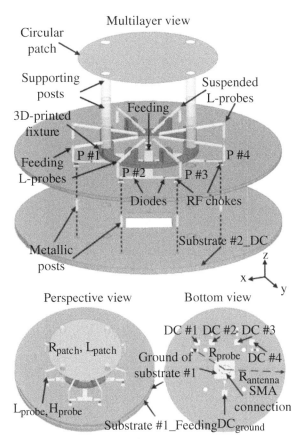

Figure 4.27 Multi-polarization reconfigurable circular patch antenna with eight L-shaped probes. *Source*: Lin and Wong [25]. © 2017, IEEE.

A reconfigurable feed network was also used to realize a multi-polarization reconfigurable circular patch antenna with eight L-shaped probes [25]. This design could switch between 0°, +45°, −45°, and 90° LPs which makes it suitable to mitigate the polarization mismatch problem in complex wireless channels. This polarization reconfigurability is realized by utilizing PIN diode switches into the feed network. As shown in Figure 4.27, this design consists of three main parts: a top circular radiating patch, a reconfigurable network of eight L-shaped probes, and an additional substrate considered for the bias lines. A circular patch was used which makes it possible to create LP along any arbitrary line of symmetry and by appropriate feeding. Four pairs of suspended L-probes were used to feed the circular patch using coupling mechanism. Each pair is responsible to generate LPs (0°, +45°, −45°, and 90°). A pair was used instead of a single probe for

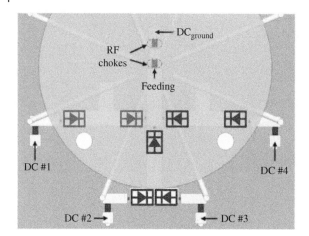

Figure 4.28 Arrangement of the PIN diode switches. *Source*: Lin and Wong [25]. © 2017, IEEE.

Table 4.6 Polarization states by status of the DC biases.

Polarization	DC #1	DC #2	DC #3	DC #4
0°	3	0	0	0
45°	0	3	0	0
−45°	0	0	3	0
90°	0	0	0	3

symmetrical excitation which would improve the polarization purity by reducing the cross-polarization level. A switching mechanism was used to select the appropriate probe. This circuitry is shown in Figure 4.28. In this feed network, seven PIN diode switches were used with their bias line located on substrate #2. Table 4.6 shows the applied DC voltage at each point on the feed network to realize a specific polarization.

An array of stacked patch antennas composed of four LP patches is used to realize a quad polarization supporting vertical LP, horizontal LP, LHCP, and RHCP [26]. In the design, as shown in Figure 4.29, a double-layer structure consists of a driven substrate integrated waveguide (SIW) cavity and four parasitic patches are used to develop the wideband antenna element. The parasitic radiation patches are excited by etching a ring slot on the top surface of the SIW cavity, using two diagonal ports. A 180° phase shifting is used for polarization reconfigurable

(a)

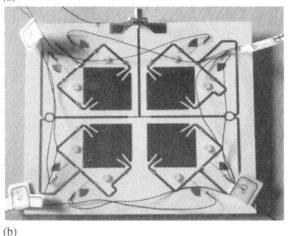

(b)

Figure 4.29 Photographs of the fabricated quad polarization reconfigurable antenna array. (a) Front and (b) bottom. *Source*: Hu et al. [26].

applications by individually feeding the antenna element with the two diagonal ports. The schematic structure of the proposed design is given in Figure 4.30. This antenna has two major parts: four LP radiating elements and a reconfigurable feeding network. The antennas are arranged in a 90° rotation in turn. Each antenna has two switchable input/output ports along its diagonal line. Due to the symmetry in the design, individual excitation at the two diagonal ports can achieve a 180° phase-shifting. For the reconfigurable feeding network, a one-to-four-way power divider alongside of four SPDT switches and two 90° phase

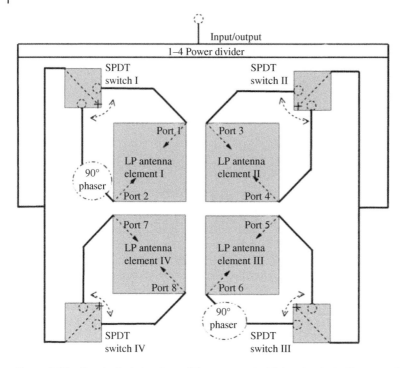

Figure 4.30 Topological structure of the proposed wideband polarization reconfigurable patch antenna array. *Source*: Hu et al. [26]. © 2017, IEEE.

Table 4.7 Polarizations by different feeding ports of the four LP antenna elements.

	Antenna I		Antenna II		Antenna III		Antenna IV	
Polarization	Port 1	Port 2	Port 3	Port 4	Port 5	Port 6	Port 7	Port 8
V-LP	ON	OFF	ON	OFF	ON	OFF	ON	OFF
H-LP	ON	OFF	OFF	ON	ON	OFF	OFF	ON
LHCP	OFF	ON	OFF	ON	OFF	ON	OFF	ON
RHCP	OFF	ON	ON	OFF	OFF	ON	ON	OFF

shifters were designed. By properly controlling the states of the four SPDT switches, quad-polarization states can be realized. As an example, if ports 1, 3, 5, and 7 are selected while other ports are left open, a vertical LP radiation will be obtained. Table 4.7 summarizes the state of the switches and their corresponding polarizations.

4.6 Polarization Reconfigurable Antennas Using Metasurface

In designing a polarization reconfigurable antenna system, as mentioned previously, a switching mechanism is involved. The switching mechanism redirects the current or waves to create different polarization in far field. Instead of changing the radiation polarization of the actual antenna, some studies have used a metasurface as an alternative method to reconfigure the radiation polarization of the antenna system.

In electromagnetic field, a metasurface refers to a periodic sub-wavelength structure behaving as an artificial material with properties different from those found in naturally occurring materials. MSs can block, absorb, enhance, or bend electromagnetic waves which allow achieving benefits that go beyond what is possible with natural materials [27]. MSs have been given several names such as artificial magnetic conductor (AMC), electromagnetic band gap (EBG), frequency selective surfaces (FSS), and metamaterial structures. They have been used for several purposes such as high impedance surface (HIS) reflectors, planar lens, vortex generator, beam deflector, radiation-absorbent material (RAM), and spectrum filters. MSs have also been used as polarization converters which lead to designing antenna systems with polarization reconfiguration.

Polarization converters are designs which control and manipulate the polarization state of electromagnetic waves. Polarization converters were initially introduced in optics and are being adopted in electromagnetic field by using MSs. Manipulating the polarization of an incident wave at will is highly desired and finds several applications. LP converter working in transmission mode is perhaps the most considered convertor. This type of converter is used to align the polarization orientation wave based on the unknown orientation of the source. A LP converter usually rotates the polarization direction of a LP electromagnetic wave by 90°, making it perpendicular to the original one. In satellite communications, polarization converters can be used to countervail the effect of Faraday rotation caused by the ionosphere. Polarization converters have also been used for polarization diversity purposes. To obtain polarization diversity, reconfigurable converters have been considered which would create a polarization reconfigurable antenna system. The configuration occurs by applying a mechanical change to the orientation or structure of the metasurface or by adding a switch which could modify the metasurface structure.

A planar polarization-reconfigurable antenna using metasurface has been proposed and designed in [28]. This antenna consists of a planar metasurface placed on top of a planar slot antenna. Both the antenna and the metasurface are in direct contact which makes it a compact and low-profile design. By mechanically

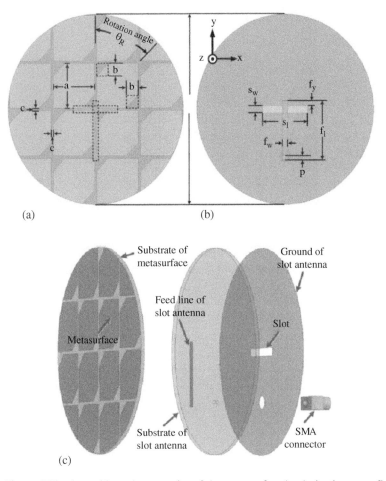

Figure 4.31 Assembly and geometries of the metasurfaced polarization reconfigurable antenna: (a) metasurface, (b) slot antenna, and (c) assembly. *Source:* Zhu et al. [28]. © 2014, IEEE.

rotating the metasurface around the center with respect to the slot antenna, the polarization of the antenna system can be reconfigured between LP, LHCP, and RHCP. This design is shown in Figure 4.31 and dimensions can be found in [28].

The polarization-reconfigurable metasurface (PRMS) antenna uses a slot antenna as the source and a metasurface superstrate. The metasurface is composed of corner truncated square unit cells as shown in Figure 4.31a which has been designed on a single-sided substrate. A Sub-Miniature Adapter (SMA) connector is used to feed the slot antenna using a transmission line. An equivalent

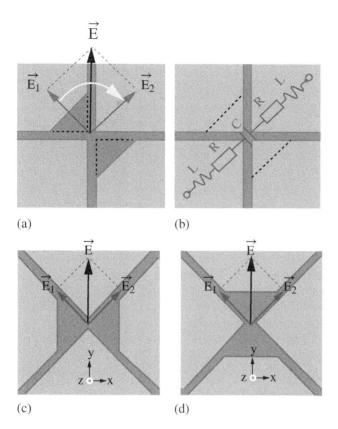

Figure 4.32 (a) E-fields on a unit cell with diagonal corner truncated at $\theta_r = 0°$, (b) equivalent circuit without truncation, (c) unit cell at $\theta_r = 45°$, and (d) unit cell at $\theta_r = 135°$. *Source:* Zhu et al. [28]. © 2014, IEEE.

circuit can be used to explain why the proposed PRMS antenna can have different polarizations by rotating the metasurface structure. As shown in Figure 4.31a, the slot antenna has a LP along the y-axis. Placing the metasurface on top of the slot antenna, the radiating field can be resolved into two orthogonal components as shown in Figure 4.32a. Considering Figure 4.32b, if the diagonal corners of the unit cell are not truncated, the two orthogonal fields at $\vec{E}_1$ and $\vec{E}_2$ would see an identical RLC circuit shown in Figure 4.32b with an equivalent impedance of

$$Z = 2R + j\omega(2L) + \frac{1}{j\omega C} = R' + jX'.$$ In this equation, C is the capacitance caused by

the gap between the patches. Truncating the corners of the patches as shown in Figure 4.32a, two different impedances of Z_1 and Z_2 would be seen by the $\vec{E}_1$ and $\vec{E}_2$ fields, respectively. Due to the cut orientation in Figure 4.32, the Z_1 would be less capacitive than the Z_2, which would cause a phase difference between the $\overline{E}_1$ and $\overline{E}_2$ fields. But carefully designing the truncation size, one can get $|Z_1| = |Z_2|$ while $\angle Z_1 - \angle Z_2 = 90°$ which would in turn give $|E_1| = |E_2|$ and $\angle E_1 - \angle E_2 = 90°$. As a result, the E-field through the metasurface would be LHCP. For the RHCP case, the metasurface needs to be rotated by $\theta_r = 90°$. If we rotate the metasurface by $\theta_r = 45°$ or $135°$, as shown in Figure 4.32c and d, the unit cell becomes symmetric which gives $\overline{E}_1$ and $\overline{E}_2$ an identical impedance. In this case, the radiated waves through the metasurface would have a LP. Simulation and measured results show that this PRMS antenna achieves a fractional operating bandwidth of 11.4%, a boresight gain of above 5 dBi, and high-polarization isolation of larger than 15 dB in the CP mode. In the LP mode, the antenna achieves a gain of above 7.5 dBi with cross-polarization isolation larger than 50 dB.

Another type of metasurface was used with the same slot antenna and same concept for frequency reconfiguration [29]. In this design, instead of corner truncated patches, a patch with unequal arm cross slot was used. The geometry of the antenna plus the metasurface is shown in Figure 4.33. Same as the design mentioned in Figure 4.31, the unequal arms of the slot in Figure 4.33 expose the orthogonal E-fields into two different impedances. This would generate a LHCP case for the orientation shown in Figure 4.33c. Rotating metasurface 90° would generate a RHCP case while 45° rotation creates a symmetric design which in turn creates a LP wave.

Mechanical changes are not popular for creating polarization reconfigurable antennas due to low reliability and hardship of implementing. Therefore, studies have been done to change the state of the metasurface polarization converters using switches. Specially, PIN diodes have been considered in several research works. An active reconfigurable polarization converter using PIN diode switches has been reported in [30]. This active metasurface consists of two identical layers of elliptic split rings loaded with PIN diodes on top of a dielectric layer. No bias network is needed for the PIN diodes as they are biased through the interconnected elliptic split rings. The active metasurface induces a linear to circular polarization conversion when the diodes are OFF, while no conversion takes place when the diodes are ON.

Studies show that meandered line metal traces formed as semi-ellipses can cause phase difference between the two components of the incident E-field vector, and the transmission wave would become a CP wave. As shown in Figure 4.34, assuming an incident E-field vector with a 45° tilt angle with respect to the axis of the metasurface plane (E_{inc}), the transmission field through the arrays can be decomposed into two orthogonal polarizations (E_x and E_y) with equal amplitude (E_0). The electric field of the transmitted wave can be expressed as $\vec{E} = \left(T_{xx}\vec{x} \pm T_{yy}\vec{y} \right)E_0$. The T_{xx} and T_{yy} are the transmission coefficients of E_x and E_y, respectively. If the diodes are OFF, there is no connection between the meander

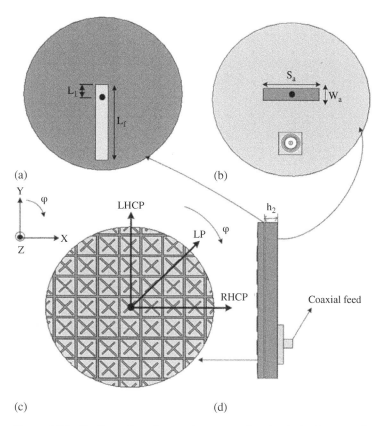

Figure 4.33 Configuration of cross slot metasurfaced polarization reconfigurable antenna: (a) slot antenna top, (b) slot antenna bottom, (c) cross slot metasurface, and (d) side view. *Source*: Kandasamy et al. [29]. © 2015, IEEE.

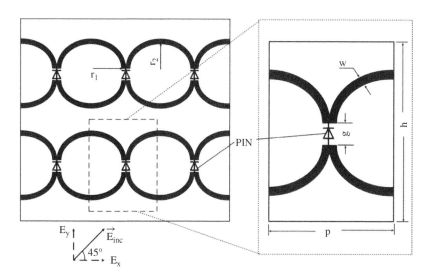

Figure 4.34 Schematic of the active metasurface and enlarged unit cell. *Source*: Li et al. [30]. © 2016, IEEE.

Figure 4.35 Photograph of the horn antenna with the proposed active polarization reconfigurable converter. *Source*: Li et al. [30].

lines and they behave as shunt inductors and capacitors for E_x and E_y, respectively. A phase difference of $\pm\pi/2$ can be induced between T_{xx} and T_{yy} which would consequently generate a RHCP or LHCP wave, respectively. By turning the switches ON, the meander lines can be connected causing a decrease in the capacitance for the E_y component, while the component E_x is not affected and sees a shut inductance. When the proper parameters of the converter are chosen, the phase difference decreases to 0°, and the transmitted wave becomes a co-polarized LP wave. A Ku-band LP horn antenna was used with the proposed reconfigurable polarization converter to achieve a polarization reconfigurable horn antenna. Figure 4.35 shows the photograph of the horn with converter. The antenna is located at $d = 5\,mm$ away from the converter. According to the above-mentioned analysis, the converter should rotate at a 45° angle with respect to the center axis of the horn. The horn with the proposed converter has a partially reflecting polarization reconfigurable feature, which allows for the horn polarization to be adjusted between LP and CP by changing the state of the PIN diodes [30].

4.7 Other Methods to Create Polarization Reconfigurable Antennas

To create a polarization reconfigurable antenna, the mechanical methods are the least favorable method due to their issues such as speed, reliability, and complexity of the design. However, mechanical methods have their own advantages and found application in reconfigurable antenna design area. Three major benefits of the mechanical methods over semiconductor and MEMS switches are given below:

1) The high linearity and low loss. This feature is beneficial especially in the case of extremely wideband application or high-power RF application. The mechanical methods are capable of handling high power without showing any nonlinear behavior over a large range of frequency.
2) As mentioned earlier, the solid-state switches or the MEMS design needs an extra circuitry for biasing or actuating the mechanism, thus increasing the complexity of the designs. Meanwhile, the self-resistance of the component used in the bias network could affect the efficiency of antennas.
3) Using mechanical methods, one can employ soft materials fabricated using common technologies. Using such material in antenna fabrication process can result in flexible and transparent design which is preferable for wearable applications.

Besides mechanical switches, different other mechanical methods such as piezo-electric transducer (PET) and fluid conductor reconfigurable antenna have been investigated which we will also review in this chapter.

A novel reconfigurable microstrip antenna with switchable circular polarization using a PET was presented in [31]. Two dielectric perturbers attached to the PETs are used to create RHCP or LHCP using an appropriate perturbation. The geometry of the proposed reconfigurable antenna is shown in Figure 4.36. The radiating element is a truncated corner square single-fed microstrip patch. As mentioned earlier, truncating the corners of a square patch antenna is a typical method to generate CP, which can be explained by the cavity model method. Lifting up both perturbers, a pair of degenerated modes (TM_{01} and TM_{10}) are excited at the same frequency. When one of the perturber is covering one corner of the patch (pulled down), this perturbation will cause two degenerate modes to split unequally. With a proper amount of perturbations, the two modes will have slightly different frequencies of resonance with equal radiation field magnitude but 90° out of phase. The dielectric constant and thickness of the perturber are the main design parameters which would affect the amount of perturbations. Studies show that the perturber's dielectric constant and thickness need to be greater than those of the patch substrate to create a sufficient amount of perturbations. Also, the dielectric perturbers are used as they induce lower loss compared to the metallic one. The PETs and perturbers are deflected vertically in the z-direction to yield the required perturbations. Disconnecting the DC source from the right PET while the left one is "ON," the top right corner of the patch will be perturbed and LHCP will be generated. RHCP can be generated in the reverse situation. Due to the symmetry of the design, the RHCP and LHCP should have the same performance. This is a desired feature for the polarization diversity applications.

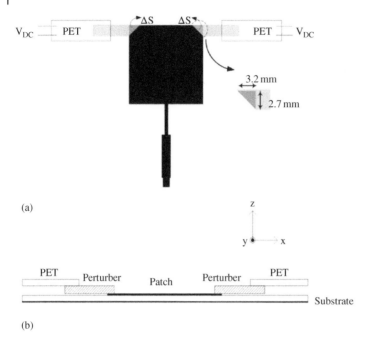

(a)

(b)

Figure 4.36 Geometry of the reconfigurable microstrip antenna with switchable circular polarization using a PET: (a) top view and (b) side view. *Source*: Hsu and Chan [31]. © 2007, IEEE.

A polarization-reconfigurable antenna was proposed using a pressure-driven fluidic loading network [32]. The proposed antenna consists of two independent, colocated, and orthogonal (crossed) narrow microstrip patch antennas. Using the fluidic loading network with repeating and alternating high-low dielectric constant fluids, the patches can be excited independently through a capacitive coupling mechanism. The geometry of this design is shown in Figure 4.37. In this geometry, the radiating element consists of two orthogonally crossed colocated narrow microstrip patches. The patches are electrically isolated at their overlap using four gaps. An SMA feed located at the center of the crossed patches is used to excite the antenna. Two mechanisms can be used to reconfigure the antenna polarization. One method is by changing the coupling level between the arms. Another method is using a switching mechanism to connect or disconnect the patch along the x- and y-directions to the excitation point.

A surface integrated fluidic network (SIFN) was used to capacitively load the gaps with dielectric fluids in an alternating "high-low" dielectric and a rotational arrangement around the central feed location as illustrated in Figure 4.38. Using this mechanism, the gaps can be loaded appropriately to independently activate

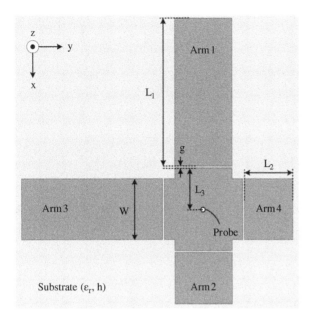

Figure 4.37 Geometry of the antenna with dual polarization capability consisting of decoupled narrow patch antenna. *Source*: Barrera and Huff [32]. © 2014, IEEE.

the two antennas. The fluids used here have different complex dielectric constants which are housed in a calculated dimension. Using a 3D-printed acrylonitrile butadiene styrene (ABS) tube, the arrangement of the fluids can be displaced. The tubes have an overlay with the gaps in the arms. The dielectric constant of the fluid strongly impacts the gap coupling capacitance C_g. This has been analytically studied in [31]. As the capacitive coupling increases with larger ε_r, the antenna resembles like a normal patch. Therefore, the operating frequency will shift toward the resonant frequency of a traditional patch (gaps replaced by solid metallization) with similar dimensions, since a short begins to occur across the gap. To minimize the gap resistance (reduce the Ohmic loss), a low-loss liquid needs to be used. Thus, the drop in the radiation efficiency from Ohmic losses stays low for low-loss fluids. Room temperature water with $\varepsilon_r = 7.69(1 - j0.127)$ was used as the high dielectric fluid for a proof-of-concept. The high dielectric water loads the gaps to excite the antenna near the resonant frequency of a traditional patch with similar dimensions. In this design, the lossy dielectric causes a significant effect on the antenna efficiency. On the other hand, air was used as the second dielectric which would appropriately isolate the arms from the central feed section. A peristaltic pump was connected to both ends of the ABS interface creating a closed loop within the SIFN. This applies pressure-driven forces to push

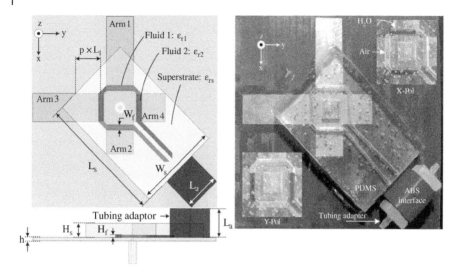

Figure 4.38 Crossed patches with a SIFN for polarization reconfiguration purpose. *Source*: Barrera and Huff [32]. © 2014, IEEE.

the two water strips into X-pol and Y-pol configurations as shown in Figure 4.38. Each of the antennas can be turned OFF by pushing air over its corresponding gaps. Therefore, the polarization can be reconfigured between X-pol and Y-pol by appropriately putting the air and water dielectrics at the gaps of the patch antennas.

Polarization reconfiguration was implemented for the same antenna geometry using RF PIN diode switches and compared to the fluidic approach discussed above. Using RF PIN diode switches, an electrical connection can be established instead of capacitively loading the gaps. The diodes are oriented in opposite directions as shown in Figure 4.39. The switches can electrically connect or disconnect appropriate arms to the feed pin and independently excite different antenna polarizations. To use switches, appropriate bias circuit was designed using RF choke lines for the DC ground. An external bias tee was used to apply the DC voltage to PIN diode switches. Studies shows that X-pol and Y-pol antennas can be excited independently using the switching mechanism [31].

Using both SIFN as well as the PIN diode switches facilitate reconfigurable polarization. However, some differences regarding antenna performance can be observed between the two methods. In particular, electrical size, radiation efficiency, radiation patterns, and switching speed are examined. Studies shows that PIN diodes tend to load the antenna more severely compared to the SIFN. Therefore, the physical size of the antenna for a specific design frequency would be smaller if using PIN diodes rather than the SIFN. Smaller physical size would give the

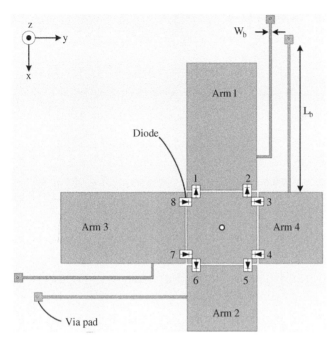

Figure 4.39 Eight RF PIN diodes used to create a polarization reconfigurable antenna using two crossed patch antennas. *Source*: Barrera and Huff [32]. © 2014, IEEE.

antenna a smaller aperture which in turn would give lower radiation efficiency [31]. Also, studies show that the SIFN coupling mechanism is less sensitive to Ohmic losses from nonideal components (dielectric fluids in this case) compared to the PIN diode switches. These two factors make the overall efficiency of the SIFN design to be placed above the PIN mechanism. The control lines for the SIFN have minimal effect on the antenna radiation pattern. The circuit feeding the peristaltic pump to the SIFN is far removed from the radiating surface of the antenna. Thus, no DC interference to the radiation patterns is contributed in the SIFN model. As for the PIN diode switches, the coupling between the antenna and the bias lines cause higher cross polarization. Regarding the switching speed, the electrically driven RF PIN diode model can be switched in order of $1\,\mu s$ whereas the SIFN will have inherently slower switching speeds due to the kinetic operation of a pump.

A circularly polarized water-based spiral antenna with polarization reconfigurability has been designed and tested in [33]. This reconfigurable antenna is basically an Archimedean spiral antenna with two water arms fed by a parallel stripline. To create a directional pattern, a conducting plate is used as a reflector.

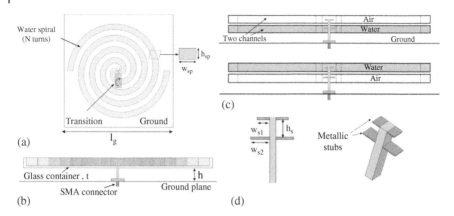

Figure 4.40 Reconfigurable Archimedean spiral antenna using water. *Source*: Hu et al. [33] © 2017, IEEE.

By controlling the water flow among two water channels, upper and lower, the polarization of the antenna can be reconfigured between LHCP and RHCP, respectively. As shown in Figure 4.40a, the Archimedean spiral antenna made of pure water is composed of two water channels with a specific cross section. A square conducting plate was used underneath the Archimedean for unidirectional radiation. A wideband transition from the coaxial line to the parallel stripline is used to feed the antenna. Separated by distance h, a container of thickness t is used to hold the water (Figure 4.40b). Pure water, which is basically a dielectric material of high permittivity, is used as the arms to guide the electromagnetic waves. The water arm acts as a leaky waveguide at frequencies where the power cannot be well confined to the arm resulting in radiation. Circular polarization is expected to occur at the place where the circumference is around one wavelength, which is as per the operating principle of a conventional metallic spiral antenna. Traditional metallic spiral antennas have a good CP performance in a wide frequency range. Circularly polarized pattern is fixed based on the arms' curving direction. In this polarization reconfigurable design, two water channels, one on top and one on the bottom side, are made to support an RHCP and LHCP water spiral, respectively, as shown in Figure 4.40c. The parallel stripline with two pairs of short metallic stubs is connected to both channels for excitation of electromagnetic waves. By controlling the water flow to fill only one channel at a time, the polarization reconfigurability is achieved. A good impedance match can be obtained by optimizing the lengths and positions of the metallic stubs (Figure 4.40d). For the fabricated prototype, measured results show an overall bandwidth ($|S11| < -10\,\mathrm{dB}$, $AR < 3\,\mathrm{dB}$) of 40% for the two polarization states.

4.8 Conclusion

In this chapter, various polarization reconfigurable antennas have been presented and discussed. Using these designs, different methods of creating polarization reconfiguration have been evaluated and presented. These methods include RF switches (RF relays, solid-state switches, and MEMS), switching feeding network, metamaterials, MSs, and fluid driven with micro-pumps. These methods have their own advantages and disadvantages such as the antenna efficiency, bandwidth, reconfiguration speed, power handling, cost, and implementation complexity. With the advancement in technology, new areas have emerged especially for the high frequency and millimeter waves. Millimeter wave is the key for the new generation of wireless communications (5G) to achieve extremely high throughput. The 5G systems at millimeter frequencies also need to maintain a decent diversity to support massive and multi-user multiple-input multiple-output (MU MIMO) to achieve the maximum capacity of the system, especially in non-line of sight (NLOS) communication situation. Finally, polarization reconfigurable antennas at millimeter wave frequencies can serve as the means of creating polarization diversity, and can find great applications.

References

1 Fang, S.-T. (2000). A novel polarization diversity antenna for WLAN applications. *2000 IEEE AP-S Dig.*, Salt Lake City, Utah, USA (16–21 July 2000), 282–285.

2 Hettak, K., Delisle, G.Y., and Stubbs, M.G. (2000). A novel variant of dual polarized CPW fed patch antenna for broadband wireless communications. *2000 IEEE AP-S Dig.*, Salt Lake City, Utah, USA (16–21 July 2000), 286–289.

3 Boti, M., Dussopt, L., and Laheurte, J.-M. (2000). Circularly polarized antenna with switchable polarization sense. *Electron. Lett.* 36 (18): 1518–1519.

4 Yang, X.-X. and Zhong, S.-S. (2000). Analysis of two dual-polarization square-patch antennas. *Microwave Opt. Technol. Lett.* 26 (3): 153–156.

5 Huang, J. (2001). Miniaturized UHF microstrip antenna for a Mars mission. *2001 IEEE AP-S Dig.* 4: 486–489.

6 Yang, F. and Rahmat-Samii, Y. (2002). A reconfigurable patch antenna using switchable slots for circular polarization diversity. *IEEE Microwave Wirel. Comp. Lett.* 12 (3): 96–98.

7 Sung, Y.J., Jang, T.U., and Kim, Y. (2004). A reconfigurable microstrip antenna for switchable polarization. *IEEE Microwave Wirel. Comp. Lett.* 14 (11): 534–536.

8 Chen, W.-S., Wu, C.-K., and Wong, K.-L. (2001). Novel compact circularly polarized square microstrip antenna. *IEEE Trans. Antennas Propag.* 49: 340–342.

9 Khidre, A., Lee, K., Yang, F., and Elsherbeni, A.Z. (2013). Circular polarization reconfigurable wideband E-shaped patch antenna for wireless applications. *IEEE Trans. Antennas Propag.* 61 (2): 960–964.

10 Guterman, J., Moreira, A., Peixeiro, C., and Rahmat-Samii, Y. (2009). Reconfigurable E-shaped patch antenna. *Proc. IEEE IWAT Int. Workshop on Antenna Technology Symp. Digest,* Santa Monica, CA, USA (2–4 March 2009),1–4.

11 Chen, R. and Row, J. (2008). Single-fed microstrip patch antenna with switchable polarization. *IEEE Trans. Antennas Propag.* 56 (4): 922–926.

12 Lin, W. and Wong, H. (2015). Polarization reconfigurable wheel-shaped antenna with conical-beam radiation pattern. *IEEE Trans. Antennas Propag.* 63 (2): 491–499.

13 OMRON Corporation. Technical explanation for general-purpose relays. http://www.ia.omron.com/data_pdf/guide/36/generalrelay_tg_e_10_3.pdf.

14 Liu, Y., Bey, Y., and Liu, X. (2016). Extension of the hot-switching reliability of RF-MEMS switches using a series contact protection technique. *IEEE Trans. Microwave Theory Tech.* 64 (10): 3151–3162.

15 Rebeiz, G.M. (2003). *RF MEMS: Theory, Design, and Technology.* New York, NY: Wiley.

16 Jaafar, H., Beh, K.S., Yunus, N.A. et al. (2014). A comprehensive study on RF MEMS switch. *Microsyst. Technol.* 20 (12).

17 Tilmans, H.A.C. Fullin, E., Ziad, H. et al. (1999). A fully-packaged electromagnetic microrelay. *Technical Digest. IEEE International MEMS 99 Conference. Twelfth IEEE International Conference on Micro Electro Mechanical Systems (Cat. No.99CH36291),* Orlando, FL (21–21 January 1999), 25–30.

18 Lee, H.-C., Park, J.-Y., and Jong-Uk, B. (2005). Piezoelectrically actuated RF MEMS DC contact switches with low voltage operation. *IEEE Microwave Wirel. Comp. Lett.* 15 (4): 202–204.

19 Mahameed, R. and Rebeiz, G.M. (2010). A high-power temperature-stable electrostatic RF MEMS capacitive switch based on a thermal buckle-beam design. *J. Microelectromech. Syst.* 19 (4): 816–826.

20 Cetiner, B.A., Jafarkhani, H., Qian, J.-Y. et al. (2004). Multifunctional reconfigurable MEMS integrated antennas for adaptive MIMO systems. *IEEE Commun. Mag.* 42 (12): 62–70.

21 Jung, T.J., Hyeon, I., Baek, C., and Lim, S. (2012). Circular/linear polarization reconfigurable antenna on simplified RF-MEMS packaging platform in K-band. *IEEE Trans. Antennas Propag.* 60 (11): 5039–5045.

22 Cui, Y.H., Zhang, P.P., and Li, R.L. (2018). Broadband quad-polarisation reconfigurable antenna. *Electron. Lett.* 54 (21): 1199–1200.

23 Babakhani, B., Sharma, S.K., and Labadie, N.R. (2016). A frequency agile microstrip patch phased array antenna with polarization reconfiguration. *IEEE Trans. Antennas Propag.* 64 (10): 4316–4327.

24 Quirarte, J.L.R. and Starski, J.P. (1993). Novel Schiffman phase shifters. *IEEE Trans. Microwave Theory Tech.* 41 (1): 9–14.

25 Lin, W. and Wong, H. (2017). Multipolarization-reconfigurable circular patch antenna with L-shaped probes. *IEEE Antennas Wirel. Propag. Lett.* 16: 1549–1552.

26 Hu, J., Hao, Z., and Hong, W. (2017). Design of a wideband quad-polarization reconfigurable patch antenna array using a stacked structure. *IEEE Trans. Antennas Propag.* 65 (6): 3014–3023.

27 Sievenpiper, D. (2008). Artificial impedance surfaces for antennas. In: *Modern Antenna Handbook* (ed. C. Balanis), 737–778. Wiley. ISBN: 978-0-470-03634-1.

28 Zhu, H.L., Cheung, S.W., Liu, X.H., and Yuk, T.I. (2014). Design of polarization reconfigurable antenna using metasurface. *IEEE Trans. Antennas Propag.* 62 (6): 2891–2898.

29 Kandasamy, K., Majumder, B., Mukherjee, J., and Ray, K.P. (2015). Low-RCS and polarization-reconfigurable antenna using cross-slot-based metasurface. *IEEE Antennas Wirel. Propag. Lett.* 14: 1638–1641.

30 Li, W., Xia, S., He, B. et al. (2016). A reconfigurable polarization converter using active metasurface and its application in Horn antenna. *IEEE Trans. Antennas Propag.* 64 (12): 5281–5290.

31 Hsu, S. and Chang, K. (2007). A novel reconfigurable microstrip antenna with switchable circular polarization. *IEEE Antennas Wirel. Propag. Lett.* 6: 160–162.

32 Barrera, J.D. and Huff, G.H. (2014). A fluidic loading mechanism in a polarization reconfigurable antenna with a comparison to solid state approaches. *IEEE Trans. Antennas Propag.* 62 (8): 4008–4014.

33 Hu, Z., Wang, S., Shen, Z., and Wu, W. (2017). Broadband polarization-reconfigurable water spiral antenna of low profile. *IEEE Antennas Wirel. Propag. Lett.* 16: 1377–1380.

5

Liquid Metal, Piezoelectric, and RF MEMS-Based Reconfigurable Antennas

Jia-Chi S. Chieh and Satish K. Sharma

5.1 Introduction

This chapter will discuss reconfigurable antennas that utilize unique material properties. Topics include antennas and arrays developed using liquid metal, piezo-electric material, and RF MEMS.

5.2 Liquid Metal – Frequency Reconfigurable Antennas

Often times pattern and frequency reconfigurable antennas require switches to activate different modes. Researchers have also long envisioned antennas which are reconfigured based on their physical attributes using liquid metal. Although not commonly used, it remains a topic of high interest in the research community.

In [1], researchers developed and demonstrated a liquid metal monopole antenna using micro-fluidic channels. Figure 5.1 shows a stack-up of the proposed antenna. The researchers used mercury as the liquid metal, which is enclosed inside a 2 mm micro-fluidic channel made of PDMS ($\varepsilon_r = 2.8$, tan $\delta = 0.02$). The micro-fluidic channel is sealed with 1-mil liquid crystal polymer (LCP) ($\varepsilon_r = 2.9$, tan $\delta = 0.0025$). A 62-mil Rogers RT5880 ($\varepsilon_r = 2.2$, tan $\delta = 0.0009$) is then laminated to the LCP layer. The Duroid layer carries the 50 Ω feed line as well as the ground plane. The feed line couples to the liquid metal through the thin LCP layer. A small bidirectional pump is then used to lengthen or shorten the liquid-metal slug inside the micro-fluidic channel. The length of the liquid-metal monopole determines the frequency of operation. The width of the micro-fluidic channel has a great impact on the operation of the antenna. Figure 5.2 shows a test of various micro-fluidic channel widths. As the channel gets wider, the liquid

Multifunctional Antennas and Arrays for Wireless Communication Systems, First Edition.
Edited by Satish K. Sharma and Jia-Chi S. Chieh.
© 2021 John Wiley & Sons, Inc. Published 2021 by John Wiley & Sons, Inc.

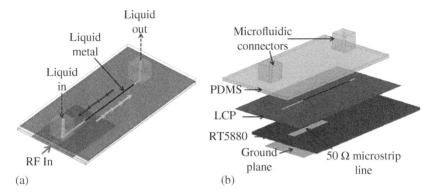

Figure 5.1 Microfluidically reconfigured wideband frequency-tunable antenna. *Source*: Dey et al. [1]. © 2016, IEEE.

Figure 5.2 Various micro-fluidic channel width testing with liquid metal. *Source*: Dey et al. [1]. © 2016, IEEE.

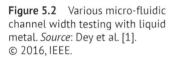

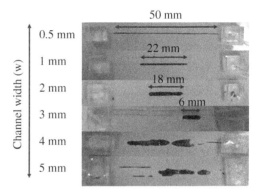

metal starts to take shapes that did not cover the entire channel length, and therefore movable/tunable metal-slugs were not attainable.

Figure 5.3 shows the practical implementation of the liquid-metal frequency reconfigurable antenna. The antenna requires bidirectional pumps which control the flow of the liquid metal. Figure 5.4 shows examples of various liquid-metal slug lengths, which correspond to various frequencies (1.29, 2.48, 3.53, and 5.17 GHz). The measured reflection coefficient magnitude (S_{11}) is also shown in Figure 5.4. The radiating lengths are 50, 25, 20, and 10 mm, which correspond to 1.29, 2.48, 3.53, and 5.17 GHz, respectively. The measured normalized radiation patterns of the antenna at 2.4 and 4.8 GHz are shown in Figure 5.5. The cross-polarization levels were less than 10 dB in both E- and H-planes. Figure 5.6 shows the measured realized gain and efficiency of the antenna over frequency. From 1.5 to 5 GHz, the measured efficiency of the antenna is >80%.

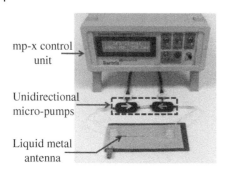

Figure 5.3 Liquid-metal flow control system. *Source*: Dey et al. [1].

mp-x control unit

Unidirectional micro-pumps

Liquid metal antenna

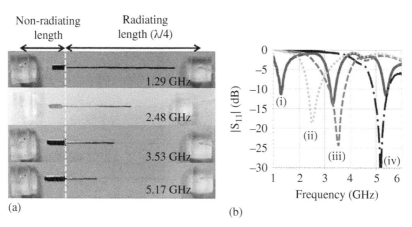

(a)

(b)

Figure 5.4 Liquid-metal frequency tuning. *Source*: Dey et al. [1]. © 2016, IEEE.

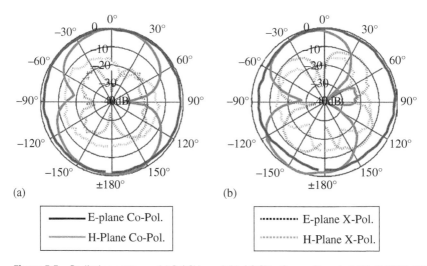

(a)

(b)

| ——— E-plane Co-Pol. | ·········· E-plane X-Pol. |
| ——— H-Plane Co-Pol. | ·········· H-Plane X-Pol. |

Figure 5.5 Radiation patterns: (a) 2.4 GHz and (b) 4.8 GHz. *Source*: Dey et al. [1]. © 2016, IEEE.

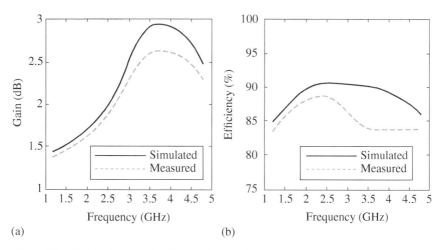

Figure 5.6 Realized gain and efficiency. *Source*: Dey et al. [1]. © 2016, IEEE.

A similar concept is presented in [2], in which an inverted-F-type antenna was realized utilizing liquid metal. Figure 5.7 illustrates this concept. In this realization, a Teflon tube is used to pump in the liquid Galinstan metal. Galinstan maintains a liquid state over a wide temperature range (from −19 to 1300 °C). The antenna is realized on a FR-4 substrate, with the upper arm implemented with the liquid metal. A Teflon tube is filled with Galinstan metal, and controlled using a vacuum pump. The liquid metal is connected to the copper using a shorting pin, which is inserted into the Teflon tube. Figure 5.8 shows the experimental setup for the automatic tuning of this antenna. Control system feedback is used to tune the length of the antenna automatically to the desired frequency of operation. Figure 5.9 shows the measured and simulated reflection coefficient magnitude of the antenna. The antenna was tuned from 698 to 746 MHz, and designed for LTE applications in mind.

5.3 Liquid Metal – Pattern Reconfigurable Antennas

Researchers have also been looking at using liquid metal to realize pattern reconfigurable antennas. In [3], researchers propose a pattern reconfigurable liquid metal antenna through the use of a main cone antenna and reflectors which can be activated by filling rods with metal. Figure 5.10 illustrates this concept. An aluminum cone is the driven antenna element, which is on a circular ground plane. Eight parasitic poles serve as reflector elements. The eight poles are hollow sleeve structures made from polymethyl methacrylate (PMMA), with a relative

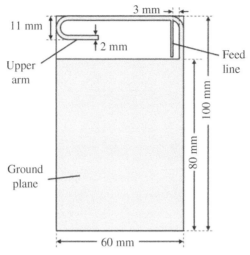

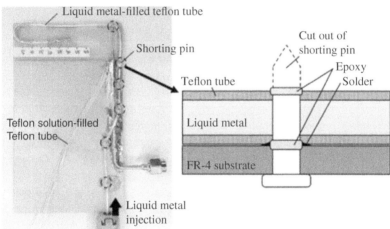

Figure 5.7 Liquid metal inverted-F-type antenna. *Source*: Ha and Kim [2]. © 2016.

permittivity of 3.7, which is nonconductive and transparent to RF. The poles are placed in 45° intervals. The whole antenna structure is mounted on a FR-4 dielectric substrate.

Since the current distribution of the cone is a traveling wave, the bandwidth is not affected by the radiation or impedance characteristics. The concept of pattern reconfigurability is similar to that of a Yagi-Uda antenna, which can switch its beam by controlling the directors and reflectors. The eight poles/sleeves act as reflectors, and are inductive in nature. The phase of the inducted electromagnetic wave is

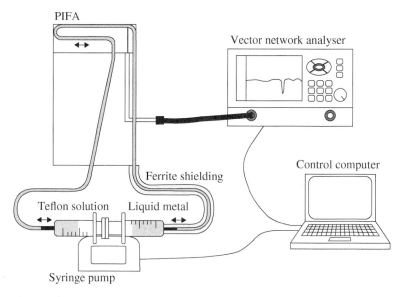

Figure 5.8 Automatic frequency tuning setup. *Source*: Ha and Kim [2]. © 2016.

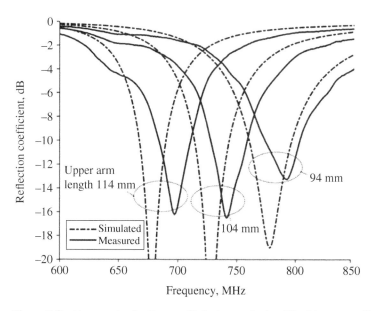

Figure 5.9 Measured reflection coefficient magnitude of liquid antenna. *Source*: Ha and Kim [2]. © 2016.

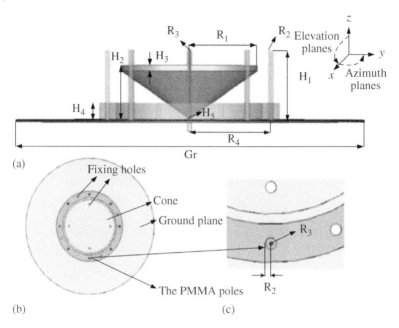

Figure 5.10 Pattern reconfigurable liquid metal antenna. *Source*: Bai et al. [3]. © 2018, IEEE.

opposite of the main reflector, and therefore the radiation pattern will be deflected in the opposite direction to the metal probes. The eight metal poles result in four antenna modes, and a total of 21 different beam patterns. When all poles are empty, the beam is omnidirectional and known as mode 1. When one pole is activated, this is known as mode 2. Mode 3 is realized as a dual-beam configuration and is activated using two poles in diagonal directions. Finally, mode 4 is when adjacent poles are used to realize a narrow beam. All of the possible modes are shown in Table 5.1.

In mode 1, since the poles/sleeves are not filled with liquid metal, the dielectric posts have little effect on the omnidirectional radiation pattern. In mode 2, the effects of the height of liquid metal poles do make a difference. As the liquid metal poles get higher, the reflection coefficient gets better at lower frequencies, but the direction of the radiation pattern degrades. Radiation patterns as well as the beam width and reflection coefficient of mode 2 are shown in Figure 5.11. The beam width is broad as expected since only a single reflector is used. Mode 3 exhibits a bidirectional pattern as two liquid metal poles are used, which are opposite each other. Radiation patterns, beam widths, and the reflection coefficient are shown in Figure 5.12 for mode 3. Finally, mode 4, adjacent metal posts are used, and therefore the pattern is more directional than in mode 2. Radiation patterns, beamwidth, and the reflection coefficient for mode 4 are shown in Figure 5.13.

Table 5.1 Possible radiation pattern modes from [3] showing states of the liquid metal, the beam width, and the main-lobe direction at 2.2 GHz.

Mode	M_1	M_2	M_3	M_4	M_5	M_6	M_7	M_8	Beam width	Beam direction
1	0	0	0	0	0	0	0	0	360°	NA
2	1	0	0	0	0	0	0	0	238.3°	0°
	0	1	0	0	0	0	0	0		+45°
	0	0	1	0	0	0	0	0		+90°
	0	0	0	1	0	0	0	0		+135°
	0	0	0	0	1	0	0	0		+180°
	0	0	0	0	0	1	0	0		+225°
	0	0	0	0	0	0	1	0		+270°
	0	0	0	0	0	0	0	1		+315°
3	1	0	0	0	1	0	0	0	90.3°	0°
	0	1	0	0	0	1	0	0		+45°
	0	0	1	0	0	0	1	0		+90°
	0	0	0	1	0	0	0	1		+135°
4	1	1	0	0	0	0	0	0	200.9°	+20°
	0	1	1	0	0	0	0	0		+70°
	0	0	1	1	0	0	0	0		+110°
	0	0	0	1	1	0	0	0		+160°
	0	0	0	0	1	1	0	0		+200°
	0	0	0	0	0	1	1	0		+250°
	0	0	0	0	0	0	1	1		+290°
	1	0	0	0	0	0	0	1		+340°

Source: Bai et al. [3]. © 2018, IEEE.

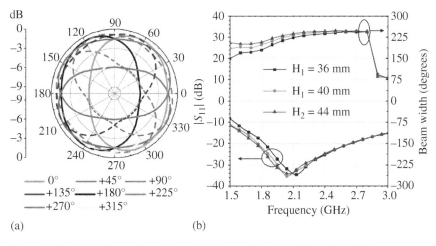

(a)　　　　　　　　　　　(b)

Figure 5.11 Radiation pattern of mode 2. *Source*: Bai et al. [3]. © 2018, IEEE.

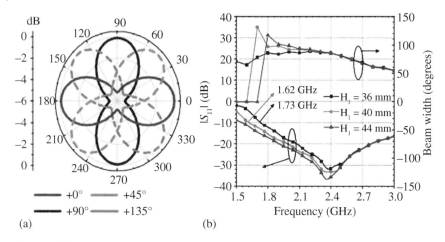

Figure 5.12 Radiation pattern of mode 3. *Source*: Bai et al. [3]. © 2018, IEEE.

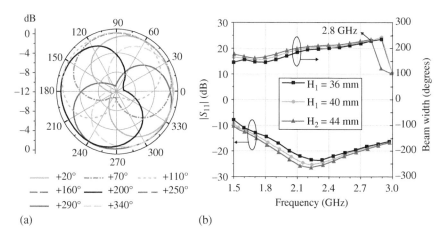

Figure 5.13 Radiation pattern of mode 4. *Source*: Bai et al. [3]. © 2018, IEEE.

The realized prototype antenna from [3] is shown in Figure 5.14. Liquid silver is used ($\sigma = 3.46 \times 10^6$ S/m). Measured radiation patterns for mode 1, mode 2, mode 3, and mode 4 at 1.7 and 2.7 GHz are shown in Figure 5.15. The fabricated antenna prototype shows 21 different beams steering over 360° of coverage. The antenna operates from 1.7 to 2.7 GHz, with a fractional bandwidth of 45.5%, with a maximum gain of 6.7 dBi. Figure 5.16 shows the measured gain over frequency for the various modes.

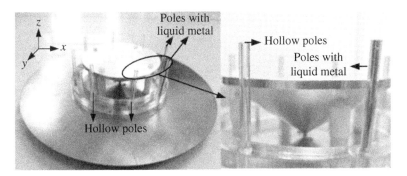

Figure 5.14 Fabricated prototype. *Source*: Bai et al. [3]. © 2018, IEEE.

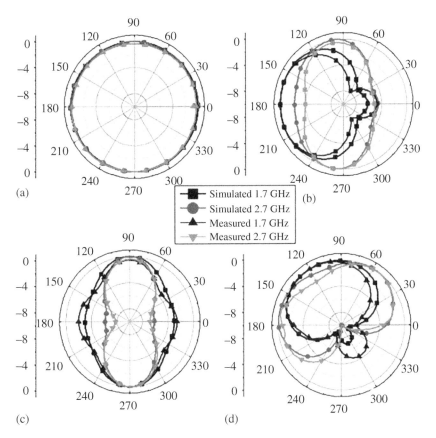

Figure 5.15 Measured radiation patterns. *Source*: Bai et al. [3]. © 2018, IEEE.

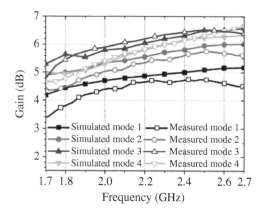

Figure 5.16 Measured gain vs. frequency. *Source*: Bai et al. [3]. © 2018, IEEE.

5.4 Liquid Metal – Directivity Reconfigurable Antennas

Liquid metal antennas also enable directivity reconfigurable antennas. In [4], researchers have designed a two-arm Archimedean spiral antenna on a stretchable elastomer. The directivity of the antenna can be adjusted based on inflating the elastomer to form a dome-shaped antenna of various heights. Microelectromechanical microblowers pneumatically control the shape of the antenna.

The antenna concept is shown in Figure 5.17. The antenna is fabricated on a SU-8 photoresist. A silicon Ecoflex is casted onto the SU-8 to form spiral channels. Gallium-indium eutectic (EGaIn) is then injected into the channel. MEMS microblowers are then used to pneumatically push the elastomer into a dome-like cap (Figure 5.18). This type of antenna is reliant on stretching the elastomer on which the antenna resides, which means that the conductor needs to also stretch with the elastomer. For this reason, liquid metal is vital to the design. Figure 5.18 shows fabricated prototype as well as a plot of the antenna height versus actuation voltage.

Figure 5.19 shows the simulated and measured time-lapsed profiles of the liquid metal antenna at various inflated heights, from 0 to 21.8 mm. Figure 5.20 shows the measured directivity patterns at 8.5 and 9.5 GHz on both vertical and horizontal polarizations. The directivity of the antenna at 8.5 GHz with 0, 10.9, and 21.8 mm are 5.39, 7.24, and 7.74 dBi, respectively. At 9.5 GHz, with 0, 10.9, and 21.8 mm, the directivities are 7.57, 8.92, and 9.90 dBi, respectively. The antenna achieves over 58% increase in directivity from the uninflated state to the fully inflated state.

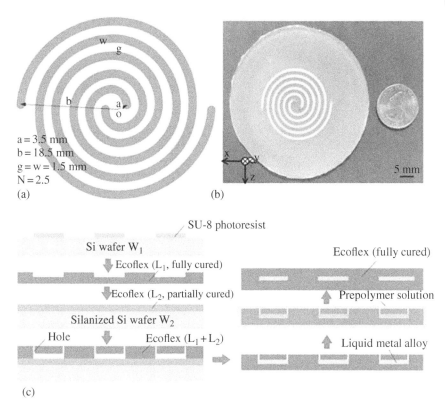

Figure 5.17 Inflatable elastomer-based liquid metal antenna with directionality re-configurability. *Source*: Liu et al. [4]. © 2017, IEEE.

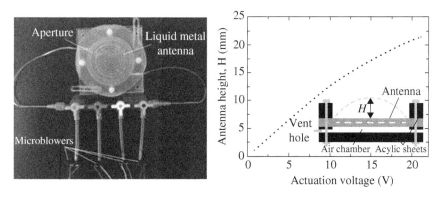

Figure 5.18 Fabricated prototype. *Source*: Liu et al. [4]. © 2017, IEEE.

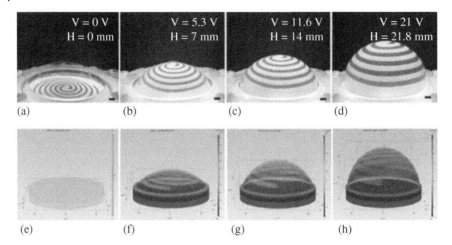

Figure 5.19 Simulated and measured time-lapse images of various inflation heights. *Source*: Liu et al. [4]. © 2017, IEEE.

5.5 Piezoelectric – Pattern Reconfigurable Array

Phased arrays are a subset of pattern reconfigurable antennas. Typically, they are implemented by using solid-state phase shifters/time delays, or through RF MEMs phase shifters. Historically, phase shifters have been the main bottleneck for widespread adoption of phased array antennas. For this reason, researchers have tried using piezoelectric materials to implement low-cost phased array antennas.

In [5], a piezoelectric transducer (PET) is used. The PET is a piezoelectric ceramic, which is deflected by applying a voltage. A dielectric perturber is attached to the PET, and moves vertically on microstrip lines. The dielectric perturbation changes the propagation constant and therefore the phase. This method is especially appropriate for linear phased array antennas. This method is also broadband because the perturbations on the transmission line change the phase but does little to the impedance mismatch and the insertion loss. Figure 5.21 illustrates this concept.

In [5], DC bias voltages of 0 and 90 V are used to deflect the PET. The dielectric perturbation affects the distributed capacitance on the transmission line below, changing the phase. The phase shift is proportional to the length of the perturbed line, and therefore, a triangular-shaped perturber is adopted. The dielectric used has a $\varepsilon_r = 10.8$ and a thickness of 25 mm. The PET has a size of $2.75 \times 1.25 \times 0.085\,\text{in}^3$ and is composed of lead zirconate titanate. Figure 5.22 shows the simulated and measured phase shift over frequency using this method.

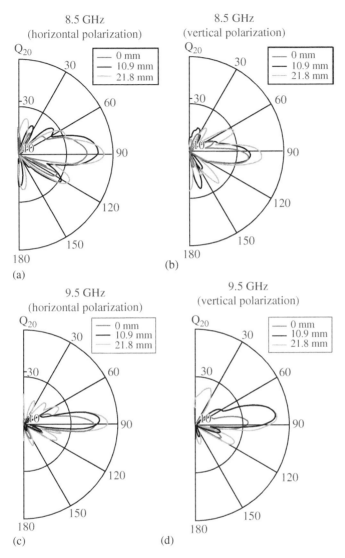

Figure 5.20 Measured directivity patterns for various heights at 8.5 and 9.5 GHz. *Source*: Liu et al. [4]. © 2017, IEEE.

An E-plane linear phased array was built and demonstrated. The test configuration is shown in Figure 5.23. The fabricated prototype is shown in Figure 5.24. Figure 5.25 shows the measured maximum beam scan patterns. A scanning angle of $-16°$ and $+17°$ was achieved at 10 GHz (Figure 5.26).

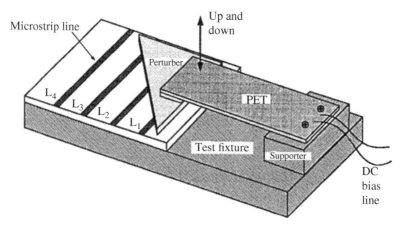

Figure 5.21 Piezoelectric transducer (PET) phase shifter. *Source*: Yun and Chang [5]. © 2001, IEEE.

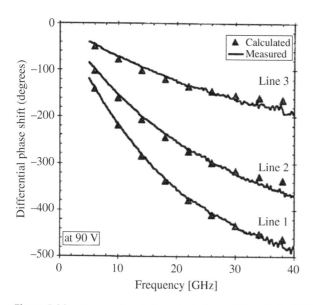

Figure 5.22 Measured and simulated phase shifter of the PET. *Source*: Yun and Chang [5]. © 2001, IEEE.

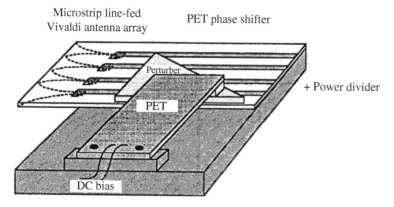

Figure 5.23 E-plane PET-based phased array. *Source*: Yun and Chang [5]. © 2001, IEEE.

Figure 5.24 E-plane PET-based phased array prototype. *Source*: Yun and Chang [5]. © 2001, IEEE.

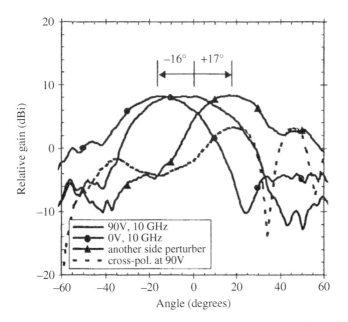

Figure 5.25 E-plane PET-based phased array scan patterns. *Source*: Yun and Chang [5]. © 2001, IEEE.

Researchers in [6] used a similar method to develop a system that operates at 60 GHz, in the millimeter-wave spectrum. This offers a low-cost beam scanning array at high frequencies. Figure 5.27 shows the measured beam patterns at 62 GHz with max perturbation in the negative and positive angles. This

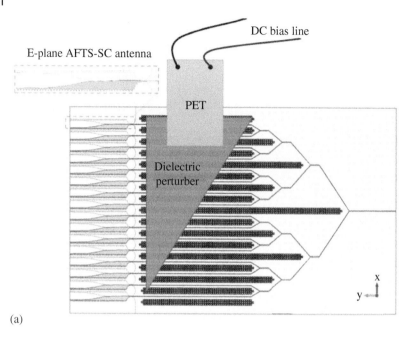

(a)

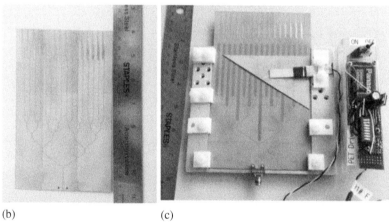

(b)   (c)

Figure 5.26 E-plane V-Band PET-based phased array prototype. *Source*: Briqech et al. [6]. © 2017, IEEE.

method allows for ±2° of beam deflection at 62 GHz. Although interesting, fully integrated silicon beamforming chipsets will most likely replace this PET method, especially as 5G rolls out and phased array antennas become ubiquitous.

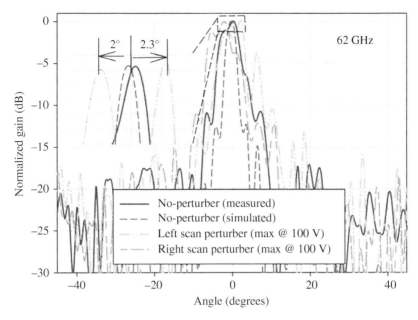

Figure 5.27 Beam scanning patterns. *Source*: Briqech et al. [6]. © 2017, IEEE.

5.6 RF MEMS – Frequency Reconfigurable

Micro-electro-mechanical systems (MEMS) have shown a lot of promise as MEMS switches have incredibly low insertion loss as well have the ability to reach millimeter-wave frequencies. As such, they are often used for reconfigurable antennas.

In [7], researchers use MEMS switches to create a pixelated reconfigurable patch antenna to operate from 1.13 to 1.7 GHz. The antenna topology is that of a slot-fed patch antenna. However, instead of a solid metal patch, the patch is composed of small squares which can be joined or disconnected through a MEMS switch. This concept is shown in Figure 5.28. The authors use the RMSW101 RF MEMS switch, which has a 0.2 dB insertion loss with a 20 dB isolation. The patch layer, slot layer, and feed line layers are fabricated on separate FR-4 substrates,

Figure 5.28 System Level Concept for MEMs pixelated reconfigurable antenna. *Source*: Wright et al. [7]. © 2018, IEEE.

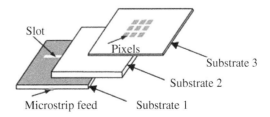

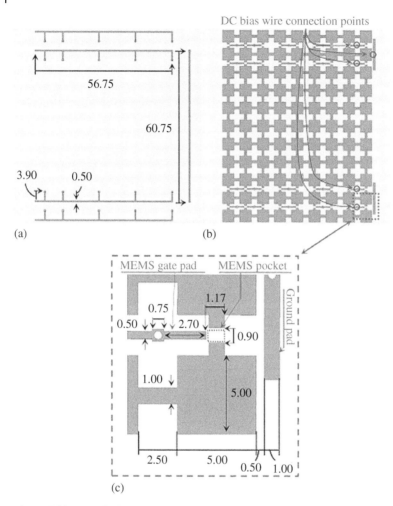

Figure 5.29 MEMS pixelated reconfigurable antenna concept element level and array level. *Source*: Wright et al. [7]. © 2018, IEEE.

with a ROHACELL foam in between the slot and the patch layers. Various columns can be activated/addressed, yielding a metal patch that can have varying electrical sizes. Figure 5.29 shows how the pixels are connected and how the MEMS switches are integrated onto the printed circuit board. The authors were able to preset three different modes/sizes, and the corresponding broadside gain of the pixelated patch antenna is shown in Figure 5.30. Figure 5.31 shows the measured broadside gain in the three modes. As expected, at higher frequencies the physical aperture is larger and so the gain increases as the frequency increases.

 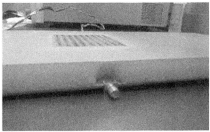

Figure 5.30 Prototype MEMs pixelated reconfigurable antenna. *Source*: Wright et al. [7].

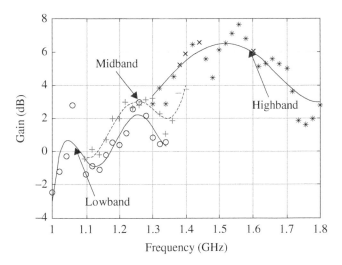

Figure 5.31 Measured broadside gain of the pixelated patch in three different modes. *Source*: Wright et al. [7]. © 2018, IEEE.

5.7 RF MEMS – Polarization Reconfigurable

In [8], researchers realized a polarization reconfigurable antenna using RF MEMS switches. The antenna has a 10 dB impedance bandwidth of 22.9% in linear polarized mode, and 28.43% in circular polarized mode. The 3 dB axial ratio bandwidth is 13.07% with measured gain at 21 GHz of 2.63 dBi (LP) and 3.90 dBi (CP).

Figure 5.32 shows the proposed antenna topology. The antenna is a circular patch excited using a coupled ring slot aperture feed. A stub, within the slot, is used to generate the orthogonal electric field versus the feed line, to generate the circular polarized radiation. The antenna is fed via finite coplanar waveguide

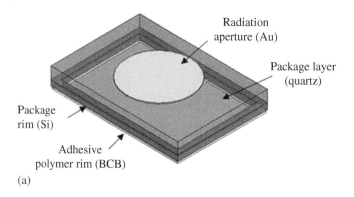

Radiation aperture (Au)

Package layer (quartz)

Package rim (Si)

Adhesive polymer rim (BCB)

(a)

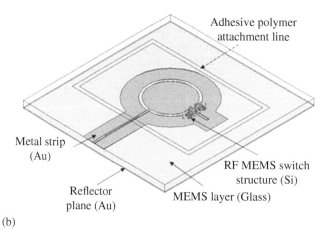

Adhesive polymer attachment line

Metal strip (Au)

RF MEMS switch structure (Si)

Reflector plane (Au)

MEMS layer (Glass)

(b)

Figure 5.32 Polarization reconfigurable antenna using RF MEMS switch. *Source:* Jung et al. [8]. © 2012, IEEE.

(F-CPW). The antenna is realized using a multilayer approach with substrates including Si, BCB, and Quartz. The RF MEMS switch is implemented on the bottom layer. A schematic of the switch is shown in Figure 5.33. The switch membrane and mechanical springs are made of single crystal silicon (SCS). When a bias voltage is applied between the membrane and the bottom electrode, the membrane descends and makes contact between the ring slot and the stub. Figure 5.34 shows the simulated reflection coefficient magnitude of this antenna in both linear and circular polarization modes, as well as the axial ratio in both modes.

The fabricated prototype of this antenna is shown in Figure 5.35. The measured axial ratio is shown in Figure 5.36. In the CP state, the axial ratio is lower than 3 dB from 19.3 to 22 GHz, which corresponds to a 13.07% axial ratio bandwidth. In the LP state, the impedance bandwidth is 22.9%.

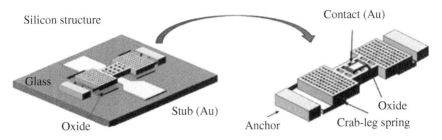

Figure 5.33 RF MEMS switch. *Source*: Jung et al. [8]. © 2012, IEEE.

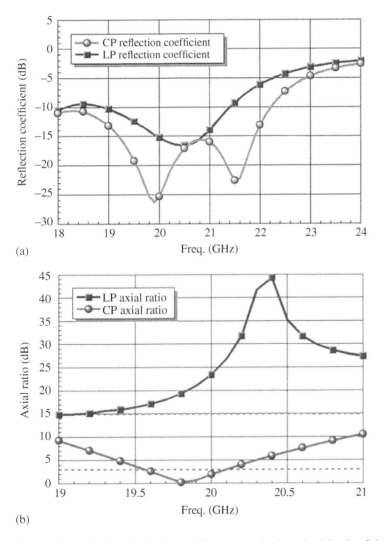

Figure 5.34 Simulated reflection coefficient magnitude and axial ratio of the two polarization modes. *Source*: Jung et al. [8]. © 2012, IEEE.

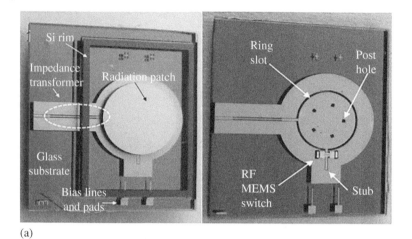

(a)

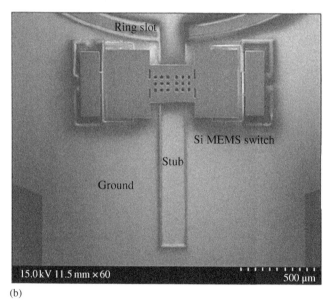

(b)

Figure 5.35 Fabricated prototype. *Source*: Jung et al. [8].

5.8 RF MEMS – Pattern Reconfigurable

In [9], researchers sought to design and realize a pattern reconfigurable antenna using RF MEMS switches. The antenna is a single-turn microstrip square spiral (Figure 5.37). To change the standing electric field distribution, and reconfigure the radiation pattern, two in-line switching elements are incorporated into the

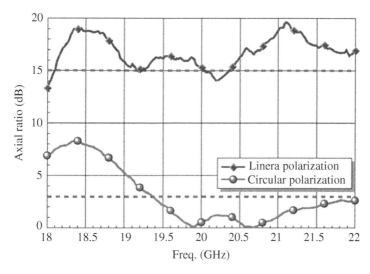

Figure 5.36 Measured axial ratio. *Source*: Jung et al. [8]. © 2012, IEEE.

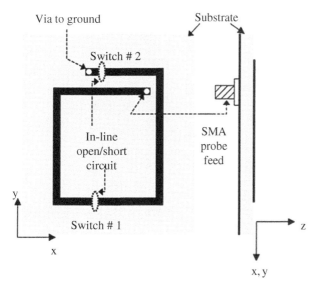

Figure 5.37 Pattern reconfigurable antenna using RF MEMS switch. *Source*: Huff et al. [9]. © 2006, IEEE.

spiral antenna. Mode 1 is the end-fire configuration (Figure 5.38), where $\theta = 90°$ and $\varphi = 90°$, switch #1 is closed (ON), and switch #2 is open (OFF). For broadside radiation (Figure 5.39), $\theta = 0°$, switch #1 is open (OFF), and switch #2 is closed (ON). Two Radant MEMS SPST-RMSW100 reflective RF switches are used to

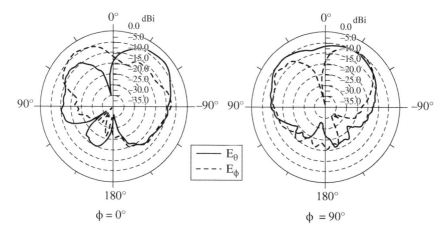

Figure 5.38 Measured end-fire radiation pattern. *Source:* Huff et al. [9]. © 2006, IEEE.

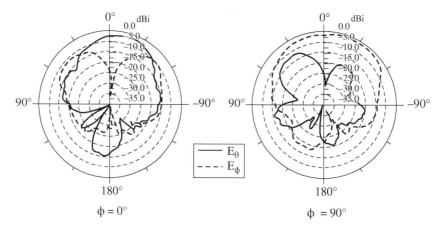

Figure 5.39 Measured broadside radiation pattern. *Source:* Huff et al. [9]. © 2006, IEEE.

reconfigure the antenna. The actuation voltage is 0 and 90 V, with 0.225 dB insertion loss and 16 dB of isolation. In the actuated (ON) state, the device behaves as a 50 Ω transmission line. In the OFF state, the device mainly looks capacitive.

5.9 Conclusion

In this chapter, we have discussed reconfigurable antennas that utilized liquid metal, piezoelectric materials, and RF MEMs. These antennas can be reconfigured in frequency, polarization, pattern, and directivity. Specific applications will influence adoption of reconfigurable antennas utilizing the discussed techniques. Practical, simple, and cost-effective implementations will ultimately dictate how widely adopted these technologies will be used for reconfigurable antennas. Future technologies may combine many of these approaches together. As an example, there has been many efforts to integrate RF MEMs with silicon CMOS processes in order to integrate the high performance of RF MEMs with the low cost and economies of scale of silicon CMOS. A recent example of this is the Diverse Accessible Heterogeneous Integration (DAHI) program that DARPA has initiated. There have also been efforts to co-integrate ferromagnetic materials onto semiconductor processes (Magnetic Miniaturized and Monolithically Integrated Components M3IC) from DARPA. Research, such as this, will enable multi-material techniques to enable future high-performance low-cost reconfigurable antennas and RF systems.

References

1 Dey, A., Guldiken, R., and Mumcu, G. (2016). Microfluidically reconfigured wideband frequency-tunable liquid-metal monopole antenna. *IEEE Transactions on Antennas and Propagation* 64 (6): 2572–2576.

2 Ha, A. and Kim, K. (2016). Frequency tunable liquid metal planar inverted-F antenna. *Electronics Letters* 52 (2): 100–102.

3 Bai, X., Su, M., Liu, Y., and Wu, Y. (2018). Wideband pattern-reconfigurable cone antenna employing liquid-metal reflectors. *IEEE Antennas and Wireless Propagation Letters* 17 (5): 916–919.

4 Liu, P., Yang, S., Wang, X. et al. (2017). Directivity-reconfigurable wideband two-arm spiral antenna. *IEEE Antennas and Wireless Propagation Letters* 16: 66–69.

5 Yun, T.-Y. and Chang, K. (2001). A low-cost 8 to 26.5 GHz phased array antenna using a piezoelectric transducer controlled phase shifter. *IEEE Transactions on Antennas and Propagation* 49 (9): 1290–1298.

6 Briqech, Z., Sebak, A., and Denidni, T.A. (2017). Low-cost wideband mm-wave phased array using the piezoelectric transducer for 5G applications. *IEEE Transactions on Antennas and Propagation* 65 (12): 6403–6412.

7 Wright, M.D., Baron, W., Miller, J. et al. (2018). MEMS reconfigurable broadband patch antenna for conformal applications. *IEEE Transactions on Antennas and Propagation* 66 (6): 2770–2778.

8 Jung, T.J., Hyeon, I., Baek, C., and Lim, S. (2012). Circular/linear polarization reconfigurable antenna on simplified RF-MEMS packaging platform in K-band. *IEEE Transactions on Antennas and Propagation* 60 (11): 5039–5045.

9 Huff, G.H. and Bernhard, J.T. (Feb. 2006). Integration of packaged RF MEMS switches with radiation pattern reconfigurable square spiral microstrip antennas. *IEEE Transactions on Antennas and Propagation* 54 (2): 464–469.

6

Compact Reconfigurable Antennas

Sima Noghanian and Satish K. Sharma

6.1 Introduction

Compact antennas have been a topic of research for many years. The reduction of antenna size is limited by physical limits [1, 2] while with the advances in miniaturizing the electronics systems, the demand for compact antennas is increasing. The reduction of the physical size of the antenna causes challenging problems such as gain and efficiency reduction, higher cross-polarization levels, and reduction of impedance bandwidth [3, 4].

In this effort, the use of reconfigurable design has an important role. Reconfigurable antennas offer another way of reducing the antenna size. Since most of the modern systems call for different antennas with different functionalities, the reduction of size can be considered when one antenna can be reconfigured to be used for different subsystems. The antenna occupies less volume and size than having several miniaturized antennas each performing one function.

Although most of the reconfigurable antennas in this chapter cannot be considered as compact antennas, if they are used for a single function and at a single frequency, but they are considered compact antennas since the combination of the functionality, frequency, pattern, or polarization is used to have multiple antennas in one physical location. This makes the antenna small, in comparison to equivalent non-reconfigurable antennas with the same functionalities.

There have been various methods used to achieve this goal. Some of these methods are the use of metasurfaces, liquid metal, ferrite, and pixel antennas. Some examples and practical methods are explained in this chapter.

Multifunctional Antennas and Arrays for Wireless Communication Systems, First Edition.
Edited by Satish K. Sharma and Jia-Chi S. Chieh.
© 2021 John Wiley & Sons, Inc. Published 2021 by John Wiley & Sons, Inc.

6.2   Reconfigurable Pixel Antenna

Pixel antenna or matrix array is based on the concept of dividing a conductor part of the antenna into several smaller pixels that are connected through switches. The reconfigurable part of the pixel antenna can be the radiator, a reflecting surface, or a superstrate. Mostly MEMS or PIN diode switches are used to turn ON or OFF the connections. These types of antennas are categorized as small antennas since, with the flexibility of design, it is possible to have multiple elements within the same physical space or volume. The simplest design is based on the uniform shape of cells.

In [5], a planar design is proposed that consists of an array of phase-agile reflection cells on a thin substrate. The reflection phase of each cell is controlled by a bias voltage. The concept is shown in Figure 6.1. The main beam is reflected by the superstrate by $R_1 \angle \phi_1$ and gets reflected from the ground plane by $R_2 \angle \phi_2$. The height of the superstrate L_r and the phase of reflection coefficients will determine the resonance frequency. $\angle \phi_1 + \angle \phi_2 - 2k = -2m\pi$, where k is the wave number. $\angle \phi_2$ is controlled by the cells. L_r can be as small as $\lambda/6$.

The proposed phase-agile cells were designed on the substrate R04230 ($\varepsilon_r = 3.0$, tan $\delta = 0.0023$) with a thickness of 1.524 mm and an area of 240 mm × 240 mm, as shown in Figure 6.2. The main microstrip patch antenna was attached to a SubMiniature version A (SMA) connector. The parasitic patch was placed 10 mm on top of the source patch to improve tracking and better coupling to the resonant modes inside the cavity. The phase-agile cells consist of a microstrip patch of 14 mm × 17 mm with a 1 mm air gap between two halves patch with a tunable resonance frequency (Figure 6.3a). The bias network consists of a line (connected to the ground through a bypass capacitor of 2.2 pF). The two halves are attached through two varactor diodes (Aeroflex Metallics MGV 125-20-0805-2) with the junction capacitance ranging from 0.1 pF to 1.0 pF as the bias voltage changes from 2 V to 20 V. A series RLC equivalent circuit is shown in Figure 6.3b.

The superstrate consisted of metallic patches on the lower side of FR4 ($\varepsilon_r = 4.4$, tan $\delta = 0.018$) with a thickness of 0.8 mm. The square patches had a

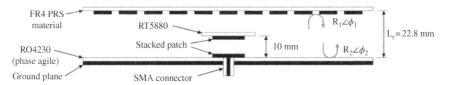

Figure 6.1   Schematic of reconfigurable pixel antenna consisting of a stacked patch antenna, phase-agile cells, and periodic structure on top. *Source*: Weily et al. [5]. © 2008, IEEE.

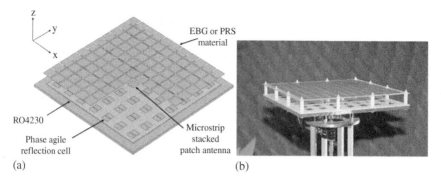

(a) (b)

Figure 6.2 (a) Layers reconfigurable antenna and (b) photo of the fabricated antenna. *Source*: Weily et al. [5]. © 2008, IEEE.

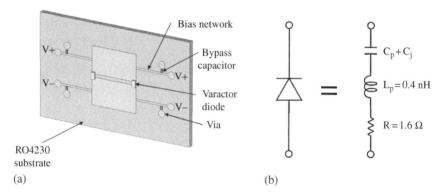

(a) (b)

Figure 6.3 (a) Frequency-agile cells consisting of two halves of a patch, bias network, and two varactor diodes and (b) equivalent circuit of the varactor diode. *Source*: Weily et al. [5]. © 2008, IEEE.

size of $18\,\text{mm} \times 18\,\text{mm}$ and a spacing of $2\,\text{mm}$ in both directions. Nylon spacers were used to keep the superstrate at the height of $L_r = 22.2\,\text{mm}$. The patch antenna was $14.5\,\text{mm} \times 14.5\,\text{mm}$, and the probe location of $0.13\,\text{mm}$ is from the edge. The parasitic patch had the dimension of $16.5\,\text{mm} \times 16.5\,\text{mm}$, printed on RT/Duroid 5880 ($\varepsilon_r = 4.4, \tan \delta = 0.018$) with a thickness of $0.254\,\text{mm}$ and $40\,\text{mm} \times 40\,\text{mm}$ area. By changing the bias voltage, maximum directivity is moved from 5.2 to 5.95 GHz. Table 6.1 and Figure 6.4 show details of this variation.

In a similar design, in [6], a patch antenna was placed on a tunable electromagnetic band gap (EBG) structure, consisting of a mushroom-type electronic band-gap structure (Figure 6.5). The patch is fed through proximity-fed microstrip line. The EBG cells are connected to the ground through vias that are controlled by switches. When the switches are in the ON state, the vias are connected, the

Table 6.1 Directivity, gain, antenna efficiency, and aperture efficiency for different reconfigurable antennas.

Bias voltage (V)	Operating frequency (GHz)	Directivity (dBi)	Realized gain (dBi)	Antenna efficiency (%)	Aperture efficiency (%)
6.49	5.2	13.8	10.0	42	11
7.13	5.3	14.8	11.3	45	13
7.92	5.4	15.0	11.3	43	13
8.90	5.5	15.6	11.5	39	15
10.80	5.6	16.1	11.2	32	16
11.49	5.7	16.8	11.6	30	18
12.95	5.775	17.7	13.0	34	22
15.22	5.875	19.0	15.0	40	29
18.50	5.95	19.8	16.4	46	34

Source: Based on Weily et al. [5].

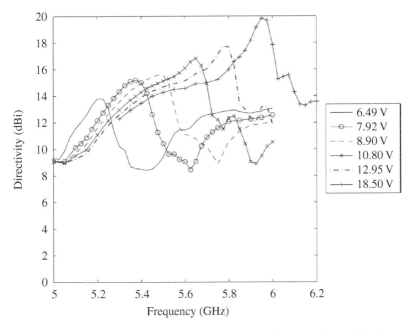

Figure 6.4 Measured directivity versus frequency for different values of diode voltage. *Source*: Weily et al. [5]. © 2008, IEEE.

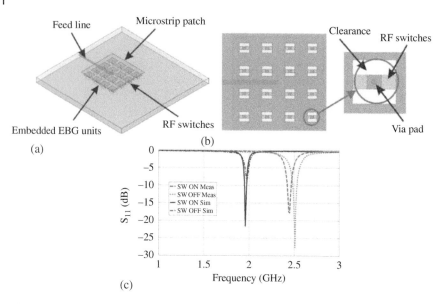

Figure 6.5 (a) Microstrip antenna on EBG structure, (b) EBGs are connected to the ground through vias controlled by switches, and (c) reflection coefficients for two modes of operation. *Source*: Liang and Yang [6]. © 2009, IEEE.

frequency is shifted down, and when the switches are in the OFF state, the EBG becomes ineffective. The patch antenna has a size of 28 mm × 28 mm, operating at 1.86 GHz when the switches are ON and at 2.52 GHz when the switches are OFF. The 4×4 element EBG was designed on FR4 with a height of 31 mils ($\varepsilon_r = 4.2$, tan $\delta = 0.02$). The width was 7 mm and the EBG patch size was 5 mm, with 0.2 mm vias. Although the principal of the design was shown through fabricating the ON and OFF switch states, the switches were not fabricated. Therefore, the results shown do not include the switch loss.

The same concept can be applied to different size patches to create the radiating or parasitic elements with various shapes. The original pixel antenna design uses uniform small patches to generate various shapes by connecting them. In [7], to reduce the number of switches, different size patches were suggested to generate various antenna-like planar monopole, various patch antennas, and planar dipole (Figure 6.6). These antennas consist of two parts. The parts closer to the excitation port have more effects on the antenna performance and they are made of smaller patch sizes and used as the radiator part. The parts farther away from the port are made of larger patches. An example prototype was made with $\lambda_0/20$ size patches. The parasitic patches have the dimensions of $0.15\lambda_0$. The patches are connected using PIN diodes with an insertion loss of 0.3 and 12 dB isolation at 2.45 GHz. The ON state of the switch was modeled as a 3.5 Ω series resistance and the OFF state

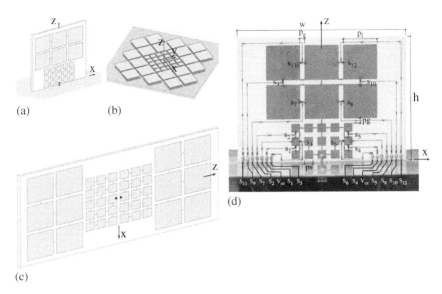

Figure 6.6 Pixel antennas with reduced number of switches and nonuniform patch sizes: (a) planar monopole, (b) patch, (c) planar dipole, and (d) fabricated planar monopole antenna. *Source*: Rodrigo and Jofre [7]. © 2012, IEEE.

is modeled as a shunt resistance of 2.6 KΩ and 0.17 pf capacitor. The optimization was done to maximize the realized gain at different directions of arrival and the reflection coefficient is minimized at the center frequency. A vertical symmetry was enforced, and 12 switches were equally distributed between the radiating and parasitic elements. The substrate was R04003 with the thickness of 0.81 mm, and the ground plane of 160 mm × 200 mm. By various combinations of ON and OFF switches, the main beam is switched toward different angles (Figure 6.7).

At the next step, the realized gain and reflection coefficient were optimized for multiple frequencies. Also, the antenna can be configured to have an omnidirectional pattern like a monopole (Figure 6.8). It can be noticed that as the frequency is increased, more switches are in the OFF state to create smaller size antennas.

The authors of [7] also proposed a pixel superstrate design driven by a patch antenna [8]. The design is illustrated in Figure 6.9. The superstrate is 6 × 6 pixels connected with 60 PIN diode switches. The overall size of the antenna is 87 mm × 87 mm. The pixel size of the superstrate patches was λ/10 at 2.5 GHz. This design is capable of having maximum beam at 30° and −30° in E- and H-planes and switches between x-linear, y-linear, left-hand circular polarization (LHCP), and right-hand circular polarization (RHCP). The configuration of different switches was optimized over six frequencies, four polarization, and five directions of arrivals. The antenna can steer the beam in XZ- and YZ-planes, while

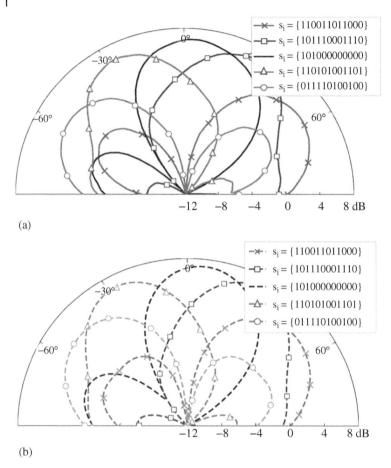

(a)

(b)

Figure 6.7 Realized gain for different beam angles at −60°, −30°, 0°, 30°, and 60° at 2.45 GHz (a) measurement and (b) simulation. *Source*: Rodrigo and Jofre [7]. © 2012, IEEE.

the beam-steering capability is better in XZ-plane, which is the same as H-plane of the driven patch. The patch is Y-polarized, and the maximum gain is also achieved for that polarization.

Table 6.2 summarizes the best gains achieved for different frequencies, direction of arrival (DoA), and polarizations for the distance (d) of 3 mm between the patch antenna and the superstrate patches. The advantage of this design is that the bandwidth of the original antenna is preserved in different configurations while frequency, radiation pattern, and polarization can be reconfigured.

The last design to be discussed in this section is the one which uses micro-electro-mechanical system (MEMS) switches in a pixel antenna. In [9], a reconfigurable pixel antenna operating in the band of 4–7 GHz is discussed. Frequency,

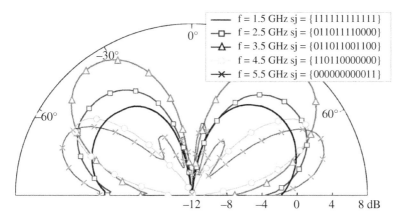

Figure 6.8 Realized gain for omnidirectional configurations at different frequencies. *Source*: Rodrigo and Jofre [7]. © 2012, IEEE.

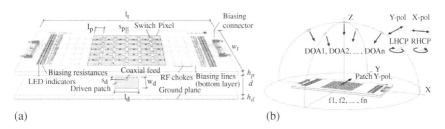

(a) (b)

Figure 6.9 (a) Schematic of patch-driven superstrate pixel antenna and (b) representation of reconfiguration parameters. *Source*: Rodrigo et al. [8]. © 2014, IEEE.

Table 6.2 The optimum gain achieved for different frequencies, degree of arrival (DOA), and polarization.

Frequency (GHz)	Optimum gain (dBi)	DOA	Optimum gain (dBi)	Polarization	Optimum gain (dBi)
2.4	3.1	XZ − 30°	4.5	Y	5.6
2.5	3.5	XZ + 30°	4.4	X	2.6
2.6	4.4	0	4.5	LHCP	4.2
2.7	4.4	YZ − 30°	3.0	RHCP	4.0
2.8	4.8	YZ + 30°	4.1		
2.9	4.3				

Source: Rodrigo et al. [8]. © 2014, IEEE.

pattern, and polarization reconfigurations are provided. There are three possible modes for this antenna: (i) multimode that allows different radiation patterns and polarizations working at the same frequency, (ii) multifrequency tunable mode, and (iii) polarization agile with applications in polarization-sensitive applications.

In diversity mode, the signals received by different patterns should be uncorrelated. Orthogonal radiation patterns can be used for this purpose. In multimode operation, pixel antenna's fundamental and higher-order modes are excited using different circular arrangements of pixels. The circular patches with different radii can generate TM_{n10}^z, where n is determined by radius a of the patch. Whereas in the case of the multifrequency tunable mode, the tuning is achieved by the size of the overall patch; therefore, the frequency range is limited by the size of the matrix.

In polarization-agile operation mode, any of TM_{n10}^z can be rotated by rotating mapped circular patches around the feeding location (in other words, while the feed location is fixed, but turning the pixels on and off, the polarization is rotated). Circular polarization can be generated by turning on the connections of the corresponding patches.

To excite TM_{n10}^z, the relationship between the patch radius (a) and resonance frequency (f_0) is given by [10]:

$$f_0 = \frac{\alpha_{mn}}{2\pi\sqrt{\varepsilon_r}\left(a\left(1 + \frac{2}{\pi a \varepsilon_r}\left(\ln\left(\frac{\pi a}{2h}\right) + 1.7726\right)^{\frac{1}{2}}\right)\right)} \tag{6.1}$$

where h is the thickness of the substrate, ε_r is the substrate dielectric constant, and α_{mn} is the mth zero of the derivative of the Bessel function of order n. The relative position ρ of the coaxial feed is given by $R\left(\frac{J_n(k\rho)}{J_n(ka)}\right)^2 = 50\,\Omega$, where R is the total resistance of the antenna due to dielectric and Ohmic losses. Since the antenna is made of pixels, the exact location of the feed is not possible; therefore, the resonance frequency will be shifted based on the number of pixels per dimension. To study the error caused by this, 8 different configurations with 3–17 pixels per dimension were simulated, as shown in Figure 6.10. The simulation results shown in Figure 6.11 show a reduction of resonance frequency and efficiency of the antenna. The simulation results showed that this design increases the number of pixels to more than seven per dimension which does not significantly change the frequency and efficiency. However, one should keep in mind that this number depends on the overall size of the antenna and a similar study should be repeated for different antenna sizes. The antenna in this simulation was fed in at a location of 0.762 mm above the ground plane and with a width of 1.78 mm at a 5.5 mm distance from the edge of the

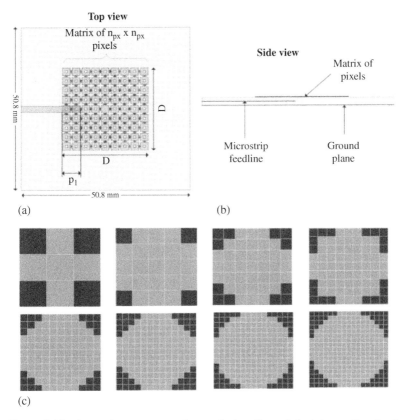

Figure 6.10 A sample antenna used to study the effect of pixel size on the circular patch performance, (a) top view, (b) side view, and (c) different pixel size (from 3 to 17 per dimension), to create a circular patch shape. The location of the microstrip feed line is calculated based on resonance frequency. Black pixels are not connected; gray pixels are connected. *Source*: Grau Besoli and De Flaviis [9]. © 2011, IEEE.

pixel's surface. The substrate was assumed to be Rogers TMM3 with a dielectric constant of 3.78 and loss tangent of 0.002 and 17 μm-thick copper. TM_{110}^z was considered to be the excited mode and the radius of the patch was assumed to be half of each side (D), where D was 25.43 mm. Figure 6.11 includes two graphs for the pixel antenna. The MEMS 3D represents the actual 3D integrated MEMS switches, compared with the ideal case switches. For a small number of pixels ($n_{px} < 7$), the MEMS switches do not show much difference with the ideal case, but as the number of pixels and the number of switches are increased, the difference is more. For $n_{px} > 7$, the resonance frequency is about 4.6% lower than the hardwired case. Also, these results do not include the effects of biasing lines.

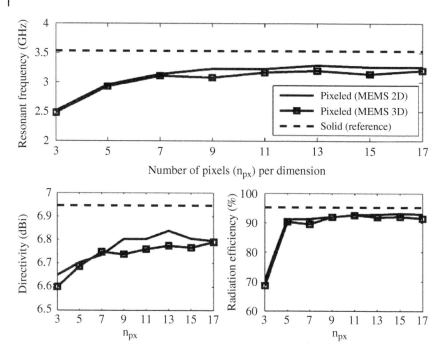

Figure 6.11 Effect of pixel size on the sample antenna resonance frequency, directivity, and radiation efficiency. *Source*: Grau Besoli and De Flaviis [9]. © 2011, IEEE.

A 9×9 matrix of pixels on a Rogers TMM3 substrate with 1.5875 mm thickness was fabricated and reported in [5]. Each pixel had a lateral dimension of 1.43 mm. The side of the overall antenna D was 17.43 mm. Two modes of TM^z_{n10} with $n = 1$, $\varnothing_0 = 0°$ and $n = 1$, $\varnothing_0 = 90°$ were excited. The range of target frequency was 4.5–7 GHz. There were 80 biasing wires for DC control. The MEMS switches in the ON state had a capacitance of 4 pF and in the OFF state capacitance of around 0.05 pf. Each pixel had four switches to connect it to the neighboring pixels, except for the feeding pixel that was connected directly to the adjacent pixels. Figure 6.12a and b depicts the simulated and measured reflection coefficients for two modes of operation. Figure 6.12c shows the simulated reflection coefficient for the multifrequency mode of operation. The pixel representations of these two modes of operation are shown in Figure 6.13. The reflection coefficient of $TM^z_{110} \varnothing_0 = 0°$ mode is 0.4 GHz off, which was found to be due to a defect in one of the MEMS switches. The radiation patterns of the two modes were measured in an anechoic chamber. Figure 6.14a shows the fabricated antenna and Figure 6.14b shows the measurement setup. The results of the pattern measurements are shown in Figure 6.15, which show good agreement with the simulated patterns.

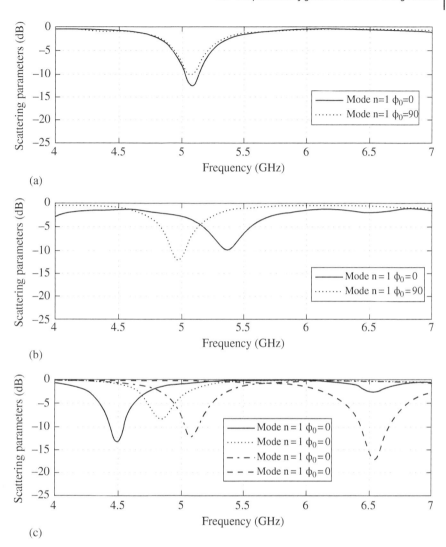

Figure 6.12 (a) Simulated reflection coefficient for two modes in multimode operation, (b) measured reflection coefficient for two modes in multi-mode operation, and (c) simulated reflection coefficient for multifrequency operation mode. *Source*: Grau Besoli and De Flaviis [9]. © 2011, IEEE.

6.3 Compact Reconfigurable Antennas Using Fluidic

Liquid metal alloys (e.g. Galinstan) are a choice for designers of reconfigurable antennas. Galinstan consists of three elements: gallium (Ga), indium (In), and tin (Sn). This metal is insoluble in water and organic solvents. Galinstan (68.5% GA,

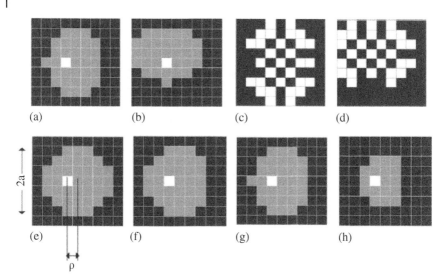

Figure 6.13 (a, b) Mapping of the pixels for multimode operation; (c, d) the biasing map for (a, b), respectively; (e–h) multifrequency operation. Gray pixels are connected, black pixels are disconnected, and the white pixel is the feed pixel. *Source*: Grau Besoli and De Flaviis [9]. © 2011, IEEE.

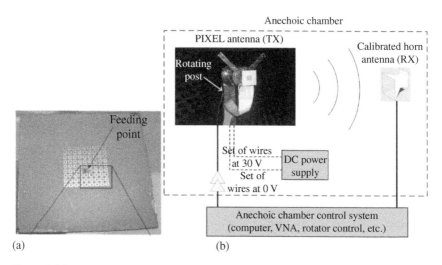

Figure 6.14 (a) Fabricated 9 × 9 pixel antenna and (b) setup of antenna pattern measurement in the anechoic chamber. *Source*: Grau Besoli and De Flaviis [9]. © 2011, IEEE.

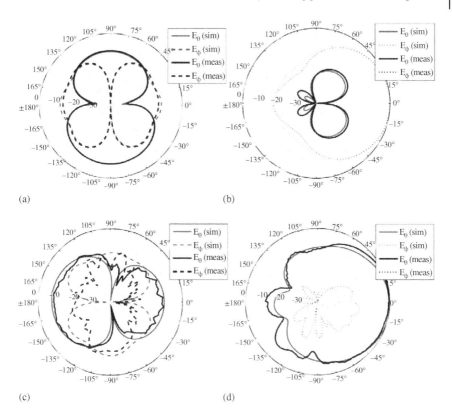

Figure 6.15 Simulated (gray) and measured (black) radiation patterns in (a, c) azimuth (xy) and (b, d) elevation (xz, $\varnothing = 0°$) planes. (a, b) are associated with TM_{110}^{z} with $\varnothing_0 = 0°$ and (c, d) are associated with TM_{110}^{z} and $\varnothing_0 = 90°$. *Source*: Grau Besoli and De Flaviis [9]. © 2011, IEEE.

21.5% In, 10% Sn) has a conductivity of 3.46×10^{6} S/m at 20 °C. Galinstan is liquid at room temperature and has a melting temperature of -19 °C. Galinstan is not toxic [11, 12].

As an example, an electrically actuated polarization and pattern reconfigurable dipole antenna is proposed in [13]. Using electro-capillary actuation (ECA), a DC signal was used to actuate the liquid metal into different states to change the polarization and patterns (Figure 6.16). In this design, a positive voltage is applied to electrolyte surrounding Galinstan (NaOH) to induce a surface tension gradient to generate forces, protracting Galinstan toward positive DC voltage. When the voltage is removed, Galinstan retracts to its equilibrium shape.

By actuating each dipole, the pattern will be rotated, as shown in Figure 6.17a–e. The actuation takes one to two seconds. It is repeatable. The measured and simulated (ANSYS HFSS) reflection coefficients are depicted in Figure 6.17f. There are

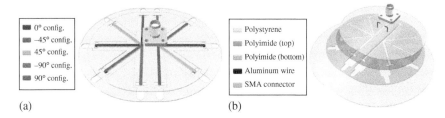

(a)

(b)

Figure 6.16 (a) Liquid metal polarization and pattern reconfigurable dipole antenna, with polyamide fixture encasing Galinstan arms; (b) layers presentation and thicknesses are polystyrenes (1 mm), top and bottom polyimide layers (0.6 mm each). *Source*: Zhang et al. [13]. © 2018, IEEE.

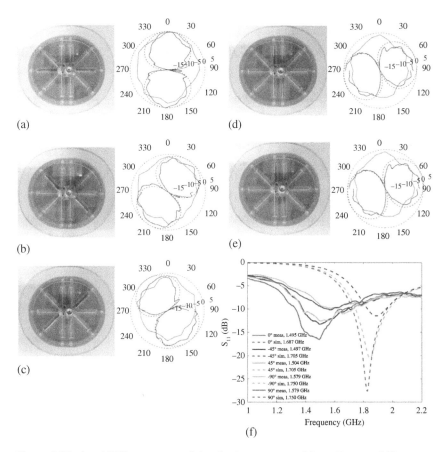

Figure 6.17 (a–e) Different states of the dipole antenna and its patterns and (f) measured and simulated reflection coefficients. *Source*: Zhang et al. [13]. © 2018, IEEE.

differences between the simulation and measurements. Several reasons for these differences including the residual NaOH around Galinstan, affecting the permittivity and shape, and possible oxidizing of Galinstan due to lack of NaOH can be mentioned. The measurements of patterns were done at 1.579 GHz. The maximum gain at 0° was around 2 dBi which was reduced to 0.58–0.75 dBi at different angles.

The method of liquid metal is applied to pixel antennas as well. In [14], a frequency-reconfigurable dipole antenna is proposed. The design started from a planar dipole antenna with a length of 64 mm and a width of 3 mm on Duroid 5880 substrate. The substrate thickness was 0.787 mm (Figure 6.18). A section of each dipole arms was divided into a four-pixel array on each side. The walls of the pixel array were made of polyimide. The top of the pixels was covered by polystyrene and the bottom was covered by polydimethylsiloxane (PDMS). The pixels were connected thorough stainless-steel walls (Figure 6.18). From the feed point, there were two pieces of copper with a length of 14 mm. These sections were soldered with stainless-steel wires to the stainless-steel walls. This is to keep the Galinstan separate from copper. Liquid metal was free to move between the top and bottom reservoirs (Figure 6.19). This happened when a voltage was applied to it. When the liquid was moved to the top side, it acted as ON switch, connecting the pixels to the copper part (Figure 6.20). Similar to [13], in this design, Galinstan was in a NaOH solution. A voltage applied to the surface of this combination causes pressure differential and actuates the liquid metal. Each pixel was 3 mm × 3 mm. The actuation voltage was 4 V square wave with 1 V DC. Table 6.3 shows the measured and simulated frequency of the antenna (using ANSYS HFSS simulation). The antenna maintains a similar radiation pattern and efficiency for all four frequencies as shown in Table 6.4. The measured radiation patterns were similar to the baseline antenna. It is observed that the shift in frequency decreases as the antenna becomes longer. This is proportional to the inverse square of the antenna length:

$$\frac{df}{dl} = -\frac{u}{2l^2} \tag{6.2}$$

where f is the resonance frequency, l is the antenna length, and u is the velocity of the wave in the dielectric.

6.4 Compact Reconfigurable Antennas Using Ferrite and Magnetic Materials

In this section, a few methods of using magnetized ferrite materials will be considered. Ferrites provide a means of reconfiguring antennas using a low frequency or DC magnetic field. Therefore, the control signal does not interfere with the

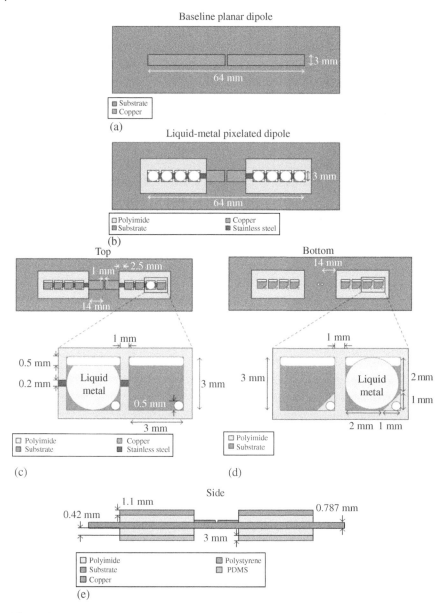

Figure 6.18 (a) Baseline antenna with copper arms, (b) pixel antenna, (c) top, (d) bottom, and (e) side view of the antenna.

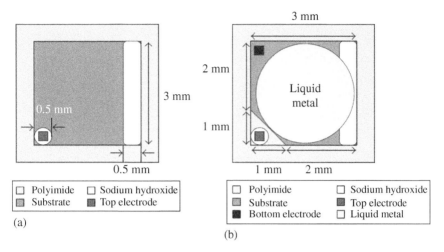

Figure 6.19 The layout of the pixel in the "OFF" state: (a) top and (b) bottom sides.

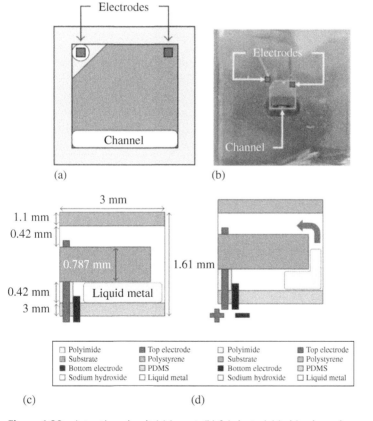

Figure 6.20 Actuation circuit (a) layout, (b) fabricated, (c) side view when metal is residing on the bottom side, and (d) application of voltage has moved the liquid metal to the top side.

Table 6.3 The resonance frequency of pixel antenna (GHz).

ON state pixels per arm	1	2	3	4
Measured	2.51	2.12	1.85	1.68
Simulated	2.43	2.08	1.88	1.78

Table 6.4 Bandwidth and efficiency of the pixel antenna in different states versus those of baseline antenna.

Parameter	Bandwidth (%)	Efficiency (%)
Baseline antenna	12.1	79.5
4 Pixels in ON state	17.9	72.6
3 Pixels in ON state	21.6	75.4
2 Pixels in ON state	17.5	72.6
1 Pixels in ON state	13.6	70.2

electromagnetic radiated fields. The advantage of using these materials is that they may be activated using a magnet and without the need for wires. The main disadvantage is the frequency dispersion and losses of magnetic materials that can cause a reduction in antenna efficiency. A good review of the types of ferrite materials suitable for use in reconfigurable antennas can be found in [15]. The miniaturization of antennas on dielectric or magnetic layers is proportional to $\frac{1}{\sqrt{\mu_r \varepsilon_r}}$. Usually, the magnetic material of interest has a moderate relative permeability (μ_r) and low magnetic losses. Sometimes, designs (Figure 6.21) are using permanent magnets for biasing versus electromagnets.

Ferrites are commonly used as a choice of magnetic materials in microwave and antenna designs. Ferrites are characterized by the saturation magnetization (M_s), ferromagnetic resonance linewidth (ΔH), and permeability (μ).

Polder [16] derived a microwave permeability tensor for ferrites assuming a "uniform" biasing magnetic field, assuming all the spinning electrons in the ferrite are aligned. This means that the ferrite material is in a saturated state. Assuming the biasing field is applied in the z-direction, the anisotropic behavior of biased ferrites is described by:

$$\vec{B} = \begin{pmatrix} \mu & j\kappa & 0 \\ j\kappa & \mu & 0 \\ 0 & 0 & 1 \end{pmatrix} \vec{H}, \tag{6.3}$$

(a)   (b)

Figure 6.21 (a) Use of electromagnet and (b) permanent magnet used for reconfigurability. *Source*: Andreou et al. [15].

$$\mu = \mu_0 \left(1 + \frac{\omega_0 \omega_m}{\omega_0^2 - \omega_m^2} \right), \tag{6.4}$$

$$\kappa = \mu_0 \frac{\omega \omega_m}{\omega_0^2 - \omega^2} \tag{6.5}$$

where $\omega_0 = \gamma \mu_0 H_0$, $\omega = 2\pi f$, $\omega_m = \gamma \mu_0 M_s$, $\gamma = g_l \gamma_e$, and γ_e is half of the electron charge to mass ratio, g_l is the Lande g factor (typically between one and two), and H_0 is a bias field.

One way to incorporate magnetic materials is by using them as a substrate material in planar antennas. In [17], a microstrip patch antenna is proposed. The magnetic material in this design operates in the saturated magnetized state. A custom magnetic ink made of iron oxide (Fe_3O_4) nanoparticles in SU-8 epoxy resin for substrate was made. The nanoparticles were in 50 : 50 wt% and had an average particle size was 20 nm. The viscosity of the ink was 37.8 cP. A metallic ink made of silver-organo-complex (SOC) was used for conducting patch and ground plane. The ink was printed using a manual screen-printing method. An FR-4 board with a sacrificial paper on the backside was used as a support. The board is needed because the magnetic ink is initially liquid and needs the support layer until it solidifies after curing. A slot was created in the support layer, the ink is applied with a thickness of 1.5 mm, and cured until it is solid. Then the support material is cut away and the sacrificial paper is removed. A 10 μm smoothening layer of 3-D Vero black plus material was printed on the top of the substrate and finally, the conductive material was deposited on top and bottom, for the patch and ground, respectively. Eight layers of SOC were printed for each side. Figure 6.22 shows these steps.

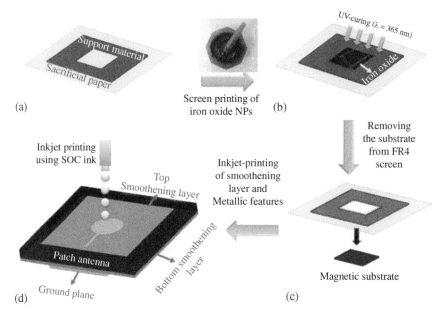

Figure 6.22 The fabrication process of the magnetic substrate: (a) FR-4 support with the sacrificial paper, (b) screen printing of the magnetic ink, (c) cutting the solid magnetic substrate, and (d) printing the smoothening layer and SOC ink on both sides. *Source*: Ghaffar et al. [17]. © 2018, IEEE.

The magnetic ink demonstrates a saturation flux density (Bs), coercive field (Hc), and remanent flux density (Bm) of 156 mT, 3.66 kA/m, and 35 mT. The magnetization frequency of the material was calculated as $f_m = 2.8 \times 10^{10}$ and $\mu_0 M_S = 4.37$ GHz. Frequency of operation was chosen as 6 GHz to be greater than the magnetization frequency. A coplanar ring resonator (using a silver ink) on top of the magnetic substrate was used to estimate the product of permittivity and permeability. The demagnetized permeability was estimated by:

$$\mu_0' = \frac{2}{3}\left[1-\left(\frac{f_m}{f_{res}}\right)^2\right]^{\frac{1}{2}} + \frac{1}{3} \tag{6.6}$$

The effective relative permittivity was calculated at different frequencies as shown in Table 6.5.

The circular patch antenna with a 4.1 mm radius fed by a microstrip line matched with a stub tuner (3.5 mm length of the open-circuit line located 1.6 mm away from the edge of the patch). The polarization is changed by applying the bias field. The antenna has a linear polarization from 6 to 6.4 GHz, while from 5 to

Table 6.5 Permittivity and permeability of the magnetic ink.

Frequency (GHz) f_{res}	Demagnetized permeability μ'_0	Relative permittivity ε_{eff}
6.04	0.796	15.50
7.90	0.860	15.29
9.90	0.900	14.20

6 GHz, it has circular polarization. Depending on the direction of the applied magnetic field, the polarization can be right-handed or left-handed as shown in Figure 6.23. Figure 6.23 shows simulated and measured radiation patterns in (a) unbiased state, (b) in magnetic bias at 5.6 GHz, (c) in magnetic bias at 6.1 GHz are shown in [17]. The resonance frequency depends on the bias field. As shown in Figure 6.24, by changing the applied magnetic field, the resonance frequency is changed.

Another way to use ferrite material is by adding ferrite loads for tuning substrate integrated waveguide (SIW). This method can be used to create frequency reconfigurable antennas. In [18], a tunable SIW antenna using ferrite slabs was proposed. The schematics of the antenna and the fabricated antenna are shown in Figure 6.25. As can be seen, the antenna consists of a square patch surrounded by vias. The top antenna is loaded with two ferrite slabs that are located at the two sides where the magnetic field is the strongest. The loading ferrite slabs will change the resonant frequency. When the external biasing magnetic field H_0 has been applied, the permeability of the ferrite changes with the strength of this biasing field. The fabricated antenna was made on Rogers Duroid 5880 substrate with 0.8 mm thickness, size of 19.8 mm × 24.8 mm, the permittivity of 2.2, and loss tangent of 0.001.

The ferrite material yttrium iron garnet (YIG) has the saturation magnetization of $4\pi M_S = 0.1799$ T and dielectric constant of 14.3. Two ferrite slabs were embedded inside the substrate. Conductor tapes were used to cover the ferrite slabs on both sides of the cavity. A permanent magnet was used to apply the bias magnetic field. The biasing field strength was controlled by changing the distance between the magnets and the ferrite material (Figure 6.25d). Figure 6.26a shows the simulated and measured reflection coefficients. The authors increased the ground plane size to 49.5 mm × 72.2 mm and observed smaller losses and a shift in frequency. They stated that the antenna may have larger losses due to small ground plane size. Figure 6.26b shows how the change of the biasing field can change the resonance frequency. With the increase in the biasing field, the resonance frequency moves to higher frequencies.

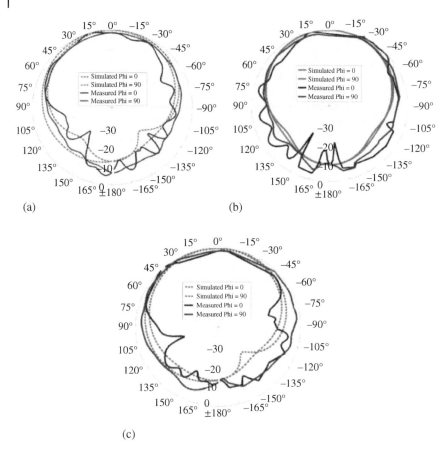

Figure 6.23 Simulated and measured radiation patterns: (a) unbiased state, (b) in magnetic bias at 5.6 GHz, and (c) in magnetic bias at 6.1 GHz. *Source:* Ghaffar et al. [17]. © 2018, IEEE.

The radiation patterns were measured (Figure 6.27) and showed small differences under biasing fields, the only change is in the total gain and frequency as listed in Table 6.6.

In [19], a similar design with a dual-band operation band is presented. This SIW antenna is utilizing a cross-slot to generate multimode and create dual-band frequency reconfigurable patterns. The design is based on multimode characteristics of the SIW resonant cavity. The cavity is loaded with two ferrite slabs. The two modes correspond to TE_{120} and TE_{220} modes. The ferrite slabs in a bias state cause the resonance frequency and polarization of the antenna to change. Figure 6.28 shows the design and two sides of the antenna. The dimensions are given in the figure caption. The antenna was fabricated on a 0.8 mm-thick F4BM-2 substrate

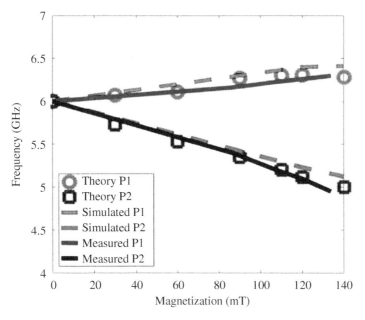

Figure 6.24 Frequency of the circular patch due to the applied magnetic bias. *Source:* Ghaffar et al. [17]. © 2018, IEEE.

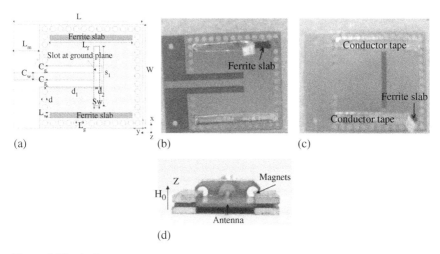

Figure 6.25 (a) Schematic of ferrite-loaded SIW antenna; (b) top side of fabricated antenna; (c) bottom side, conductor tapes are covering the ferrite slab; (d) direction of biasing field, dimensions (all in millimeters) $L_m = 5, L_w = 1, L_g = 1.5, L_f = 15.8, L = 24.8$, $C_w = 1.45, C_g = 1.3, d_1 = 10, d_2 = 1, d = 1, S_w = 1, S_1 = 11.3, W = 19.8$. *Source:* Tan et al. [18]. © 2013, IEEE.

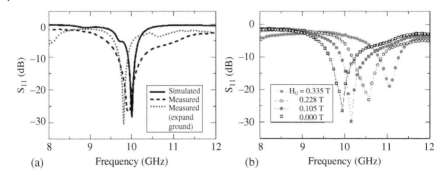

Figure 6.26 Reflected coefficient: (a) simulated and measured at zero biasing fields and (b) measured at various biasing field strengths. *Source*: Tan et al. [18]. © 2013, IEEE.

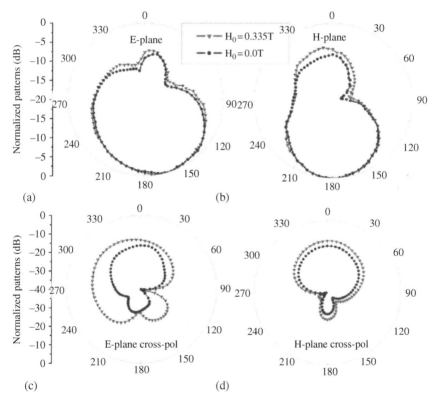

Figure 6.27 Normalized radiation patterns with and without application of the biasing magnetic fields: (a, b) co-polarization fields and (c, d) cross-polarization fields. *Source*: Tan et al. [18]. © 2013, IEEE.

Table 6.6 Gain and resonance frequency changes due to different levels of biasing field strength for the ferrite-loaded SIW antenna.

H_0 (T)	Frequency (GHz)	S_{11} (dB)	Gain (dBi)
0.000	9.95	−26.6	5.60
0.105	10.16	−29.80	5.30
0.147	10.30	−25.50	5.21
0.228	10.57	−20.13	5.13
0.265	10.71	−20.79	5.08
0.335	11.06	−19.00	5.01

Source: Tan et al. [18]. © 2013, IEEE.

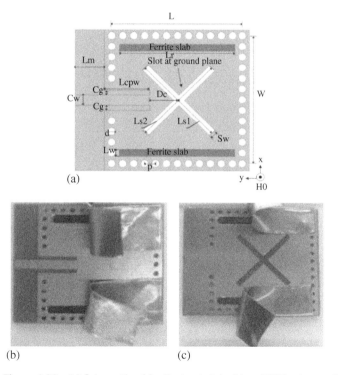

(a)

(b)   (c)

Figure 6.28 (a) Schematic of ferrite-loaded dual-band SIW antenna; (b) front; (c) back side of the fabricated antenna, dimensions (all in millimeters) L = 23.8, L_{s1} = 13, L_{s2} = 12.7, L_f = 15.8, L_w = 1, L_{cpw} = 6.2, L_m = 4, C_w = 1.45, C_g = 0.7, d = 1, p = 1.5, d = 1, S_w = 1, D_c = 4, W = 19.8.

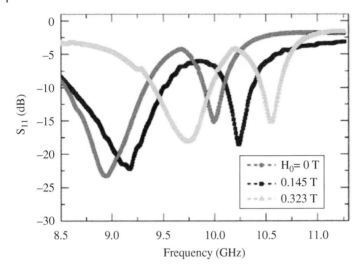

Figure 6.29 Measured reflection coefficients of cross-slot ferrite-loaded SIW antenna under various magnetic bias fields.

($\varepsilon_r = 2.2$, $\tan \delta = 0.001$). Two permanent magnets were used to provide the biasing field. Figure 6.29 shows the reflection coefficients measured for the antenna under various biasing fields. The biasing field was tuned by moving the magnets and changing their distance from the ferrite slabs. Table 6.7 shows the measured antenna parameters under different biasing field strengths. When there is no biasing field present, the second resonance frequency provides axial ratio (AR) equal to 3.4 dB, while the first frequency provides a linear polarized field. The field at both frequencies is linearly polarized for other biasing fields. Since the position of magnets is away from the radiating slots, there is little effect on the radiation fields. Below $H_0 = 0.06$ T, antenna radiation is circularly polarized at low frequency and linearly polarized at high frequency.

6.5 Metamaterials and Metasurfaces

Throughout this book, there have been many examples of the use of metamaterials and metasurfaces in reconfigurable antennas. In this chapter, one example will be explained in detail. A MEMS based reconfigurable antenna is proposed that is inspired by the earlier design by the same authors (presented in [20]). The original design consists of a coplanar waveguide (CPW)-fed antenna. To adjust the frequency and create double frequency operation, a T-shaped slot is cut out of a rectangular patch as shown in Figure 6.30. The T-shaped slot is loaded with an

Table 6.7 Measured antenna performance at two frequencies of operation of the cross-slot ferrite-loaded SIW antenna under different magnetic bias fields (AR: axial ratio, Eff: radiation efficiency).

H_0 (T)	f_1 (GHz)	S_{11} (dB)	Gain (dBi)	AR (dB)	Eff. (%)	f_2 (GHz)	S_{11} (dB)	Gain (dBi)	AR (dB)	Eff. (%)
0.000	8.98	−22.73	6.34	3.4	95.1	9.98	−15.15	5.98	32.5	93.2
0.145	9.18	−22.23	6.20	23.1	92.2	10.24	−18.53	5.77	9.4	90.7
0.323	9.74	−18.12	6.13	21.2	87.6	10.54	−15.19	5.10	23.1	81.7

interdigital capacitor. The dimension of the monopole and T-shaped slot is shown in Figure 6.30a. The size of a monopole at 2.45 GHz is $1/13.3 \lambda_0 \times 1/21.4 \lambda_0$. The overall size of the antenna (including the ground planes) is 32 mm × 24 mm. The monopole works at 5.5 GHz band, while the metamaterial-inspired reactive loading of the slot and interdigital capacitor provides resonance at 2.46 GHz (Figure 6.31). This dual-band antenna has a measured radiation efficiency

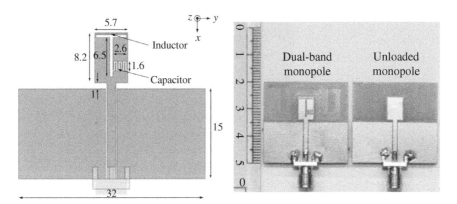

Figure 6.30 (a) The geometry of dual-band metamaterial-inspired antenna and (b) photograph of the fabricated antenna. *Source*: Zhu and Eleftheriades [20]. © 2009, IEEE.

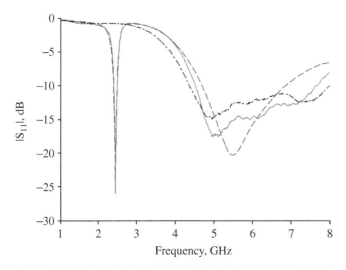

Figure 6.31 Simulated and measured S_{11} for the metamaterial-inspired dual-band monopole antenna, solid line: measured dual-band antenna, dash line: simulated dual-band antenna, dash dot line: measured single band monopole. *Source*: Zhu and Eleftheriades [20]. © 2009, IEEE.

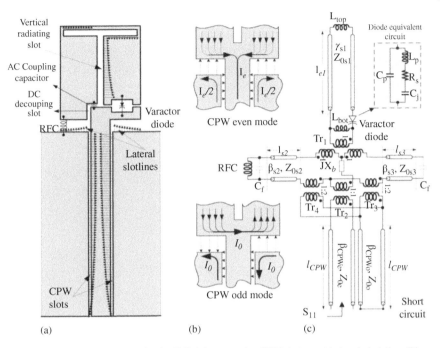

Figure 6.32 (a) Asymmetry in the E-field across the CPW slots and lateral slot-line, (b) currents and E-field along with the vertical radiating slot, and (c) circuit model of the CPW and radiating slot. *Source*: Mirzaei and Eleftheriades [21]. © 2011, IEEE.

of 89.2% at 5.5 GHz and 64.0% at 2.46 GHz. The measured gain values are 1.53 and 0.71 dBi at 5.5 and 2.46 GHz, respectively.

The authors of [21] investigated adding the reconfigurability to this antenna. First, a circuit model was developed. Figure 6.32a shows the E-field across all slots. The top horizontal slot does not contribute to the radiation due to symmetry; therefore, it acts as a loading inductor. The vertical slot is equivalent to a magnetic current and radiates. The E-field across two lateral slots and CPW slots is asymmetric, as shown in Figure 6.32b. The currents under CPW are separated as even and odd modes. The slot across the patch that is loaded with a vertical slot is modeled as an inductor, and the varactor is the radiating element and is modeled as a lossy transmission line (Figure 6.32c). The even and odd modes of the feed section are modeled as separate parts with different ports. Each of the lateral slots is modeled as a transmission line. C_f represents the fringing fields and X_b is to model CPW bridges to the lateral slot lines. L_{bot} models the current exchange between two sides of the vertical slot through the substrate.

The interdigital capacitor here is replaced by a varactor. This varactor (MGV125-08-0805-2 by Aeroflex/Metallics) is modeled with C_j that can be tuned between 0.075 and 0.6 pF (for bias voltage of 20–2 V), and parasitic elements as

$L_p = 0.4\,nH$, $C_p = 0.06\,pF$, and $R_s = 2.65\,\Omega$ (at 4 V 50 MHz). The antenna was fabricated on FR4 substrate with a thickness of 1.59 mm, with a dielectric constant of 4.35, and a loss tangent of 0.017 (Figure 6.33). The DC bias for varactor diode was applied through the AC input line using a bias-T. The DC return pass is provided by a radio-frequency choke (RFC) which is an inductor of 39 nH. The simulated and measured reflection coefficient magnitudes of the antenna with capacitance variation are shown in Figures 6.34 and 6.35, respectively. It can be

Figure 6.33 Fabricated monopole antenna. *Source*: Mirzaei and Eleftheriades [21]. © 2011, IEEE.

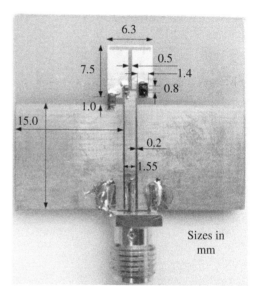

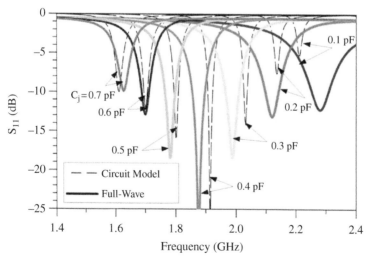

Figure 6.34 Simulation results showing the changes in S_{11} by changing the varactor capacitor. *Source*: Mirzaei and Eleftheriades [21]. © 2011, IEEE.

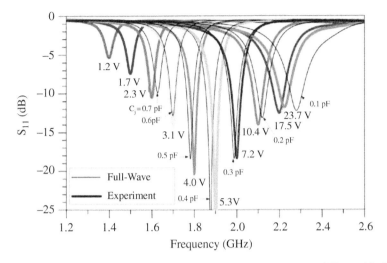

Figure 6.35 Experimental results showing S_{11} variation due to different biasing voltage. *Source*: Mirzaei and Eleftheriades [21]. © 2011, IEEE.

observed that for capacitance variation between 0.1 pF and 0.7 pF, frequency variation is approximately between 2.3 and 1.60 GHz considering acceptable matching criteria.

6.6 Conclusion

In this chapter, we looked at the concept of compact antennas from the point of multi-functionality of reconfigurable antennas. Four methods of pixel antenna, use of fluidic to reconfigure the conducting area, and use of magnetic materials and metasurfaces in small antennas were reviewed through some examples. While the reconfiguration of the radiating surface provides a method to be used in systems with multiple functions and having tight areas and volumes, the method of reconfiguring the antenna has been subject to much creative thinking and new materials and technology. The combination of the advanced materials and fabrication technology will provide antenna designers with an array of new choices in designing compact multifunctional antennas.

References

1 Collin, R.E. and Rothschild, S. (1964). Evaluation of antenna Q. *IEEE Trans. Antennas Propag.* 12 (1): 23–27. https://doi.org/10.1109/TAP.1964.1138151.

2 Hansen, R.C. (1981). Fundamental limitations in antennas. *Proc. IEEE* 69 (2): 170–182.

3 Best, S.R. (2006). Bandwidth and the lower bound on Q for small wideband antennas. *IEEE Antennas Propag. Soc. AP-S Int. Symp.* 1 (781): 647–650. https://doi.org/10.1109/APS.2006.1710607.

4 Harrington, R.F. (1960). Effect of antenna size on gain, bandwidth, and efficiency. *J. Res. Natl. Bur. Stand. Sect. D Radio Propag.* 64D (1): 1. https://doi.org/10.6028/jres.064d.003.

5 Weily, A.R., Bird, T.S., and Guo, Y.J. (2008). A reconfigurable high-gain partially reflecting surface antenna. *IEEE Trans. Antennas Propag.* 56 (11): 3382–3390. https://doi.org/10.1109/TAP.2008.2005538.

6 Liang, J. and Yang, H.Y.D. (2009). Microstrip patch antennas on tunable electromagnetic band-gap substrates. *IEEE Trans. Antennas Propag.* 57 (6): 1612–1617. https://doi.org/10.1109/TAP.2009.2019928.

7 Rodrigo, D. and Jofre, L. (2012). Frequency and radiation pattern reconfigurability of a multi-size pixel antenna. *IEEE Trans. Antennas Propag.* 60 (5): 2219–2225. https://doi.org/10.1109/TAP.2012.2189739.

8 Rodrigo, D., Cetiner, B.A., and Jofre, L. (2014). Frequency, radiation pattern and polarization reconfigurable antenna using a parasitic pixel layer. *IEEE Trans. Antennas Propag.* 62 (6): 3422–3427. https://doi.org/10.1109/TAP.2014.2314464.

9 Grau Besoli, A. and De Flaviis, F. (2011). A multifunctional reconfigurable pixeled antenna using MEMS technology on printed circuit board. *IEEE Trans. Antennas Propag.* 59 (12): 4413–4424. https://doi.org/10.1109/TAP.2011.2165470.

10 Derneryd, A.G. (1979). Analysis of the microstrip disk antenna element. *IEEE Trans. Antennas Propag.* 27 (5): 660–664. https://doi.org/10.1109/TAP.1979.1142159.

11 Liu, T., Sen, P., and Kim, C.J. (2012). Characterization of nontoxic liquid-metal alloy galinstan for applications in microdevices. *J. Microelectromech. Syst.* 21 (2): 443–450. https://doi.org/10.1109/JMEMS.2011.2174421.

12 Geratherm Medical AG (2004). Galinstan fluid safety data sheet, 2–5 [Online]. http://baisd-mi.safeschoolssds.com/document/repo/445ac403-dce0-428a-9541-c07daac255bc

13 Zhang, G.B., Gough, R.C., Moorefield, M.R. et al. (2018). A liquid-metal polarization-pattern-reconfigurable dipole antenna. *IEEE Antennas Wirel. Propag. Lett.* 17 (1): 50–53. https://doi.org/10.1109/LAWP.2017.2773076.

14 Sarabia, K.J., Yamada, S.S., Moorefield, M.R. et al. (2018). Frequency-reconfigurable dipole antenna using liquid-metal pixels. *Int. J. Antennas Propag.* 2018: 1–6. https://doi.org/10.1155/2018/1248459.

15 Andreou, E., Zervos, T., Alexandridis, A.A., and Fikioris, G. (2019). Magnetodielectric materials in antenna design: exploring the potentials for reconfigurability. *IEEE Antennas Propag. Mag.* 61 (1): 29–40. https://doi.org/10.1109/MAP.2018.2883029.

16 Polder, D. (1949). On the theory of ferromagnetic resonance. *London Edinburgh Dublin Philos. Mag. J. Sci.* 40 (300): 99–115. https://doi.org/10.1080/14786444908561215.

17 Ghaffar, F.A., Vaseem, M., Roy, L., and Shamim, A. (2018). Design and fabrication of a frequency and polarization reconfigurable microwave antenna on a printed partially magnetized ferrite substrate. *IEEE Trans. Antennas Propag.* 66 (9): 4866–4871. https://doi.org/10.1109/TAP.2018.2846796.

18 Tan, L.R., Wu, R.X., Wang, C.Y., and Poo, Y. (2013). Magnetically tunable ferrite loaded SIW antenna. *IEEE Antennas Wirel. Propag. Lett.* 12: 273–275. https://doi.org/10.1109/LAWP.2013.2248113.

19 Tan, L.R., Wu, R.X., and Poo, Y. (2015). Magnetically reconfigurable SIW antenna with tunable frequencies and polarizations. *IEEE Trans. Antennas Propag.* 63 (6): 2772–2776. https://doi.org/10.1109/TAP.2015.2414446.

20 Zhu, J. and Eleftheriades, G.V. (2009). Dual-band metamaterial-inspired small monopole antenna for WiFi applications. *Electron. Lett.* 45 (22): 1104–1106. https://doi.org/10.1049/el.2009.2107.

21 Mirzaei, H. and Eleftheriades, G.V. (2011). A compact frequency-reconfigurable metamaterial-inspired antenna. *IEEE Antennas Wirel. Propag. Lett.* 10: 1154–1157. https://doi.org/10.1109/LAWP.2011.2172180.

7

Reconfigurable MIMO Antennas

Kumud R. Jha and Satish K. Sharma

7.1 Introduction

Reconfigurable antennas are indispensable to the modern wireless communication systems whose genesis can be traced from the seminal work of Harrington [1], where the radiation characteristics of the reactively controlled directive arrays were investigated. Due to the adaptive properties of the reactively controlled arrays, the basic unit of the array is referred to as the reconfigurable element which may be in the form of switched element, reconfigurable apertures, evolutionary antennas, and self-structuring antennas [2–4]. Although these antennas refer to different structures and the operating mechanism, they have the same basic goal to operate over the multiple band of the electromagnetic spectrum.

In an antenna, by purposefully altering the radiated field from the effective aperture which redistributes the surface current density at the radiating edge of antenna, the configurability is achieved [5]. The redistribution of the current density at the edge of the aperture causes the change in the antenna functionalities which enables the versatile use of the reconfigurable antennas for various wireless communication systems.

It is expected from the reconfigurable antennas to meet the certain characteristics such as reasonable gain, stable radiation pattern, and a good impedance matching over the desired band of operation in addition to the reconfiguration property. Thus, the design of a reconfigurable antenna depends on the multiple factors and the technique to achieve these goals is determined before the design process. Reconfigurable antennas are smart to adapt to the operating environment, and besides a classical communication system, it has also yielded the improvement in the data communication at the physical layer [6–10]. Thus, the reconfigurable antennas are also playing a significant role in the diversity technique of the

Multifunctional Antennas and Arrays for Wireless Communication Systems, First Edition.
Edited by Satish K. Sharma and Jia-Chi S. Chieh.
© 2021 John Wiley & Sons, Inc. Published 2021 by John Wiley & Sons, Inc.

ad-hoc networks, multiple-input multiple-output (MIMO) ad-hoc networks, antenna beamforming, or the next-generation communication systems.

In the communication systems, to reduce the interference of adjacent channels and nodes, to maximize the overall network throughput, various forms of directional and phased array antennas find their application. The MIMO antennas comprising of a number of such antennas operating using spatial multiplexing (SM) and diversity techniques, are used to increase the network spectral efficiency [11]. However, these methods find a difficulty in being integrated on compact portable devices due to the space constraint. Thus, to overcome the space limitations and extract the benefits of MIMO SM/diversity techniques with those of directional antennas, the reconfigurable antennas may be used as a key element of the MIMO transceivers. The use of this type of antennas increases the channel capacity while reducing the space constraints on the communication devices. There are a number of reconfiguration techniques like frequency, pattern, and polarization, which are efficiently chosen to meet a specific requirement.

Due to the adaptability, in addition to the classical communication systems, the reconfigurable antennas also find their potential application in multi-link MIMO ad-hoc network where at each node, types of reconfigurable antennas are selected to meet the specific design requirements. Normally, directional antennas mitigate interference by estimating the direction of the incoming signals at the receiver. Since, reconfigurable MIMO antennas are capable to radiate more than one radiation pattern at different frequencies and polarization, it helps in enhancing the channel capacity and throughput. Adaptability due to the reconfiguration also helps in achieving high system gain and the security also which makes it superior to the non-reconfigurable MIMO antenna systems.

The reconfigurable antennas designed to efficiently meet the radiation characteristics over the operating frequency band find multiple applications including MIMO systems. With the implementation of multiple antennas, in comparison to single-input-single-output (SISO), the MIMO systems' capacity increases and the bit error decreases at the cost of the fabrication complexity. The complexity further increases with the use of the reconfiguration technique of these antennas which operate over multiple bands.

These MIMO antennas are also called as smart antennas which enable the signals to be processed adaptively in the mobile channel at the transmitter, receiver, or both sides of the communication link. In these systems, the spatial dimension of the communication channel is exploited to increase the channel capacity by mitigating the number of impairments, such as multipath fading and co-channel interference and increasing the data rate by streaming over multiple channels [12]. In this chapter, we discuss the various methods to design the reconfigurable antennas, application of the reconfigurable antenna in MIMO systems, and various MIMO performance indices related to these antennas.

7.2 Reconfigurable Antennas for MIMO Applications

Antenna in MIMO systems consists of a number of transmitting and receiving antenna elements to improve the performance of the communication system [13]. The wireless channels using these multiple antennas in fact exploit the spatial properties of the multipath channels. Multipath fading is mitigated with the help of diversity technique. These diversity techniques are:

1) Frequency diversity: The same set of information is transmitted at various widely separated frequencies to enhance the reliability of reception.
2) Time diversity: The information is transmitted at the different intervals of time.
3) Space diversity: In this technique, multiple antennas are employed at the transmitter and receiver end.

In the early days, the space diversity technique could not draw much attention of researchers as the distance between multiple radiating elements was considered as a limiting factor in addition to increased number of antennas and cost. However, with the advancement in the wireless technology, the need of the higher data rate is steadily increasing and all possible solutions have been explored in the past one decade. In general, the channel capacity is governed by Shannon's law and given as (7.1) [13]:

$$C = \log_2(1 + SNR|H|^2)\ b/s/Hz. \tag{7.1}$$

In (7.1), C and SNR are the channel capacity and signal-to-noise-ratio, respectively. To enhance the channel capacity in a simplified case, bandwidth is one of the options. The electromagnetic frequency spectrum is precious whereas current wireless communication bands are completely occupied; therefore, scope of increasing the bandwidth is not possible. The SNR can be affected by the transmitted power but cannot be increased beyond certain value depending on the wireless communication application and the state regulations in effect. Secondly, when transmitting power is increased, the channel noise also increases and thus these two methods find themselves unsuitable to mitigate the channel fading and in boosting the reliability of the wireless communication chain.

Due to the limitations of the SISO system, MIMO systems have been attractive to combat the multipath fading and to enhance the channel capacity by the parallel multichannel communication. In the MIMO system, due to the multipath communication two types of gain are observed.

1) Rate gain: When multiple channels in parallel operate in MIMO implementation, it is possible to transmit different data through different parallel channels and thus increases the data transmission rate.

2) Diversity gain: In this technique, same data are transmitted over the multiple channels to counter the multipath channel fading and thus increases the reliability of the communication link.

The schematic representation of these configurations are shown in Figure 7.1 which shows the need of multiple antennas at the transmitter and receiver ends and are needed to enhance either of these two gains in the MIMO system.

When multiple antennas operate in the vicinity of each other, it becomes difficult to ascertain the system performance due to correlation between the adjacent antennas. In such a system, the channel capacity defined earlier (7.1) is modified and depends on the number of channels being paired between transmitter and receiving ends. The upper limit of the channel capacity with multiple antennas is given by (7.2) [13].

$$C_{MIMO} = \log_2\left[\det\left(I + \frac{SNR}{N}H \cdot H^{*2}\right)\right] \text{b/s/Hz.} \tag{7.2}$$

In (7.2), C_{MIMO}, N, I, H, and H* are channel capacity of MIMO channels, number of MIMO channels, identity matrix of N×N channels, channel coefficient, and the conjugate transpose of H, respectively. Thus, C_{MIMO} depends on the channel transfer function H which contains all the frequency-dependent different sub-channels between the transmitter and receiver. Since, these sub-channels are formed between transmitting and receiving antennas, the number of antennas used in the MIMO system enhances the data rate of the wireless communication

Figure 7.1 Antenna in the SISO and MIMO configuration. *Source*: Based on Kshetrimayum [13].

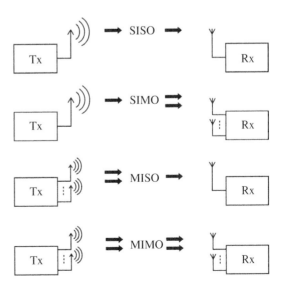

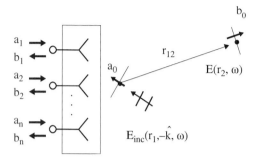

systems. From the signal analysis perspective, these antennas are considered as linear time invariant systems whose performance is evaluated in terms of the transfer function [14]. The transfer function of the antenna is related to the far-field radiation pattern and thus the radiation pattern of the antenna has a great role in the success of their MIMO system [15]. The transfer function of the antenna is also related to the port isolation of the multiport MIMO system and for a generalized multi-antenna system channel as shown in Figure 7.2, the port isolation can be determined to find the influence of one port on the other where the backscattering of the antenna array is taken into account given by (7.3) and the notations have the standard meaning [15].

In another sense, it also says that the channel matrix is related to the correlation between antennas and can be evaluated from the knowledge of the transmit antenna transfer function, channel transfer function, and the receive antenna transfer function [15]. The simplified channel matrix is written as (7.4).

$$C_{rad} = \frac{E(r_2,\omega)\hat{v}}{E(r_1-\kappa,\omega)} = f(\Theta_{TX},\Phi_{TX},\Theta_{RX},\Phi_{RX},\omega), \tag{7.3}$$

$$H = H_{TX}^T \times H_C \times H_{RX}. \tag{7.4}$$

The correlation is the function of the distance between two antennas either at the transmitting or receiving end. In the compact communication systems such as handheld devices, these antennas are closely placed and the correlation increases to prohibit antennas to operate independently, and thus the antenna design complexity increases many folds. The radiated power and the antenna gain pattern are related to the electric field (E) in the free space. Thus, the channel matrix is related to the radiated electric field from the antenna. The vector E-field contains three main elements: (a) phase, (b) shape of the pattern, and (c) polarization. Thus, three strategies, space, pattern and polarization, can be used to de-correlate the antennas.

With the advancement development in the wireless communication system, the theory and practice are moving toward cognitive radio, waveform diversity technique, and cross-layer technique for dynamic spectrum management. However, these techniques implicitly assume communication of fixed radio frequency front end. Contrary to this, the technology advancement has led to the development of the reconfigurable RF front ends and particularly in the reconfigurable antennas which may be used to optimize the performance of the wireless communication systems. The MIMO channels operate in multipath fading environments where there are a number of sparse channels. These sparse channels can be covered by the use of the adaptive antennas. Since, reconfigurable antennas are adaptive, they can play a significant role in enhancing the channel capacity of the MIMO systems. These types of antennas can enhance the link capacity by offering maximum multiplexing gain over the entire SNR range of the channel. Thus, the reconfigurable antennas find a unique position in the development of the modern MIMO systems.

7.3 Isolation Techniques in MIMO Antennas

The spectral efficiency of the MIMO communication system largely depends on the channel matrix and the de-correlation among the antenna elements. The antennas can be de-correlated by increasing the spatial distance between antennas which is against the concept of miniaturization. In another approach, the orthogonal placement of two antennas disturbs the current phase distribution on the radiating elements and the ground plane and thus the radiated field pattern becomes orthogonal which helps in minimizing the correlation. However, the large separation is still needed to achieve a reasonably good isolation between the two antennas. Since isolation is inversely proportional to the correlation, this single parameter invites attention of designers to improve the isolation between two antennas which can be improved by various methods [16]. There are a number of isolation implementation methods which have been used to de-correlate fixed band or reconfigurable band antennas.

7.3.1 Decoupling Network

The decoupling network when placed between the closely spaced antennas reduces the coupling by the means of the negative coupling. The decoupling network consists of a transmission line of specific length which changes the phase of the current between two closely spaced antennas. The decoupling theory is documented in [17] and the generalized schematic of two closely placed antennas is shown in Figure 7.3.

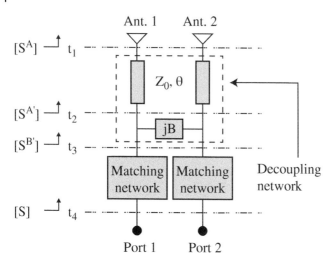

Figure 7.3 Decoupling technique for isolation improvement. *Source*: Chen et al. [17]. © 2008, IEEE.

The decoupling network can be modeled as a four-port network. The following conditions are sufficient to improve the isolation:

$$\theta = 0.5\left(\varphi \pm \frac{\pi}{2}\right) \tag{7.5}$$

$$B = \pm \frac{2\alpha}{1+\alpha^2} Y_0 \tag{7.6}$$

In Eqs. (7.5) and (7.6), θ, φ, α, Y_0, and B are electrical length of coupling network, phase angle of coupling parameter between two ports, magnitude of coupled signal, characteristic admittance of network, and substance of the passive element, respectively. The decoupling network can also be reconfigured to offer different degrees of isolation at various frequencies of operation. The reconfigurable decoupling network using a p-i-n (PIN) diode is shown in Figure 7.4.

7.3.2 Neutralization Lines

In the neutralization technique, a neutralization line (NL) is placed between two antennas. The current is tapped from minimum impedance point on one of the antenna and phase is reversed by selecting an appropriate length of the NL [19, 20]. The phase reversal of the current passing from one antenna to other through the transmission line creates an extra electromagnetic field to reduce the mutual

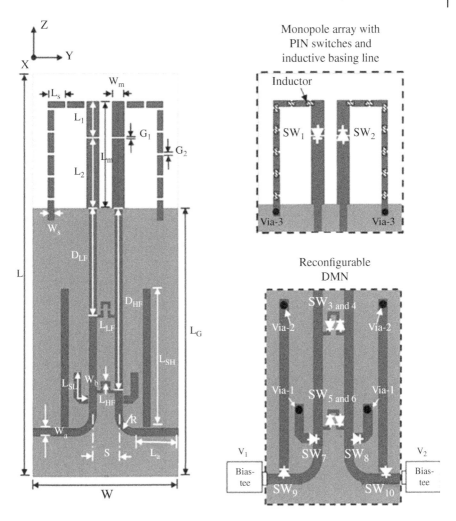

Figure 7.4 Reconfigured decoupling network with MIMO antennas. *Source*: Cai and Guo [18]. © 2011, IEEE.

coupling. Although, this technique reduces the coupling over a narrow frequency band because the phase of the current is not the same at every frequency point in the desired band, but in reconfigurable antennas, where the bandwidth is intentionally narrow, the technique finds a good application [20]. A reconfigurable antenna using the NL is shown in Figure 7.5. The antenna uses reconfigurable NL with two states. When switches are ON, it operates at around 2.4 GHz with a good isolation. When both switches are OFF, the resonant frequency shifts to higher

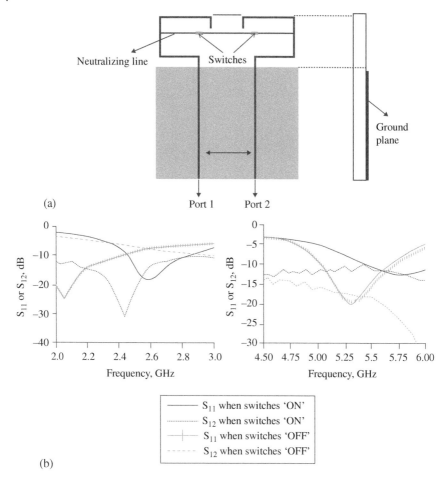

Figure 7.5 Antenna using reconfigurable neutralization line (a) structure and (b) response. *Source*: Luo et al. [20]. © 2010, The Institution of Engineering and Technology.

band. The first case describes the effect of the NL and the second one shows the effect of spacing in terms of operating wavelength.

7.3.3 Using Artificial Material

In non-reconfigurable antennas, to enhance port isolation in MIMO configuration, artificial materials are used [21, 22]. The 1-D dispersion diagram for the complementary splitring resonator unit-cell may exhibit a strong rejection depending on its size. In case of reconfigurable antennas, the use of artificial materials has limited scope

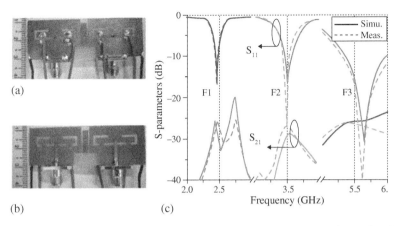

Figure 7.6 Reconfigurable antenna with artificial material loading. *Source*: Pandit et al. [23]. © 2017, IEEE.

due to the narrow band frequency response. However, a MIMO antenna using artificial material with improved isolation is shown in Figure 7.6 [23]. The antenna operates over multiple bands above 2 GHz and isolation is better than 20 dB in all cases.

7.3.4 Defected Ground Plane

The defected ground structure (DGS) finds an application in the isolation improvement of reconfigurable MIMO antennas. When a ground plane is defected, it increases the current path in the ground plane which increases the port isolation. The dimension of the DGS has a direct influence on the isolation level. Reconfigurable MIMO with DGS as an isolator is reported in [24, 25]. In general, DGS is lossy at resonant frequency and the lost energy is given by Eq. (7.7).

$$E_{Loss} = \left(1 - \left|S_{11}\right|^2 - \left|S_{21}\right|^2\right). \tag{7.7}$$

Thus, in antennas with low S_{11}, the energy lost by the DGS is related to the mutual coupling and by using appropriate dimension and shape of the DGS structure, it can be controlled.

7.4 Pattern Diversity Scheme

By dynamically changing its properties, a simple reconfigurable antenna can perform multiple tasks and in 5G communication systems, these multitasking reconfigurable antennas are being widely explored for the various applications.

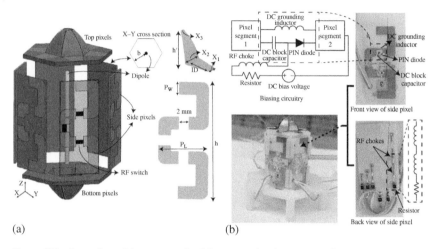

(a) (b)

Figure 7.7 Reconfigurable antenna for 5G communication system. *Source*: Hossain et al. [26]. © 2017, IEEE.

The radiation pattern is one of the key properties of an antenna which can be reconfigured to increase its utility in the small-cell 5G communication systems. A sub-6 GHz (4.8 GHz $< f <$ 5.2 GHz) antenna with the pattern diversity scheme is shown in Figure 7.7. The antenna consists of driven dipoles and surrounded by 3-D reconfigurable parasitic layers with intermediate switches (PIN) and basing circuits. The working of this antenna is similar to Yagi–Uda antenna which consists of driven, reflector, and director elements. The dipole in this case acts as a driven element and the electric length of the connected/disconnected parasitic elements of 3-D layer determines its operating characteristics. Based on the ON/OFF status of switches, the parasitic elements behave as a director or reflector. When the electric length of the pixels (elements) is small which is obtained in the OFF state of the switches, they are transparent to the radiated fields of the dipole and it generates an omnidirectional pattern. When the parasitic layer with connected pixels is electrically large, it generates an azimuthal beam-steering mode. Similarly, the pattern can also be steered in elevation plane by tilting the axis which runs from the reflector to the director. In addition, it also operates in modified dipole pattern mode by interconnecting switches of other three modes. These four modes generated by the antenna are shown in Figure 7.8 [26] and the capability may be extended to MIMO implementation.

In MIMO antenna, to reduce the envelope correlation coefficient (ECC), the pattern of various radiation elements is placed such that these are complementary to each other or they should not overlap. This property is achievable by designing the antenna in such a way that radiation patterns are at offset and the method is called the pattern reconfigurable MIMO [27].

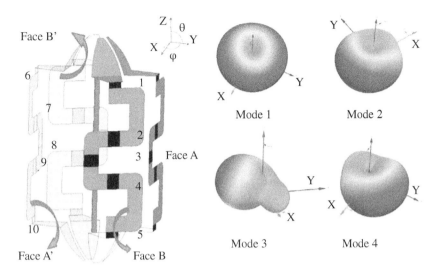

Figure 7.8 Radiation patterns of the reconfigurable antenna. *Source*: Hossain et al. [26]. © 2017, IEEE.

The isolation and ECC can be improved by diagonal placement of radiating elements. This type of antenna configuration also reduces the security threat by avoiding the radiation in unintended direction. By using different switching states, the pattern of the antenna can be controlled [28].

In conical radiation pattern, the maximum directivity is offset to boar sight and pattern shapes resemble a cone. By exchanging the mean beam direction, the average receiver SNR can also be improved [8]. However, omnidirectional antennas can receive rich multipath resulting in a low sub-channel correlation [29]. In such antennas, multiple switching posts are used to reconfigure the radiation pattern [30].

Microstrip antenna can be excited in different modes by using a switching device like PIN diode by changing the mode of operation. To resonate the antenna at the same frequency in different modes, the electric length can be tuned to about 2.5 shorter than the original. The number of switches can be approximately decided to help the antenna to resonate with multiple modes. The arrangement finds an application in MIMO-orthogonal frequency division multiplexing (OFDM) system. The low ECC (below 0.5) is a benefit of pattern reconfigurable antenna and it has been used in a MIMO-OFDM to estimate the frequency response over 117 OFDM sub-channels in a 40 MHz bandwidth with a carrier spacing of 312.5 kHz at 5.25 GHz band. The setup is shown in Figure 7.9.

The pattern reconfigurable antenna, due to the multimode operation, reduces the need of the multiple antenna, with a narrow bandwidth without any control

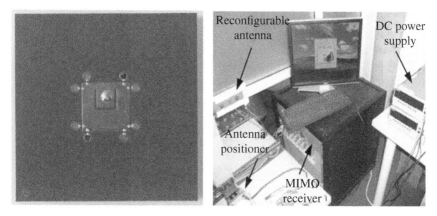

Figure 7.9 A pattern reconfigurable antenna using multiple shorting posts. *Source*: Qin et al. [30].

circuit and two independent patterns can be achieved. A pattern reconfigurable antenna operating at 1.62 GHz with a modified feed line to control the pattern reconfiguration is shown in Figure 7.10 [31].

The diversity property of the reconfigurable antenna is an attractive feature to reduce the multipath effect by avoiding the source noise coming from unknown direction by operating the same antenna in different conical modes like TM_{01} and TM_{11}, etc. Pattern reconfiguration allows the system to adapt to nonstationary propagation environment to improve SINR by directing the nulls of their radiation pattern steering, larger coverage, and data rate. The pattern can be controlled by operating the antenna in characteristics mode [32]. For satisfactory performance of MIMO, three conditions should be satisfied: (i) environment need to support multipath propagation, (ii) antenna diversity, and (iii) received signal should be powerful to overcome noise. Under these conditions, the angle diversity is well suited to the typical indoor- and outdoor-rich multipath propagation environments [33].

7.5 Reconfigurable Polarization MIMO Antenna

In general, spatial diversity technique reduces the correlation between two adjacent antennas. However, in the compact systems, to improve the data rate and channel capacity, multimode antennas offer a good solution. The multimode antennas are an alternative to the spatially separated antennas. The working of this type of antenna depends on the feeding network, mode of operation in place of using different types of antennas. In general, multimode antennas yield the combination of pattern and the polarization diversity to improve the

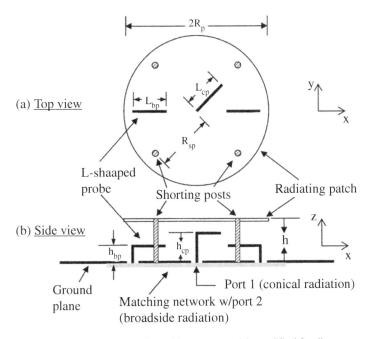

(a) Top view

(b) Side view

L-shaaped probe

Shorting posts

Radiating patch

Ground plane

Matching network w/port 2 (broadside radiation)

Port 1 (conical radiation)

Figure 7.10 Pattern reconfigurable antenna with modified feeding network. *Source*: Yang and Luk [31]. © 2006, IEEE.

uncorrelated channel impulse response for the MIMO or diversity system. The first of this kind of antenna system with multimode operation was suggested in 1965 [34] where an orthogonal azimuth pattern with switched diversity reception technique was used to uncorrelate the received signals and the correlation in terms of the electric magnetic and total energy density for vertical dipoles is shown in Figure 7.11.

Multimode antennas from the signal analysis perspective were investigated in [35], where a biconical antenna with overmoded waveguide excitation supports multimode radiation. Each mode represents a unique solution of Maxwell equations to satisfy the boundary conditions of the articular geometry. Such an arrangement of the biconical antenna is shown in Figure 7.12, where the number of modes is governed by the diameter of the feed line. In the multipath scenario, the far-field radiation patterns obtained from different modes are quite different to result in the low correlation.

In multimode antennas, in addition to uncorrelated signals due to the orthogonal patterns, the mean signal-to-noise ratio (SNR) needs to be similar to obtain a diversity gain or useful MIMO systems. Using log periodic and spiral topology, such properties can be easily achieved as shown in Figure 7.13. By selecting an approximate antenna topology, mean effective gain (MEG) which is related to the

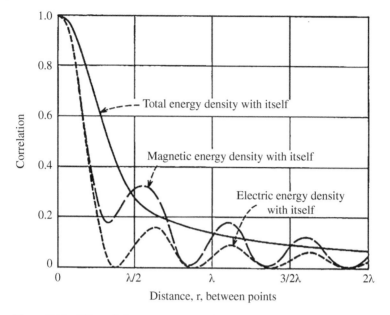

Figure 7.11 Effect of distance on correlation. *Source*: Gilbert [34]. © 1965, John Wiley & Sons.

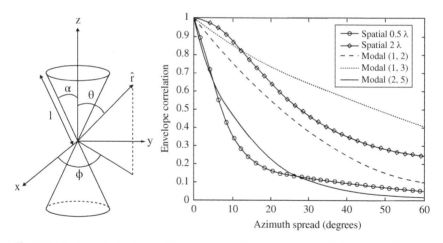

Figure 7.12 Biconical antenna (a) topology and (b) envelope correlation for multimodes. *Source*: Svantesson [35]. © 2000, IEEE.

SNR of the single modes may be controlled to enhance diversity gain. For the MIMO, the total received power or mean SNR is an important parameter for an antenna array [36] which describes the potential acceptability of multimode antenna in MIMO systems.

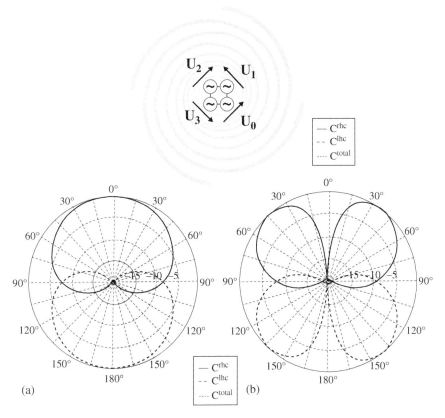

Figure 7.13 Spiral antenna (a) design and (b) radiation pattern. *Source*: Waldschmidt and Wiesbeck [36]. © 2004, IEEE.

The decoupling effects between different polarizations are a complex mechanism to be understood [37]. To de-correlate two signals in the outdoor environment, 20λ horizontal and 15λ vertical separation is needed which may be accomplished by the slant polarized antenna without any spatial separation due to the change in the polarization.

The polarization state of an antenna can be changed by pulling or pushing the resonance frequency [38]. Piezo electric transducer (PET) is used to pull up or pull down the resonance frequency which in turn controls the phase response of an oscillator. When used with a planar structure, it perturbs the electromagnetic fields on the line and changes the effective permittivity, propagation mechanism, and the phase response of the line. Owing to the phase shifting behavior, this type of transducer is used to change the phase response of the patch antenna and thus the polarization state which is dependent on the phase response of the antenna

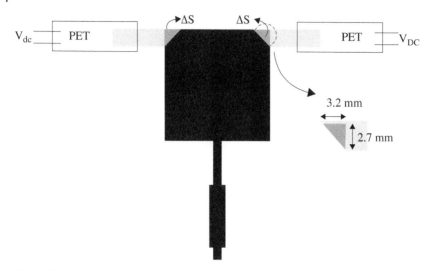

Figure 7.14 Patch antenna with PET switch. *Source*: Hsu and Chang [38]. © 2007, IEEE.

changes. Two such devices are used to operate the antenna in left-handed or right-handed circular polarization state and a PET-controlled reconfigurable antenna is shown in Figure 7.14. The method uses a truncated, nearly square patch to generate different circularly polarized state of the antenna being described by the cavity model. When both perturbs (PET) are lifted up, two degenerate modes (TM_{010} and TM_{100}) modes are generated.

When one device is pulled down, two degenerate modes are split. By controlling the response of the PET, the fixed amount of perturbation is introduced to cause a phase shift of 90° between the two modes which also depends on the host device and the antenna parameters such as dielectric constant and thickness of the substrate. To introduce the phase shift, the dielectric perturbers are introduced. The perturber's dielectric constant and thickness should be high and low with respect to the substrate, respectively. Since, this type of transducer is capable of introducing polarization reconfigurable states in an antenna, a number of such antennas may be configured to work in the MIMO implementation.

The polarization state can also be changed by using shorting pins. The shorting pin can be switched ON and OFF to connect the main radiator to the ground plane. Thus, two modes depending on the switching state, with the same amplitude and a 90° phase-shifted response, are possible to generate the circularly polarized response of the antenna. A polarization reconfigurable antenna topology is shown in Figure 7.15. The circularly patched antenna is short-circuited at its rim with a fixed angular separation with respect to the center of the patch.

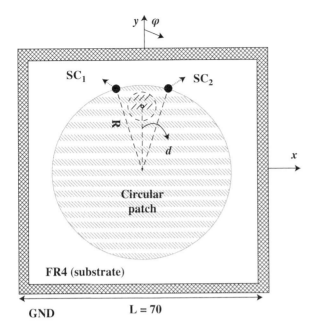

Figure 7.15 A reconfigurable polarization antenna. *Source*: Khaleghi and Kamya [39].
© 2009, IEEE.

These shorting points are selected either side of the line joining feed line to the center of the patch which can be switched ON or OFF with the suitable switching element.

The wireless channels are time-variant and thus the reconfigurable antennas with full polarization capabilities (orthogonal linear, right-hand, and left-hand) are needed for the optimum performance of the communication systems. Thus, quad-polarization comprising of two-orthogonal polarization and two circular polarization states of the antenna would help the communication systems to adapt time-varying channel environments. Such an antenna is shown in Figure 7.16 where the tunable coupler is used to change the state of antenna. In this topology, the coupler is reconfigured to change the polarization states. Thus, this type of antenna needs a reconfigurable hybrid coupler which can provide the −3 dB hybrid mode (90° phase difference between direct and coupled ports) or the uncoupled line mode. Due to the polarization agility, the antenna is suitable for the modern wireless communication systems in the multipath propagation environment.

In general, a polarization diversity antenna consists of two antennas with the orthogonal polarization with the similar radiation pattern and a high degree of polarization to maximize the capacity. In the simple way, with the two dipoles,

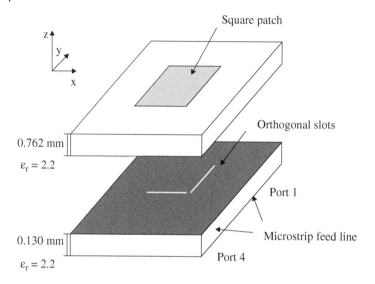

Figure 7.16 Quad-polarization state antenna. *Source*: Ferrero et al. [40]. © 2009, IEEE.

one as vertical and another as horizontal polarization, the diversity is achieved. However, at the high frequency, the polarization diversity may also be achieved by the single antenna component if one follows a certain structure pattern like Alford loop. In this kind of antenna, the surface current density on the main radiator and the ground plane maintains the same orientation to obtain the polarization diversity. Alford loop antenna to maintain the polarization diversity is shown in Figure 7.17 [41]. This kind of the antenna maintains an invariable radiation pattern in one plane and an omnidirectional radiation pattern in the orthogonal plane.

Generally, due to the miniaturization of antennas, which supports multiple frequency bands, these antennas are electrically small. When designing a reconfigurable antenna using this type of antenna at the low frequency, the gain is less which is rarely considered in the system-level design on which emphasis is put in. The antenna can be equipped with the polarization tracking capability to track the polarization orientation/matching between the transmit and receive antennas which is critical for the line of sight operation. In the non-line of sight (NLOS) application scenario, the reconfigurable antennas are used to improve the channel capacity using heterogeneous polarization configurations employing orthogonal space-time block cones (OSTBC). To meet these expectations, Grau et al. [42] have proposed an Oriol antenna as shown in Figure 7.18. It is a compact dual polarized two-port reconfigurable antenna designed on a quartz substrate. It consists of a single octagonal-shaped patch in which two orthogonal ports excite two orthogonal polarizations. The polarization states are selected using monolithically integrated

Figure 7.17 Alford loop antenna for polarization diversity. *Source*: Ahn et al. [41].

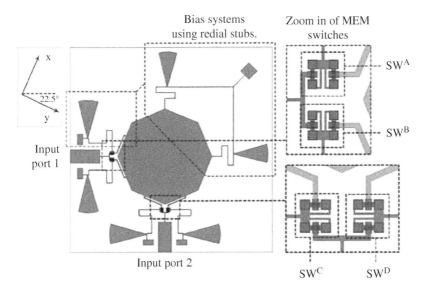

Figure 7.18 Oriol antenna. *Source*: Grau et al. [42]. © 2010, IEEE.

micro-electro-mechanical (MEM) switches. The antenna has the property to rotate its polarization states from vertical/horizontal to slant or vice versa. Since, the antenna is capable to reconfigure its polarization state, it is suitable for the polarization tracking in polarization-sensitive communication systems.

In the MIMO systems, the signal processing, coding, and radiation characteristics can be dynamically changed according to the varying channel conditions and thus it improves the channel capacity [43]. This enhancement in the channel

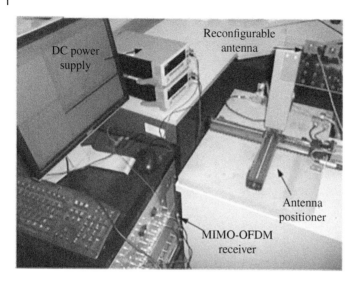

Figure 7.19 MIMO-OFDM system. *Source*: Qin et al. [43].

capacity also depends on the orientation of the transmitting and receiving antennas. In the NLOS scenario, the circularly polarized antenna is relatively more effective than the linearly polarized antennas in enhancing the channel capacity. The capacity gain of the system is significantly increased in the case of a polarization reconfigurable MIMO system. A 2×2 MIMO antenna with different polarization states were used to analyze the capacity gain by Qin et al. [43]. The measurement setup for the same is shown in Figure 7.19.

Polarization reconfigurable antennas are also called polarization agile because they alter the polarization characteristics in real time which are desirable features of wireless systems to increase frequency reuse and to provide a modulation scheme for microwave tagging systems. There are several design challenges to the polarization reconfigurable antenna which further aggravates in the MIMO implementations. The narrow bandwidth, low gain, design complexity, and the performance symmetry on actuating the switching devices are the key issues with these categories of antennas. Several efforts have been made to increase the impedance bandwidth of the circularly polarized antennas mainly by improving the impedance matching of the antenna in different polarization states. Khidre et al. [44] has proposed a simple design of E-shaped polarization reconfigurable antenna which is shown in Figure 7.20. The antenna has four different states depending on the switching states ON and OFF of the embedded diode. In this type of antenna, the circular polarization is obtained by suitable slot dimensions which are offset by the switching states to introduce the asymmetry which perturbs the field beneath the patch to produce the circular polarization. Thus, this

Figure 7.20 E-shaped circularly polarization reconfigurable antenna. *Source*: Khidre et al. [44]. © 2013, IEEE.

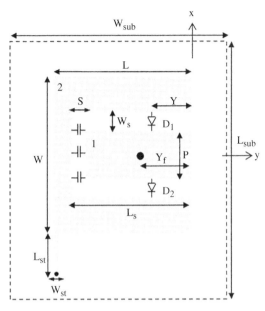

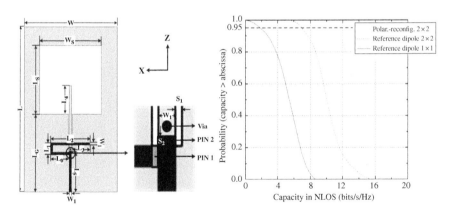

Figure 7.21 Reconfigurable antenna for indoor wireless communications. *Source*: Li et al. [45]. © 2011, John Wiley & Sons.

kind of antenna may be used for the linear as well as circular polarization. When multiple antennas are operated in tandem, they can be implemented in the MIMO configuration to enhance the channel capacity of the system.

The polarization diversity antennas are preferred in the indoor environment and their performance in the MIMO implementation using a slot antenna is shown in Figure 7.21 and Table 7.1 [45].

Table 7.1 Average and 95% outage channel capacity (bit/s/Hz).

Channel capacity	Condition	1 × 1 dipole	2 × 2 dipole	2 × 2 polar – reconfig.
Average	LOS	5	7.86	10.62
	NLOS	5	9.9	13.18
95% outage	LOS	1.75	4.91	7.11
	NLOS	1.94	6.87	11.32

Source: Li et al. [45]. © 2011, John Wiley & Sons.

7.6 MIMO Antenna Performance Parameters

In addition to the conventional characteristics of antennas, such as the bandwidth, resonance frequency, the radiation patterns, the gain, and the efficiency, the MIMO antennas are characterized by various other parameters which are entirely different from the behavior of the single antenna [46]. Reconfigurable antennas in MIMO implementation are also characterized by the various parameters used for the non-reconfigurable MIMO antennas.

7.6.1 Envelope Correlation Coefficient (ECC)

In the communication system, the correlation coefficient (σ) is the measure of the interaction of the channels. The correlation metric considers the radiation pattern of antennas and states the effect of the radiation pattern of one antenna on other when operated simultaneously. In the MIMO system, multiple antennas operate in vicinity of each other which increases the correlation among antennas. The correlation of the antennas increases the redundancy and the channel capacity is reduced. It states the effect of the radiation pattern of one antenna on the other which is measured as the square of the correlation coefficient and known as the ECC. The ECC may be calculated using scattering parameters (S-parameter) and from the 3-D radiation pattern fields. These methods are given by Eqs. (7.8) and (7.9) [47, 48].

$$\text{ECC} = \left| \frac{\left| S_{ii}^{*} S_{ij} + S_{ji}^{*} S_{jj} \right|}{\sqrt{\left(1 - \left| S_{ii} \right|^{2} - \left| S_{ji} \right|^{2} \right)\left(1 - \left| S_{jj} \right|^{2} - \left| S_{ij} \right|^{2} \right)} \eta_{\text{radi}} \eta_{\text{radj}}} \right|^{2}. \tag{7.8}$$

$$\mathrm{ECC} = \frac{\left|\iint_{4\pi} \vec{F}_i(\theta,\phi) \cdot \vec{F}_j(\theta,\phi)^* \partial\Omega\right|^2}{\iint_{4\pi}\left|\vec{F}_i(\theta,\phi)\right|^2 \partial\Omega \iint_{4\pi}\left|\vec{F}_j(\theta,\phi)\right|^2 \partial\Omega}. \tag{7.9}$$

In Eq. (7.8), S_{ij} and η_{radi} are the scattering parameter at port i due to the excitation at port j and radiation efficiency of the antenna i, respectively. Whereas, in (7.9), $\vec{F}_j(\theta,\phi)$ indicates the radiation pattern fields of the antenna j in a polar coordinate system. In these two types of measurement, the first method (S-parameter) describes the port ECC which is greatly affected by the isolation between antennas. However, in the second case, the interaction of the radiated far-field is considered as a deterministic parameter of the ECC. The S-parameter-based ECC calculated with the help of (7.8) is simple in comparison to the 3-D radiated field-based approach given by (7.9) and is considered robust. It is to note that isolation and correlation coefficient are two different entities and high isolation does not guarantee low ECC until antenna efficiency is substantially high. The high isolation and low correlation coefficient are needed for a MIMO antenna system with good diversity performance.

In the case of the reconfigurable antennas, the ECC needs to be computed for each tuning range. To measure the ECC, in this case, first of all, the two MIMO antennas are tuned to operate in the same frequency band to enable them to work as the MIMO antenna and then the ECC is calculated. In the reconfigurable antennas, the antenna efficiency (η) can be low and thus the S-parameter method may lead to erroneous value. However, for the highly efficient antennas with the assumption that $\eta \approx 1$, the method may be used to roughly estimate the ECC of the MIMO antennas.

7.6.2 Total Active Reflection Coefficient (TARC)

In the MIMO implementation, multiple antennas operate and thus scattering (S-parameter) matrix of a single antenna is not enough to describe the bandwidth and efficiency. Thus, the MIMO antennas are characterized in the form of total active reflection coefficient (TARC) which is defined as the ratio of the square root of the total reflected power divided by the square root of the total incident power in the MIMO system. The TARC is computed from the knowledge of the S-parameter using formula (7.10) which is modified as (7.11) for the two-port network [49–51].

$$\mathrm{TARC} = \frac{\sqrt{\sum_{i=1}^{N}|b_i|^2}}{\sqrt{\sum_{i=1}^{N}|a_i|^2}}. \tag{7.10}$$

$$\text{TARC} = \frac{\sqrt{\left(\left|S_{11} + S_{12}e^{j\theta}\right|^2\right) + \left(\left|S_{21} + S_{22}e^{j\theta}\right|^2\right)}}{\sqrt{2}}. \tag{7.11}$$

In (7.10)–(7.11), b_i and a_i are the reflected and incident signals. Boldfaced letters in these equations represent the vector signal.

For the N-port MIMO antenna system, the S-matrices are of the order of $N \times N$. The TARC accounts for the coupling and the random variation signal combinations between the ports. Its value lies between 0 and 1 where 0 indicates the total radiated power and 1 indicates total reflected power by the antennas. The TARC is usually presented on a dB scale. In general, the pattern of the TARC is similar to the reflection coefficient magnitude parameter of the antenna and a maximum 3 dB variation in magnitude with respect to the single antenna is allowed for the better performance of the MIMO system over the desired operating frequency band. With the increase in the number of antenna elements, the S-parameter matrix increases exponentially and thus tracking all the curves pertaining to the random variation in the phase of input signals at a port from other ports is difficult. Thus, TARC is the method of manipulating all these S-parameters for a N-port system and displaying a single curve which contains information of the S-parameters.

For the two-port network, once the S-parameters are measured or calculated, the random phase is swept between $0°$ and $180°$ to find the effect of the phase variation between the two ports on the resonance behavior of the antenna, which is used to create the corresponding TARC curves for effective bandwidth assessment. For the reconfigurable MIMO antennas, the TARC is calculated by synchronously operating the antennas over a particular band by tuning the biasing voltage. When two or more antennas operate in the same band, the TARC is calculated using (7.10).

7.6.3 Mean Effective Gain (MEG)

Normally, the radiation characteristics of a stand-alone antenna are measured in the ideal situation in the anechoic chamber which differs from the practical application scenario. The antenna in the communication systems operate in certain environment for a specific application and thus it is inevitable to know the performance of the antenna operating in a specific environment. Therefore, to know the behavior of the antenna in the specific application scenario, in one way, it has to be compared with the standard antenna whose characteristic is known in such a working condition. This is a complex process, as every time, we have to tune the antenna to specific application, it is to be fabricated repeatedly. In other words, the MEG of the antenna is calculated to take into account the environmental conditions. The solution of this problem is proposed in [52] where a probabilistic model for the environment was proposed. In this model, using 3-D radiation pattern fields along with the proposed statistical model, the mathematical expressions are solved to find the MEG of the antenna. This method allows us to find the MEG using the simulated/measured gain pattern in an ideal environment and a model of the environment suitable for the application scenario.

$$\text{MEG} = \int_0^{2\pi} \int_0^{\pi} \left[\frac{\text{XPD}}{1+\text{XPD}} \ G_\theta(\theta,\varphi)P_\theta(\theta,\varphi) + \frac{\text{XPD}}{1+\text{XPD}} \ G_\varphi(\theta,\varphi)P_\varphi(\theta,\varphi) \right] \sin\theta d\theta d\varphi. \quad (7.12)$$

In Eq. (7.12), XPD is the cross-polarization discrimination. Other terms, G and P, given in a polar coordinate system are the gain pattern and the statistical distribution of the incoming waves in environment assuming the two are uncorrelated. In addition, the following conditions of the gain pattern of the antenna and statistical distribution of the incoming wave need to satisfy (7.13)–(7.14) to be used in the calculation of the MEG.

$$\int_0^{2\pi} \int_0^{\pi} \left[G_\theta(\theta,\varphi) + G_\varphi(\theta,\varphi) \right] \sin\theta d\theta d\varphi = 4\pi, \quad (7.13)$$

$$\int_0^{2\pi} \int_0^{\pi} P_\theta(\theta,\varphi)\sin\theta d\theta d\varphi = \int_0^{2\pi} \int_0^{\pi} P_\varphi(\theta,\varphi)\sin\theta d\theta d\varphi = 1. \quad (7.14)$$

The XPD shows the distribution of the incoming power, i.e. the ratio of vertical mean incident power to the horizontal mean incident power. A more general formulation using the polarization matrix and other incoming wave distributions is discussed in [53, 54].

There are a number of channel models. A channel model suits a particular environment such as urban, rural, etc. A general channel model is presented in [55] which assumes a uniform distribution for the signals in the azimuth direction and Gaussian distribution in the elevation direction. It represents a regular Rayleigh fading channel for the cellular communication systems. To compute the MEG of the reconfigurable antenna in the MIMO system, the antenna gain patterns are measured in the horizontal plane and for the different expected value of XPD, the MEG is calculated.

7.6.4 Diversity Gain

In the MIMO system, diversity reception is a technique by which multiple versions of the transmitted signals are received through multiple channel paths. Diversity gain is a measure of the effect of diversity on the communication system.

The diversity gain is defined by (7.15) as the difference between the time-averaged SNR of the combined signals within the diversity antenna system and that of a single antenna on one diversity channel-provided SNR is above a reference level [56].

$$\text{Diversity gain} = \left[\frac{\gamma_C}{\text{SNR}_C} - \frac{\gamma_1}{\text{SNR}_1} \right]_{P\left(\gamma_C < \gamma_s / \text{SNR}\right)}. \tag{7.15}$$

With the assumption that incident signals are uncorrelated with a Rayleigh distribution, the probability of finding the instantaneous mean SNR of the diversity system less than the reference level [P($\gamma_C < \gamma_s$/SNR)] is approximated by (7.16).

$$P\left(\gamma_C < \frac{\gamma_s}{\text{SNR}} \right) = \left[1 - e^{\left(-\frac{\gamma_s}{\text{SNR}} \right)} \right]^M. \tag{7.16}$$

In (7.15)–(7.16), γ_C, SNR, γ_1, SNR$_1$, γ_s/SNR, and M are the instantaneous SNR for diversity system, mean SNR for single branch, reference SNR, and the number of antennas, respectively. The increase in the number of antennas will increase the received combined power in a diversity system (MIMO system).

7.7 Some Reconfigurable MIMO Antenna Examples

In this antenna design example, a four-port electrically small antenna designed on the handheld ground plane size is demonstrated. The example describes the measurement of various MIMO parameters of the reconfigurable antennas in lower LTE band of the electromagnetic spectrum. The MIMO antenna consists of two different sets of antennas to operate in lower bands (700–900 MHz) and the upper LTE, GSM, Wi-Fi, and Wi-Max bands (1.7–5.75 GHz), respectively. However, to understand the behavior of the reconfigurable MIMO antennas, only the characteristics of the lower LTE band antennas are described. The LTE 700–900 MHz band of the electromagnetic spectrum is covered by the reconfigurable antenna #1 and #2 operating in MIMO implementation. Antennas have been printed on a 120 mm × 65 mm × 0.8 mm FR-4 substrate ($\varepsilon_r = 4.4$, tan δ = 0.02). The fabricated antenna is shown in Figure 7.22.

The reconfigurable antenna consists of narrow strips. Since, the narrow strip monopole antennas suffer from the poor impedance matching, the operating bandwidth is enhanced by analyzing the antenna using equivalent circuit (EC) model and its equivalent lumped circuits are reported in [57]. The reconfigurable antennas operating in the lower LTE band were measured for the S-parameters with the help of vector network analyzer (VNA). To show the effect of the biasing

Figure 7.22 Four-element MIMO antenna with reconfigurable lower band. *Source*: Jha and Sharma [57].

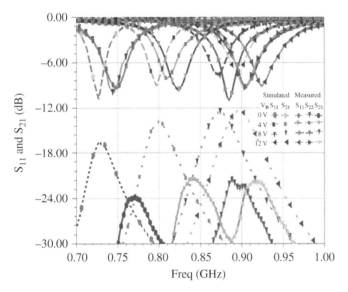

Figure 7.23 Simulated and measured S-parameters (simulated: dash, measured: solid). *Source*: Jha and Sharma [57]. 2018, IEEE

voltage on the antenna tuning range, the S_{11} and S_{22} parameters were separately measured at 0, 4, 8, and 12 V, respectively, as shown in Figure 7.23. With the increase in the biasing voltage, due to the nonlinearity of the diodes and fabrication imperfection, the resonance frequencies of Antenna #1 and Antenna #2 shift apart. For the analysis of various MIMO characteristics of reconfigurable antennas, it is required to bias the varactors in such a way that the antennas cover a common operating band and therefore, the biasing voltages are tuned to align the S_{11}, S_{22}, and S_{12} parameters of the two antennas at the center frequency of 755, 840, and 880 MHz, respectively, as shown in Figure 7.24.

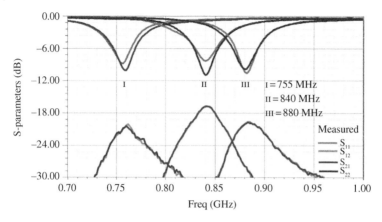

Figure 7.24 Synchronized frequency response of Antenna #1 and Antenna #2. *Source:* Jha and Sharma [57]. 2018, IEEE.

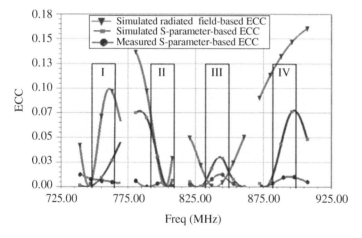

Figure 7.25 Computed ECCs of reconfigurable MIMO antennas. *Source:* Jha and Sharma [57]. 2018, IEEE.

ECC is an important parameter of a MIMO antenna and for this particular reconfigurable antenna elements, the response is shown in Figure 7.25. If we look at the computed ECC using these methods, there is a significance variation in results.

The MEG calculation of a reconfigurable antenna at different frequencies is shown in Figures 7.26 and 7.27. For the successful cellular communication, it is expected that the MEG ratio of an antenna is less than 3 dB. The measured radiation pattern of Antenna #1 and #2 in the azimuthal plane ($\theta = 90°$) is shown in Figure 7.26 [58]. The pattern shows a good de-correlation in the horizontal plane as

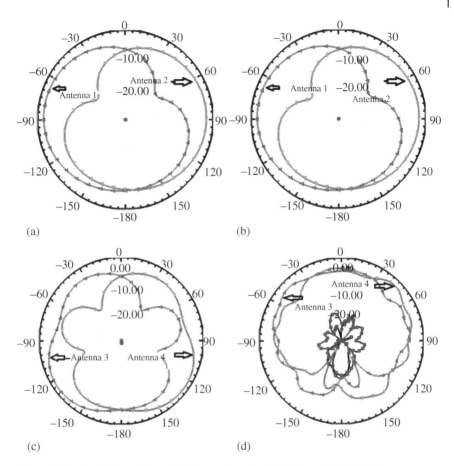

Figure 7.26 Radiation pattern of antennas in azimuthal plane at (a) 830 MHz, (b) 860 MHz, (c) 1.82 GHz, and (d) 5.3 GHz. *Source*: Jha and Sharma [58].

required to calculate the MEG. From this gain pattern, for the various possible value of XPD, the calculated MEG of the antennas and its ratio are shown in Figure 7.27.

The TARC of Antenna #1 and Antenna #2 is computed at two different frequencies, 800 and 880 MHz, for the lower band antennas which is shown in Figure 7.28. When comparing the performance of the single antenna, in the passband with the angular variation of phase, the impedance matching level shifts up or down. However, the pattern of the impedance bandwidth is identical. The variation in the magnitude accounts for the reflected power when the phase angle of incident wave at any port is randomly varying and in the worst-case scenario, it is expected that TRAC does not vary more than 3 dB for the practical cases. The other reference antenna of interest to this chapter is discussed in [57].

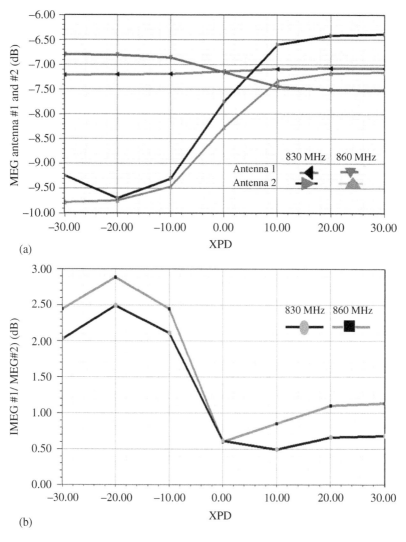

Figure 7.27 Computed (a) MEG and (b) MEG ratio for Antenna #1 and Antenna #2. *Source*: Jha and Sharma [58].

The reconfigurable MIMO antennas are also important for the cognitive and the software-defined radio applications. The front and back faces of a multiband antenna with reconfigurable LTE band is shown in Figure 7.29. The antenna is designed on a low-cost FR-4 substrate ($\varepsilon_r = 4.4$ tan $\delta = 0.02$) of dimension $65 \times 120 \times 1.6$ mm^3 thus making it compact and suitable for the wireless handheld devices. The five-port antenna contains a sensing antenna, two lower LTE band

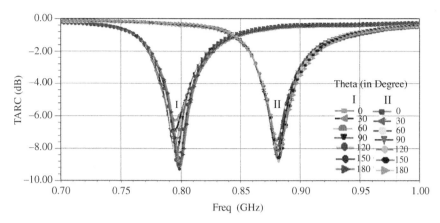

Figure 7.28 Measured TARC of reconfigurable antennas in LTE band. *Source*: Jha and Sharma [57].

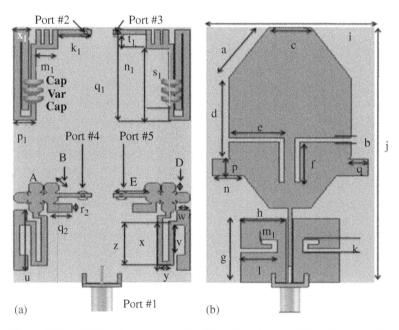

Figure 7.29 MIMO antenna system for CR: (a) top view and (b) bottom view. *Source*: Jha et al. [59].

reconfigurable antennas, and two higher band monopole antennas and the antenna system has two sets of MIMO antennas and one additional sensing antenna making it suitable for MIMO implementation with cognitive radio applications. The design parameters of these antennas are shown in Table 7.2 [59].

Table 7.2 Design parameters of MIMO antenna.

Parameters	Size (mm)	Parameters	Size (mm)	Parameters	Size (mm)	Parameters	Size (mm)
a	28.57	I	65	k1	12	q2	8.8
b	2	J	120	m1	8	r2	4
c	16	K	6	n1	25.31	u	28
d	12.5	L	14	p1	8	v	14
e	21.5	M	6	q1	42.8	w	4.8
f	19	N	12.31	s1	37.8	x	23.5
g	30	P	8	t1	6.5	y	6
h	17.65	Q	7.15	x1	5.5	z	21
A	3	B	1.63	D	2.36	E	12.10

Source: Jha et al. [59].

The lower LTE band antennas #2 and #3 are reconfigurable in nature. In the lower LTE, the maximum bandwidth requirement of a channel is 20 MHz which is covered by the reconfigurable antennas. The reconfigurable meandered antenna element is mounted on the top corner of the substrate and connected to the ground plane through a coaxial feed. Due to the small electrical size and inter-element reactance, monopole antennas suffer from the poor impedance matching and by using open or shunt impedance matching technique, the useful band of the reconfigurable antenna are achieved. In this example, the open stub matching technique is used. The frequency agility is introduced by using a varactor diode which is supported by the DC blocking capacitance. When these elements are used, along with the radiating patch, the quality factor is reduced which helps in increasing the operating bandwidth and lowering the operating frequency. To improve the impedance matching and frequency agility, SMV1281-011LF varactor diode with the varactor capacitance 13.3 pF (0 V biasing) along with two DC blocking capacitors of 15 pF each are added to the circuit. The measured response of reconfigurable antennas is shown in Figure 7.30 which indicates the high isolation >10 dB in lower LTE band as this is one of the design requirements of a MIMO antenna.

The radiation pattern behavior of the reconfigurable antenna at 800 MHz is plotted in Figure 7.31. When we observe the radiation pattern, we find the change in the orientation of the radiation pattern of two antennas which are operating in MIMO configuration.

To accurately determine the ECC of the MIMO antennas, the 3-D radiation patterns are to be measured and reconstructed. The reconstructed 3-D radiation

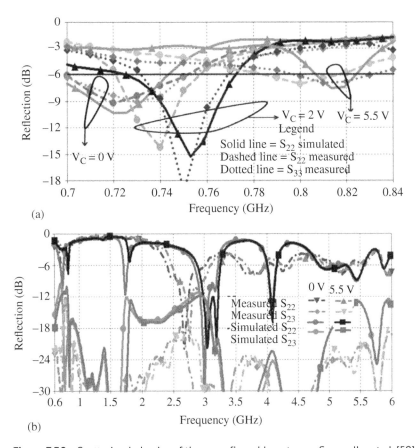

Figure 7.30 Scattering behavior of the reconfigurable antenna. *Source:* Jha et al. [59].

pattern of these antennas when measured at the interval of 22.5° in φ-plane at 2.4 GHz is shown in Figure 7.32 which shows variation in the main beam directions in two cases. Once the 3-D radiation pattern was measured, the ECC was calculated by writing a code using Eq. (7.8). In this antenna, the measured ECC of the reconfigurable antennas at 0.7 and 0.8 GHz is 0.019 and 0.196, respectively.

In the MIMO system, due to the multipath richness of the wireless propagation model, the channel capacity increases. However, the performance is limited by a number of antennas, power imbalance, and the multiple links or their correlation. The isolation enhancement techniques help in improve the performance of the MIMO antenna but do not guarantee the reduction in ECC, always. Thus, the implementation of the de-correlation enhancement technique is another

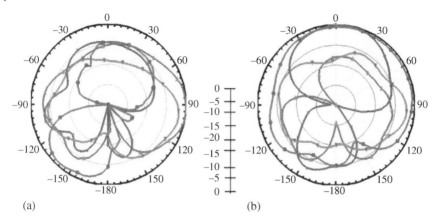

(a)   (b)

Figure 7.31 2-D radiation pattern of antenna (a) #2 and (b) #3 at 800 MHz. *Source*: Jha et al. [59].

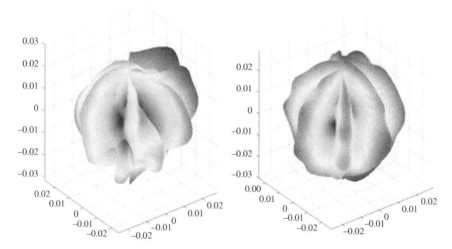

Figure 7.32 Measured correlation pattern of reconfigurable antennas at 900 MHz. *Source*: Jha et al. [59].

attractive feature of the MIMO antenna design. The correlation coefficient of the antennas is reduced by the use of a good matching network. To reduce the ECC, an antenna along with its matching network is shown in Figure 7.33 [60].

The matching network of the antenna contains two inductors and one tunable capacitor, all arranged in the pi-network configuration. The first inductor of 2.5 nH helps in impedance matching and the second inductor of 5 nH governs the initial resonance frequency of the antenna. The correlation of the MIMO antenna

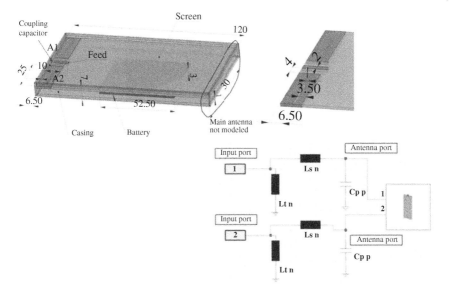

Figure 7.33 MIMO antenna with its matching network (Unit in mm). *Source*: Tatomirescu et al. [60]. © 2014, IEEE.

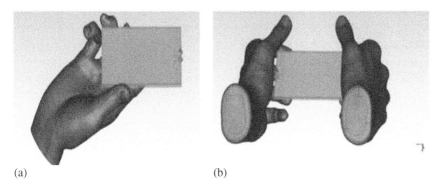

(a)　　　　　　　　　　　　(b)

Figure 7.34 The antenna in (a) free space and (b) data mode. *Source*: Tatomirescu et al. [60]. © 2014, IEEE.

has been calculated in the isotropic channel and a measurement-derived channel model with cross polarization discrimination (XPR) of 5.5 dB. The correlation of the antenna has been analyzed in the two modes: (a) free space and (b) data modes. The position of the antenna is shown in Figure 7.34.

The analysis shows that the coupling capacitor used in the matching network design helps in de-correlating the antennas. When the capacitance changes, the array factor and the gain pattern of the individual antenna changes and thus

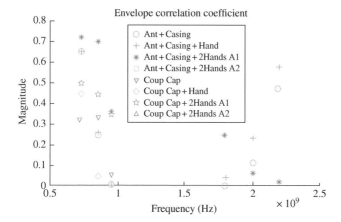

Figure 7.35 ECC of antenna with different configurations. *Source*: Tatomirescu et al. [60]. © 2014, IEEE.

de-correlation is improved. The ECC of the antenna in various simulation environments is shown in Figure 7.35. Thus, in conclusion, the use of the tunable capacitor also helps in correlation control in addition to reconfigurable antenna design.

The MIMO antennas also find their applications in the microwave sensing applications in various bands of the electromagnetic spectrum [23]. Instead of multiband, the antennas can be made reconfigurable to operate over these bands which depend upon the applications at different time slots. Due to the sensing ability of the antenna in MIMO implementation, it may also be used as the wireless sensors. A frequency reconfigurable MIMO antenna for the wireless sensor network applications is shown in Figure 7.36. The size of the antenna in this case is $0.49\lambda_0 \times 0.16\lambda_0$ at 2.45 GHz.

The antenna consists of a slot being excited by a microstrip transmission line. Two PIN diodes are used to reconfigure the antenna. The slot of half-wavelength is used in this design where PIN diodes are placed in the slot and to bias them, metal patches are created within the slot. The diodes are placed at the one end of the metallic patch and at the other end, the DC blocking capacitors have been placed and thus the diode can operate in the ON or OFF state. When two antennas are placed side by side, to maintain a low ECC, the isolation improvement becomes mandatory which is achieved by using a meandered slot line placed between these antennas. The MIMO implementation of the antenna with its S-parameter response is shown in Figure 7.36. The isolation is better than 25 dB which may help to reduce the ECC of the antenna to make it suitable for wireless sensor applications.

A four-port antenna designed on a tablet size ground plane is discussed in [61]. To utilize transmit spectral power density in optimal way and to eliminate the

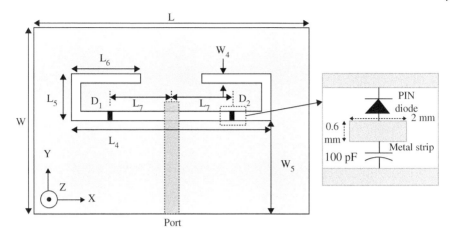

Figure 7.36 A part of the frequency reconfigurable slot antenna showing different components. *Source*: Pandit et al. [23]. © 2018, IEEE.

need of the expensive reconfigurable filters, the reconfigurable antennas are preferred in 4G communication bands and thus this antenna operating in MIMO implementation aims to provide higher data throughput. Four printed inverted F-type antennas have been placed on the four sides of the tablet ground plane as shown in Figure 7.37. The isolation-enhancing element placed between two antennas improves the isolation which also helps in improving the ECC of the antenna. Antennas are reconfigured using PIN diodes which are biased to operate in LTE 17, LTE 13, LTE 14, GSM (824–894 MHz), and EGSM (880–960 MHz) bands. While reconfiguration happens at the lower band, higher frequency bands are consistent and it covers various wireless communication bands: DCS 1800, PCS 1900, UMTS I, LTE AWS 1, and LTE 2500. The antenna has omnidirectional radiation pattern as desired for the low-frequency communication system. The antenna is shown in Figure 7.37.

Figure 7.38 shows measured reflection coefficient magnitude for the lower reconfigurable frequency band (Figure 7.38a) and higher consistent band (Figure 7.38b). It can be observed that the lower reconfigurable bands between 704 and 960 MHz cover particularly EGSM, GSM, LTE13, LTE14, and LTE17 bands as PIN diodes on the radiating element are turned ON and OFF. While the lower reconfigurable frequency bands are switching between EGSM, GSM, LTE13, LTE 14, and LTE17, the higher frequency band is consistently present between 1710 and 2690 MHz.

The measured 3-D radiation pattern of the four radiating elements is shown in Figure 7.39. A near-omnidirectional radiation pattern is observed for all the radiating elements. Additionally, pattern diversity is observed between the radiating elements which helps in improving diversity gain and ECC.

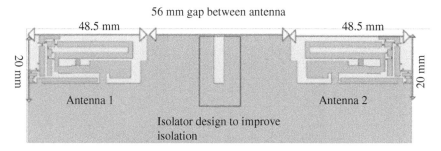

Figure 7.37 Reconfigurable modified PIFA radiating elements (Antenna #1 [top left], Antenna #2 [top right], Antenna #3 [bottom right], Antenna #4 [bottom left]) placed at four corners of the tablet size ground plane implementing four-element MIMO antenna. (a) Simulated model and (b) photograph of the fabricated antenna. *Source*: Sharma and Wang [61].

In general, MIMO antennas share a common ground plane. However, in planar design, it is also possible to use a single radiator and different ground plane to operate in MIMO configuration. A four-port MIMO antenna with coexisting ground plane and radiating element is shown in Figure 7.40. The design of the antenna begins with the analysis of a simple disc antenna which is loaded by stubs and passive elements to achieve the desired operating band. The antenna pattern is repeated to achieve the desired number of ports. The ease of fabrication and series integration of electrical components are the main advantages of this type of antenna. Due to spatial orthogonal port locations, the antenna offers pattern diversity. Generally, this type of antenna operates over multiple bands and one band can be turned on or off by using a reconfigurable reflector. The measured reflection and transmission coefficient of this type of the antenna is shown in Figure 7.41.

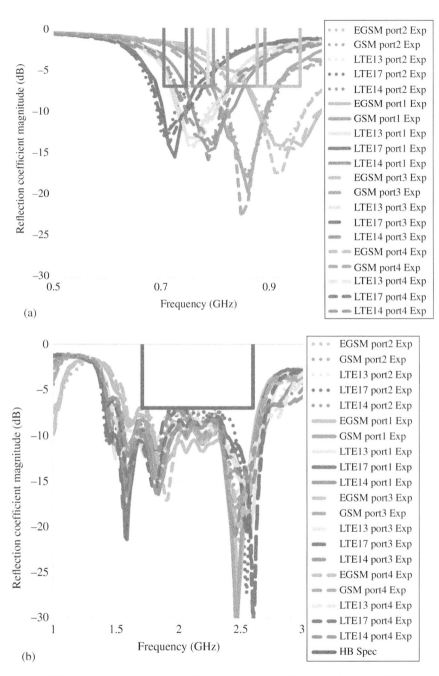

Figure 7.38 Measured reflection coefficient magnitude of the four-element MIMO antenna shown in Figure 7.37b. (a) The lower reconfigurable frequency and (b) the higher frequency band which is consistently present even when lower reconfigurable frequencies are present. *Source*: Sharma and Wang [61].

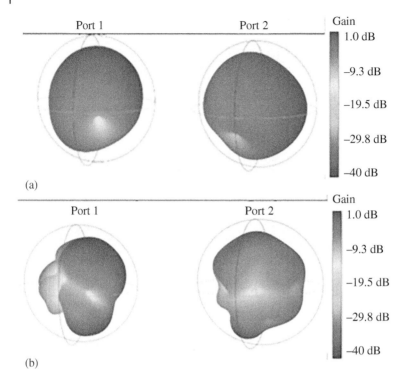

Figure 7.39 Measured 3D radiation patterns of the antennas at 780 and 1710 MHz. *Source*: Sharma and Wang [61].

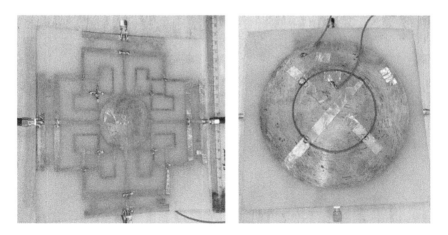

Figure 7.40 Common radiator MIMO antenna [62].

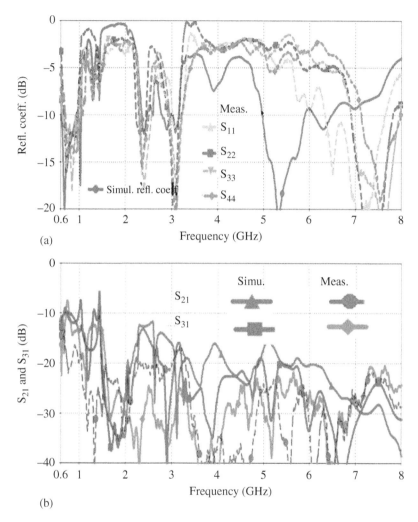

Figure 7.41 S-parameters of the common radiator antenna: (a) reflection coefficient magnitude and (b) transmission coefficient. *Source*: Jha and Sharma [62].

The numerically calculated ECC of this antenna is shown in Figure 7.42. The results show that the orthogonal antennas have good ECC in the entire band of operation but two antennas in line show the ECC above the prescribed limit and it happens due to the existence of co-phase components of the radiated electromagnetic field components.

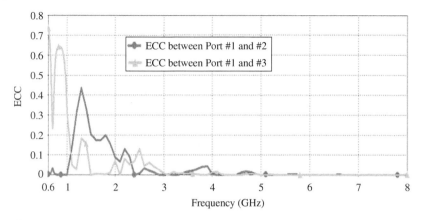

Figure 7.42 ECC of the four-port antenna. *Source*: Jha and Sharma [62].

7.8 Conclusion

In this chapter, the design of reconfigurable antennas in the MIMO system is described. To meet the different requirements, different variants of the reconfigurable antennas are used. There are different types of the reconfiguration technique which are realized using switching elements, artificial materials, and by changing the material property like magnetic materials. When the reconfigurable antennas designed using these methods are used in the MIMO implementation, the channel capacity of the communication system tremendously increases. However, this depends on the placement of antennas and its radiation capabilities. These capabilities are calculated as ECC, TARC, MEG, and diversity gain. To operate the antennas in MIMO configuration, these indices must satisfy a predefined value which depends on the environmental conditions. Although, there are various other parameters such as multiplexing efficiency, diversity gain, and channel capacity which are investigated in MIMO antenna studies but all are related to the correlation coefficient and thus by governing ECC, these parameters can also be optimized.

References

1 Harrington, R.F. (1978). Reactively controlled directive arrays. *IEEE Trans. Antennas Propag.* 26 (3): 390–395.

2 Pringle, L.N., Harms, P.H., Blalock, S.P. et al. (2004). A reconfigurable aperture antenna based on switched links between electrically small metallic patches. *IEEE Trans. Antennas Propag.* 52 (6): 1434–1445.

3 Linden, D.S. (2004). Evolving antennas in-situ. *Soft Comput.* 8 (5): 325–331.

4 Coleman, C.M., Rothwell, E., Ross, J., and Nagy, L. (2002). Self-structuring antennas. *IEEE Antennas Propag. Mag.* 44 (3): 11–23.

5 Haider, N., Caratelli, D., and Yarovoy, A.G. (2013). Recent development in reconfigurable and multiband antenna technology. *Int. J. Antennas Propag.* 2013: 14 pages, Article ID 869170.

6 Kountouriotis, J., Piazaa, D., Mookiah, P. et al. (2011). Reconfigurable antennas and configuration selection methods for MIMO ad hoc networks. *EURASIP J. Wirel. Commun. Netw.* 147 (1): 14.

7 Piazza, D., Kirsch, N., Forenza, A. et al. (2008). Design and evaluation of a reconfigurable antenna array for MIMO systems. *IEEE Trans. Antennas Propag.* 56 (3): 861–881.

8 Boerman, J.D. and Bernhard, J.T. (2008). Performance study of pattern reconfigurable antennas in MIMO communication systems. *IEEE Trans. Antennas Propag.* 56 (1): 231–236.

9 Cetiner, B.A., Jafarkhani, H., Qian, J.-Y. et al. (2004). Multifunctional reconfigurable MEMS integrated antennas for adaptive MIMO systems. *IEEE Commun. Mag.* 42 (12): 62–70.

10 Cetiner, B., Akay, E., Sengul, E., and Ayanoglu, E. (2006). A MIMO system with multifunctional reconfigurable antennas. *IEEE Antennas Wirel. Propag. Lett.* 5 (31): 463–466.

11 Blostein, S.D. and Leib, H. (2003). Multiple antenna systems: their role and impact in future wireless access. *IEEE Commun. Mag.* 41 (7): 94–101.

12 Bellofiore, S., Foutz, J., Govindarajula, R. et al. (2002). Smart antenna system analysis, integration and performance for mobile ad-hoc networks (MANETs). *IEEE Trans. Antennas Propag.* 50 (5): 571–581.

13 Kshetrimayum, R.S. (2017). *Fundamentals of MIMO Wireless Communications*. Cambridge University Press.

14 Manteuffel, D. and Kunisch, J. (2004). Efficient characterization of UWB antennas using the FDTD method. *International Symposium on Antennas and Propagation*, Monterey, 1752–1755 (20–25 June 2004).

15 Manteuffel, D. (2009). MIMO antenna design challenges. *Antennas and Propagation Conference*, Loughborough, 50–56 (16–17 November 2009).

16 Malathi, A.C.J. and Thiripurasundari, D. (2016). Review on isolation techniques in MIMO antenna systems. *Ind. J. Sci. Technol.* 9 (35): 1–10.

17 Chen, S.-C., Wang, Y.-S., and Chung, S.-J. (2008). A decoupling technique for increasing the port isolation between two strongly coupled antennas. *IEEE Trans. Antennas Propag.* 56 (12): 3650–3658.

18 Cai, Y. and Guo, Y.J. (2011). A frequency-agile compact array with a reconfigurable decoupling matching network. *IEEE Antennas Wirel. Propag. Lett.* 10: 1031–1034.

19 Zhang, S. and Pedersen, G. (2015). Mutual coupling reduction for UWB MIMO antennas with a wideband neutralization line. *IEEE Antennas Wirel. Propag. Lett.* 15: 166–169.

20 Luo, Q., Pereira, J.R., and Salgado, H.M. (2010). Reconfigurable dual-band C-shaped monopole antenna array with high isolation. *Electron. Lett.* 46 (13): 1–2.

21 Lee, H. and Lee, H. (2012). Isolation improvement technique for two closely spaced loop antennas using MTM absorber cells. *Int. J. Antennas propag.* 2012, article ID: 736065: 9.

22 Zhai, H., Xi, L., Zang, Y., and Li, L. (2018). A low-profile dual-polarized high-isolation MIMO antenna arrays for wideband base-station applications. *IEEE Trans. Antennas Propag.* 66 (1): 191–202.

23 Pandit, S., Mohan, A., and Ray, P. (2017). Compact frequency-reconfigurable MIMO antenna for microwave sensing applications in WLAN and WiMAX frequency bands. *IEEE Sensor Lett.* 2 (3): 1–4.

24 Chiu, C.-Y., Ho Cheng, C., Murch, R.D., and Rowell, C.R. (2007). Reduction of mutual coupling between closely-packed antenna elements. *IEEE Trans. Antennas Propag.* 55 (6): 1732–1738.

25 Asadallah, F.A., Costantine, J., Tawk, Y., and Christodoulou, C.G. (2017). Isolation enhancement in MIMO reconfigurable PIFAs for mobile devices. *European Conference on Antennas and Propagation*, Paris, 3803–3805 (19–24 March 2017).

26 Hossain, M.A., Bahceci, I., and Cetiner, B.A. (2017). Parasitic layer-based radiation pattern reconfigurable antenna for 5G communications. *IEEE Trans. Antennas Propag.* 65 (12): 6444–6452.

27 Shoaib, S., Shoaib, N., Shoaib, I., and Chen, X. (2017). Design and performance analysis of pattern reconfigurable MIMO antennas for mobile smartphones. *Microwave Opt. Tech. Lett.* 59 (1): 148–156.

28 Rhee, C., Kim, Y., Park, T. et al. (2014). Pattern-reconfigurable MIMO antenna for high isolation and low correlation. *IEEE Antennas Wirel. Propag Lett.* 13: 1373–1376.

29 Piazza, D., D'Amico, M., and Dandekar, K.R. (2009). Performance improvement of a wideband MIMO system by using two-port RLWA. *IEEE Antennas Wireless Propag. Lett.* 8: 830–834.

30 Qin, P.-Y., Guo, Y.J., Weily, A.R., and Liang, C.-H. (2012). A pattern reconfigurable U-slot antenna and its applications in MIMO systems. *IEEE Trans. Antennas Propag.* 60 (2): 516–528.

31 Yang, S.-L.S. and Luk, K.-M. (2006). Design of a wide-band L-probe patch antenna for pattern reconfiguration or diversity applications. *IEEE Trans. Antennas Propag.* 54 (2): 433–438.

32 Lian, R., Pan, J., and Huang, S. (2017). Alternative surface integral equation formulations for characteristic modes of dielectric and magnetic bodies. *IEEE Trans. Antennas Propag.* 65 (9): 4707–4716.

33 Chiau, C.C., Gao, Y., Chen, X., and Parini, C.G. (2005). Evaluation of indoor MIMO channel capacity with a realistic four-element diversity antenna array on a PDA terminal. *IEEE Int. Workshop on Antenna Technology: Small Antenna Sand Novel Metamaterials*, Singapore, 454–457 (7–9 March 2005).

34 Gilbert, E.N. (1965). Energy reception for mobile radio. *Bell Syst. Tech. J.* 44 (8): 1779–1803.

35 Svantesson, T. (2000). An antenna solution for MIMO channels: the multimode antenna. *Asilomar Conference on Signals, Systems and Computers Pacific Grove*, USA, 1617–1621 (29 October–01 November 2000).

36 Waldschmidt, C. and Wiesbeck, W. (2004). Compact wide-band multimode antennas for MIMO and diversity. *IEEE Trans. Antenna Propag.* 52 (8): 1963–1969.

37 Valdés, J.F.V., Fernández, M.A.G., González, A.M.M., and Hernández, D.S. (2006). The role of polarization diversity for MIMO systems under Rayleigh-fading environments. *IEEE Antennas Propag. Lett.* 5: 534–536.

38 Hsu, S.-H. and Chang, K. (2007). A novel reconfigurable microstrip antenna with switchable circular polarization. *IEEE Antennas Proag. Lett.* 6: 160–163.

39 Khaleghi, A. and Kamyab, M. (2009). Reconfigurable single port antenna with circular polarization diversity. *IEEE Trans. Antennas Propag.* 57 (2): 555–559.

40 Ferrero, F., Luxey, C., Staraj, R. et al. (2009). A novel quad-polarization agile patch antenna. *IEEE Trans. Antennas Proag.* 57 (5): 1562–1566.

41 Ahn, C.-H., Oh, S.-W., and Chang, K. (2009). A dual-frequency omnidirectional antenna for polarization diversity of MIMO and wireless communication applications. *IEEE Antennas Wirel. Propag. Lett.* 8: 966–969.

42 Grau, A., Romeu, J., Lee, M.-J. et al. (2010). A dual-linearly-polarized MEMS-reconfigurable antenna for narrowband MIMO communication systems. *IEEE Trans. Antennas Propag.* 58 (1): 4–17.

43 Qin, P.-Y., Guo, Y.J., and Liang, C.-H. (2010). Effect of antenna polarization diversity on MIMO system capacity. *IEEE Antennas Propag. Lett.* 9: 1092–1095.

44 Khidre, A., Lee, K.-F., Yang, F., and Elsherbeni, A.Z. (2013). Circular polarization reconfigurable wideband E-shaped patch antenna for wireless applications. *IEEE Trans. Antennas Propag.* 61 (2): 960–964.

45 Li, Y., Zhang, Z., Zheng, J., and Fen, Z. (2011). Channel capacity study of polarization reconfigurable slot antenna for indoor MIMO system. *Microwave Opt. Technol. Lett.* 53 (6): 1209–1213.

46 Sharawi, M.S. (2013). Channel capacity study of polarization reconfigurable slot antenna for indoor MIMO system. *IEEE Antennas Propag. Mag.* 55 (5): 281–232.

47 Hallbjorner, P. (2005). The significance of efficiencies when using S-parameters to calculate the received signal correlation from two antennas. *IEEE Antennas Wirel. Propag. Lett.* 4: 97–99.

48 Blanch, S., Romeu, J., and Corbella, I. (2003). Exact representation of antenna system diversity performance from input parameter description. *Electron. Lett.* 39 (9): 705–707.

49 Manteghi, M. and Rahmat-Samii, Y. (2005). Multiport characteristics of a wideband cavity backed annular patch antenna for multipolarization operations. *IEEE Trans. Antennas Propag.* 53 (1): 466–474.

50 Pozar, D.M. (1998). *Microwave Engineering*, 2e. New York: Wiley.

51 Manteghi, M. and Sami, Y.R. (2003). Broadband characterization of the total active reflection coefficient of multiport antennas. *IEEE Antennas and Propagation Society International Symposium*, Ohio, 20–23 (22–27 June 2003).

52 Nielsen, J.O., Pederson, G.F., Olesen, K., and Kovacs, I.Z. (1999). Computation of mean effective gain from 3D measurements. *IEEE 49th Veh. Technol. Conf.*, Texas, 787–791 (16–20 May 1999).

53 Jamaly, N., Derneryd, A. and Svensson, T. (2013). Analysis of antenna pattern overlap matrix in correlated non-uniform multipath environments. *IEEE 7th European Conference on Antennas and Propagation*, Gothenburg, 2113–2117 (8–12 April 2013).

54 Kalliola, K., Sulonen, K., Laitinen, H. et al. (2002). Angular power distribution and mean effective gain of mobile antenna in different propagation environments. *IEEE Trans. Veh. Technol.* 51 (5): 823–838.

55 Taga, T. (1990). Analysis of mean effective gain of mobile antennas in land mobile radio environments. *IEEE Trans. Veh. Technol.* 39 (2): 117–131.

56 Vaughan, R. and Andersen, J.B. (2003). *Channels, Propagation, and Antennas for Mobile Communications*. London: IET.

57 Jha, K.R. and Sharma, S.K. (2018). Combination of MIMO antennas for handheld devices. *IEEE Antennas Propag. Mag.* 60 (1): 118–131.

58 Jha, K.R. and Sharma, S.K. (2014). Combination of tunable printed monopole and elliptical monopole antennas in MIMO configurations for cell phone application. *IEEE Int. Microwave and RF Conference*, Bangalore, 194–197 (15–17 December 2014).

59 Jha, K.R., Bukhari, B., Singh, C. et al. (2018). Compact planar multi-standard MIMO antenna for IoT applications. *IEEE Trans. Antennas Propag.* 66 (7): 3327–3336.

60 Tatomirescu, A., Buskgaard, E., and Pedersen, G.F. (2014). Correlation coefficient control for a frequency reconfigurable dual band compact MIMO antenna destined for LTE. *Int. Workshop on Antenna Tech.*, Sydney, 80–83 (4–6 March 2014).

61 Sharma, S.K. and Wang, A. (2018). Two elements MIMO Antenna for tablet size ground plane with reconfigurable lower bands and consistent high band radiating elements. *IEEE Antennas and Propagation Symposium*, Boston, 25–26 (08–13 July 2018).

62 Jha, K.R. and Sharma, S.K. (2018). A novel four-port pattern diversity antenna for 4G communications. *IEEE Antennas and Propagation Symposium*, Boston, 27–28 (08–13 July 2018).

8

Multifunctional Antennas for 4G/5G Communications and MIMO Applications

Kumud R. Jha and Satish K. Sharma

8.1 Introduction

Due to the exponential growth of the data transmission rate over the wireless communication channel, the flexible and bandwidth-efficient communication systems are required worldwide. The development in the field of the communication system, especially in the coding, has enabled the scientific community to approach the Shannon's channel capacity limit with a single antenna attached to the communication system which can further be increased by using multiple-input-multiple-output (MIMO) antenna system and it is a most significant technological breakthrough in the field of the contemporary communications.

MIMO schemes are categorized as (i) diversity techniques, (ii) multiplexing schemes, (iii) multiple access methods, (iv) beamforming, and (v) multifunctional MIMO arrangements. These are shown in Figure 8.1 [1]. In the diversity scheme, the number of replica of the same transmitted signal is received at the receiver with an assumption that at least some of them are not attenuated by the multipath fading effects in the environment. The spatial diversity is achieved by employing multiple antennas at the transmitter or receiver ends. These antennas are used to transmit and receive the encoded replicas of the same information and hence it helps in improving the bit error rate (BER) of the received information.

The multiplexing techniques focus to improve the attainable spectral efficiency of the system. In this scheme, signals from each transmitter are independent of the transmitting antennas. In multiplexing and diversity schemes of MIMO antenna techniques, the antennas are placed as far as possible to enable transmitting or receiving antennas to experience the fading independently and thus highest possible diversity or multiplexing gain is achieved.

Multifunctional Antennas and Arrays for Wireless Communication Systems, First Edition.
Edited by Satish K. Sharma and Jia-Chi S. Chieh.
© 2021 John Wiley & Sons, Inc. Published 2021 by John Wiley & Sons, Inc.

MIMO techniques

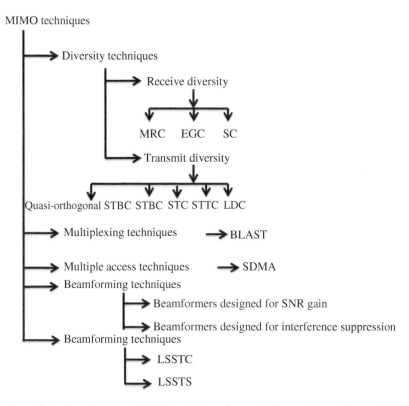

Figure 8.1 Classification of MIMO technique. *Source*: Hajjar and Hanzo [1]. © 2010, IEEE.

The use of multiple antenna system can also be beneficial to improve the signal-to-noise ratio (SNR) at the receiver of the signal-to-interference-plus-noise ratio (SNIR) in a multiuser scenario which is achieved by the beamforming technique. In this technique, the antenna gain is increased in the desired direction and reduced in the other direction to reduce the interference from other users. In a simple diversity scheme, without any loss, the bandwidth is obtained by employing multiple antennas at the receiver end where a number of methods are used to combine the independently fading signal replicas. For the narrowband frequency-flat fading, the maximum ratio combining (MRC) is used. In addition, there are other techniques such as equal gain combining (EGC) and selection combining (SC). These diversity schemes are capable to achieve full diversity order.

On the other hand, it is also possible to use several transmit-rather than receive-diversity schemes using two transmit antennas to avoid the complex joint detection of multiple symbols. Similarly, the information can be decoded at the receiver side with MRC, EGC, or SC scheme using multiple antennas. When multiple antennas are employed at either transmitter or the receiver side, it leads to the development of MIMO systems which improves the attainable spectral efficiency of the system by transmitting a number of parallel signal streams independent of each other and results in the increase in the multiplexing gain. This type of MIMO is known as Bell Labs' layered space time (BLAST) scheme and it aims to increase the throughput of the system in terms of bits per symbol to be transmitted in a given bandwidth at a given integrity. In contrast to the BLAST technique where multiple antennas are activated by single user for increasing the throughput, space division multiple access (SDMA) uses multiple antennas to support multiple users. It uses the user-specific channel impulse response of different users to differentiate different signals.

In the beamforming arrangement, typically half wavelength spaced antenna elements are used to create the spatially selective transmitter/receiver beam. The technique is employed to receive the signals from different angular directions. It is also capable of suppressing the effect of the co-channel interference and thus allowing system to connect multiple users by angularly separating them. This is possible when users are separated in terms of the angle of arrival of their beams at the receiver antennas. Finally, the term multifunctional MIMOs, as the term suggests, combines the benefits of a number of MIMO schemes including diversity gains, multiplexing gains, and beamforming gains. In addition, to make the MIMO system multifunctional, the beamforming is combined with both spatial diversity as well as spatial multiplexing techniques.

8.2 MIMO Antennas in Multifunctional Systems

A generalized multifunctional MIMO scheme is shown in Figure 8.2, which combines the benefits of the space time coding (STC), BLAST, and beamforming technique. The system's architecture shows the transmit antenna arrays consisting of N_T antenna elements which are sufficiently apart in order to experience independent fading. It causes the transmit diversity and/or multiplexing. The antenna elements are separated by half wavelength distance to achieve a beamforming gain. The STC employed can be orthogonal space-time block code (OSTBC), space-time trellis codes (STTC), or STC to attain a diversity gain. In this scheme, data transmitted from each STC equipment are independent of the data being transmitted from other STCs which results in high multiplexing gain. In the BLAST scheme, each STC is considered as a layer and thus the overall multiplexing gain is increased. It can also be helpful in attaining the beamforming gain since it uses the antenna array in the transmission.

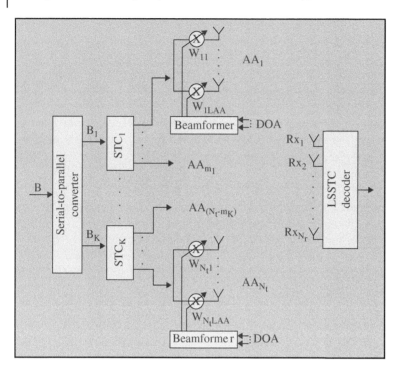

Figure 8.2 Multifunctional MIMO system block diagram. *Source*: Hajjar and Hanzo [1]. © 2010, IEEE.

A dual-functional MIMO scheme was proposed to combine the benefits of vertical-Bell laboratories layered space-time (V-BLAST) and OSTBC in [2]. The scheme considers transmissions over OSTBCs where several parallel OSTBC blocks are capable to transmit data independently. It helps in increasing the diversity gain of the OSTBC and the multiplexing gain due to the use of several independent OSTBC layers. However, this scheme suffers from the drawback that decoder implements group successive interference cancellation (GSIC) which does not take into the account of the antenna-specific received signal power at the different OSTBC layers for the sake of ordering the layers before interference cancellation. On the other hand, a dual-functional MIMO scheme which combines the STBC with beamforming benefits from the diversity gain of the STBCs and the SNR gain of the beamforming [3, 4].

In place of this, dual-functional MIMO schemes which combine STC with beamforming technique benefits from the diversity gain of the STBCs and the SNR gain of the beamformer [4, 5]. This combination technique works with an assumption that the transmitter has the partial knowledge of the channel. It is also possible to combine the twin-antenna-aided Alamouti STBC to a beamforming network with an assumption that the transmitter is aware of the channel status and also aware of signal arrival direction at the receiver [5]. Inspired by this, a tri-functional MIMO scheme which combines diversity gain with multiplexing gain and beamforming gain has also been reported [6]. The multi-functionality of the antenna largely depends on its operating bands and the radiation characteristics. For example, antenna with the unidirectional radiation patterns is most suitable for the satellite communication or point-to-point communication systems. Whereas antenna with omnidirectional radiation patterns are suitable for the global positioning systems (GPSs) or the personal mobile systems. In the multiband and multifunctional devices, one single antenna with various characteristics is useful [7]. Similarly, for the vehicular communication, aircraft communications, or wireless body communications, the antenna with the different radiation pattern in azimuth and elevation plane is required and thus, these tasks make the antenna multifunctional [8–10]. In the modern wireless system, multiband MIMO antennas are widely used in carrier aggregation technique of LTE-Advance [11]. A multimode–multifunctional antenna can meet this requirement.

In comparison to the traditional antennas, frequency reconfigurable antennas have distinct advantage in terms of size, similarity in the radiation pattern, and gain at all the desired operating frequencies. Besides them, reconfigurable antenna also operates as a multifunctional antenna which drastically reduces the number of components, size, and the hardware complexities of the wireless system. As shown in Figure 8.3, to make the antenna multifunctional, the surface current density is exploited by using switches. The antenna contains multiple switching elements. The multiple switches when incorporated in the antenna, various frequency bands can be covered. Similarly, the change in the surface current density

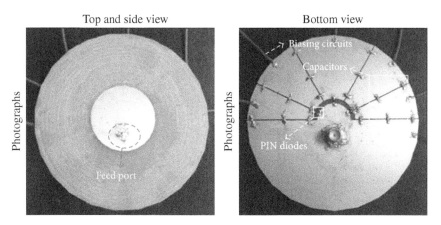

Figure 8.3 Frequency reconfigurable circular patch antenna. *Source*: Chen et al. [12].

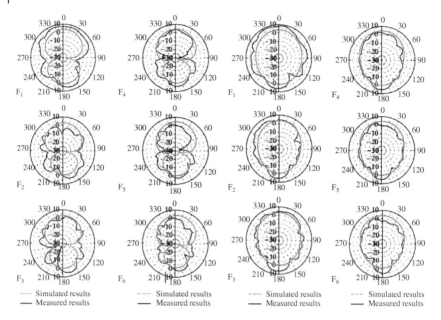

Figure 8.4 Frequency reconfigurable circular patch multifunctional antenna. *Source*: Chen et al. [12]. Licensed under CC BY 4.0.

on the surface of the antenna can operate it in the multiple modes to make it multifunctional. Due to the planar geometry, multiple radiating elements may also be used to convert this single element into the multifunctional MIMO antenna. The radiation pattern of the antenna in various switching stages is shown in Figure 8.4. These radiation patterns may be used to cover various applications in the horizontal and vertical plane of the antenna [12]. Based on the switching states, the antenna is capable to meet the DCS1800 (1.71–1.88 GHz), UMTS (2.11–2.20 GHz), WiBro (2.3–2.4 GHz), Bluetooth (2.4–2.48 GHz), and space research and earth exploration-satellite communications (2.2–2.29 GHz) services.

8.3 MIMO Antennas in Radar Systems

In recent years, the MIMO array has found a suitable application in the near-field imaging radar which has the capability of snapshot data acquisition [13]. The use of the MIMO array helps in synthesizing a large virtual aperture of the target [14]. Thus, in addition to the wideband signals, high resolutions in down-range and cross-range dimensions can be obtained. In this type of radar, the MIMO system contains both transmit and receive antenna elements. The transmit elements are divided into two groups placed at the two ends of the array, while the receive elements are uniformly located in the middle of the MIMO antenna array. For the case, there is M transmitting and N receiving elements, $M \times N$ virtual elements are synthesized. The inter-element spacing plays a vital role in the data acquisition and for the case in [14], the spacing between transmit and receive antennas are d and Md/2, respectively.

For the far-field measurement, the virtual array is equivalent to a linear equi-spaced array with a higher utilization ratio. However, for the near-field imaging applications, the array cannot be considered as a perfectly equi-spaced [15]. The down-range resolution of the radar system is determined by the frequency bandwidth of the transmitted waveform. The down-range resolution of this radar is given by (8.1).

$$\rho_r = c/2B. \tag{8.1}$$

In (8.1), c is the speed of propagation, and B denotes the frequency bandwidth. The cross-range resolution can be estimated by (8.2) [16, 17].

$$\rho_c \approx \lambda R/2L_{vir}. \tag{8.2}$$

In (8.2), λ, R, and L_{vir} are the wavelength, distance between scatterer to the center of MIMO array, and the length of the virtual aperture proportional to the length of the physical MIMO array. For the linear array, the element spacing is less than half a wavelength. For many applications, this criteria is not satisfied and thus results in the grating lobe which is attributed to the under sampling across the aperture dimension. The grating lobes of the MIMO array are determined by the virtual element spacing [18].

$$\theta_g \approx \sin^{-1}\left(\frac{\lambda}{2d_{vir}}\right). \tag{8.3}$$

In (8.3), θ_g is the position of the grating lobe relative to the main lobe and d_{vir} is the spacing between the virtual elements. The schematic of the MIMO radar is shown in Figure 8.5. The MIMO array is parallel to the X-axis. R_0 is the reference distance from the center of the MIMO array to the origin of the target. R_{Tm} denotes the distance between mth transmit element to the scatterer (σ). The R_{Rn} is the distance between nth receive element and the scatterer and the range is given by (8.4a, 8.4b) [18].

$$R = \sqrt{x^2 + \left(y + R_0\right)^2}, \tag{8.4a}$$

$$\theta = \tan^{-1}\left(x/\left(y + R_0\right)\right). \tag{8.4b}$$

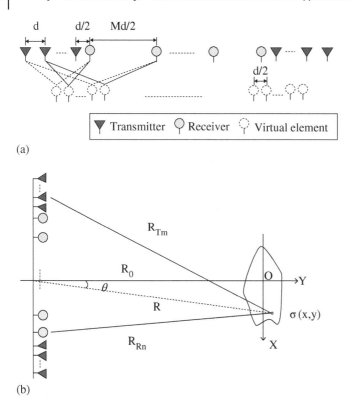

(a)

(b)

Figure 8.5 Schematic of the use of MIMO in near-field imaging radar. *Source:* Liu et al. [18]. © 2018, IEEE.

Based on the theory proposed, a near-field radar using MIMO array is shown in Figure 8.6a. The radar was used in the number of indoor and outdoor experimental verifications. This radar operates in the time division multiplexing (TDM) mode in the X-band of the electromagnetic spectrum. The MIMO array consists of 20 receive and 4 transmit antennas spaced at $d = 0.05$ m. The combinations among these elements synthesize 80 virtual channels for the data acquisition. The array is made of horn antennas in which transmit and receive antennas have the gain of 11 and 16 dBi, respectively. The −3 dB beam widths are 38° and 26° in orthogonal planes. In the X-band extending from 8 to 12 GHz, the frequency is swept in the step of 5 MHz. A radar target and its measured results for the target placed at 5 m distance are shown in Figure 8.6b and c, respectively. The analysis shows that when the received signal-form MIMO is processed appropriately, the image of the target is similar to the object. Similarly, the outdoor experiments also show the good resemblance of the target and its image acquired by the radar. In the outdoor

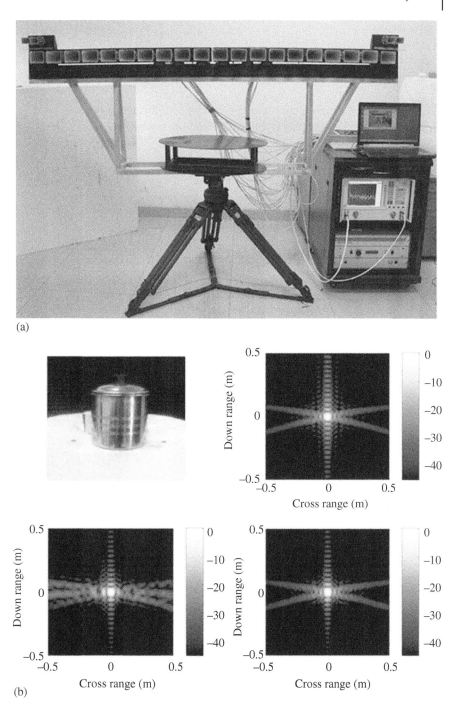

(a)

(b)

Figure 8.6 MIMO-based (a) radar and (b) radar target and its image. *Source*: Liu et al. [18].

environment, a Volvo XC 90 SUV was placed at 15 m away from the antenna and the scattering behavior was measured in 9–11 GHz frequency range at the step of 5 MHz. The target and its image at the radar are shown in Figure 8.7a and b, respectively.

Another example of the use of the MIMO system in the radar technology is described in [19] where a radar system working in L-band at 2 GHz is shown. The system contains 24 antenna elements for both transmission and reception. The antenna operates in a phased MIMO configuration in which these are divided into two groups of 12 each. The flow diagram of the signal processing and the antenna arrangements are shown in Figures 8.8 and 8.9, respectively. The transmitting antennas are further sub-divided into two subgroups and they are located at angles θ_A and θ_B with respect to the target. Both sub-arrays transmit orthogonal waveforms. During the receive mode, all 24 elements operate in the coherent mode to detect the target located an angle θ with the broadside direction of the antenna array. By transmitting orthogonal signals, the jamming probability is severely decreased. In the receive mode, the use of coherent reception helps in gaining maximum advantage of all available degrees of freedom to the receiver. Under the jammed-free conditions, the angle θ for both echo signals corresponding to each transmitted signal from each sub-array of the transmitting antennas is the same. In case of sidelobe jamming, one out of two signals identifies the jamming signals. In this radar, the threshold of 100 has been taken because most

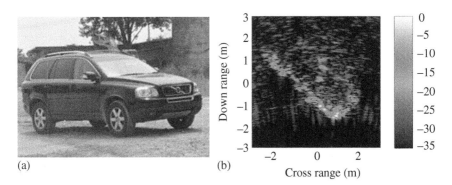

Figure 8.7 Far-field radar (a) target and (b) its image. *Source*: Liu et al. [18].

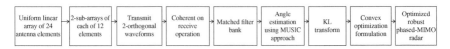

Figure 8.8 Flow chart of phased MIMO radar. *Source*: Butt et al. [19]. © 2018, IEEE.

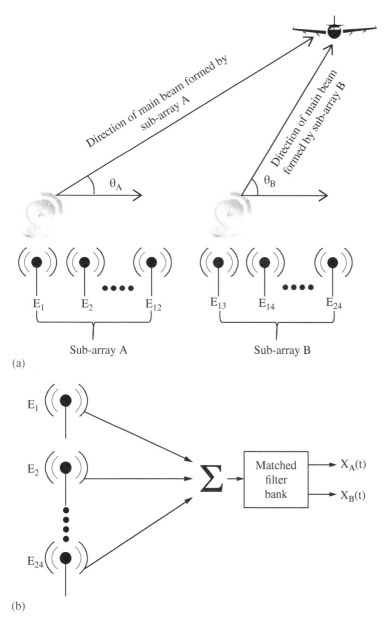

(a)

(b)

Figure 8.9 Schematic of (a) transmit mode and (b) receive mode. *Source*: Butt et al. [19]. © 2018, IEEE.

of the ECM is carried out in the sidelobes owing to inability of counter–counter measures to detect the jamming source.

When the MIMO antennas are arranged in this fashion, the power level of two transmit sub-arrays are different and the power level is compared to find out the sub-array being jammed. The sub-array with higher power level is treated as the jamming signal because the digital radio frequency memory (DRFM) signals have high power in comparison to the echo signal [20, 21]. In case of no jamming, the signals received from two sub-arrays are identical. Under the jamming condition, once the angle of arrival is identified, the transmitting antennas steer the beam in such a way that the null is placed in the direction of jamming.

8.4 MIMO Antennas in Communication Systems

The multifunctional MIMO antennas have numerous advantages as the same antenna can be used over multiple frequency bands and may be deployed to perform multiple tasks. They can be configured as a dual function antenna by controlling the current flowing through the ground plane [22]. To meet these expectations, a multifunctional inverted F-shape (IFA) 2-element MIMO antennas integrated to slotted reconfigurable antenna is designed to form a dual two-element MIMO system [23]. In this antenna, the annular slots also act as a defected ground structure (DGS). The antenna is the first dual-function reconfigurable MIMO antenna, which can be used to increase the throughput of the communication system by increasing the antenna elements. The dual function antenna designed on a substrate area of $50 \times 110 \, \text{mm}^2$ is shown in Figure 8.10. The IFA antenna operates in 2.4 GHz band where the slot antenna covers 1.73–2.28 GHz band of the electromagnetic spectrum with a minimum bandwidth of 60 MHz to support a number of wireless standard bands. The radiation pattern of the antenna shows a wide coverage in these frequency bands which are needed for the wireless communications.

8.5 MIMO Antennas for Sensing Applications

In general, MIMO antennas have widely been studied to increase the throughput of a wireless communication system. However, due to its behavior of an antenna as a sensing element, the application range of the MIMO can be extended to the wireless sensor network. These sensors have been integrated to WLAN and WiMaX networks to increase the throughput of the system as per IEEE 802.11n/ ac and IEEE 802.16m standards. WLAN operates over two bands 2.4–2.48 and 5.3–5.9 GHz. Similarly, WiMax operates in 3.4–3.6 GHz and in these bands the

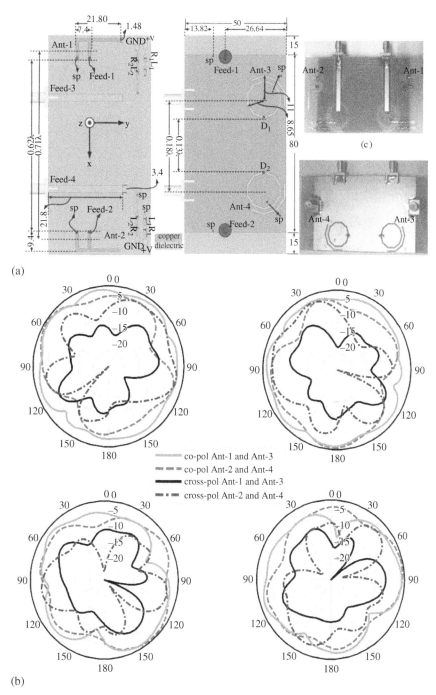

Figure 8.10 Antenna (a) configuration and (b) radiation pattern at 2.45 and 2 GHz. *Source*: Hussain et al. [23]. © 2018, IEEE.

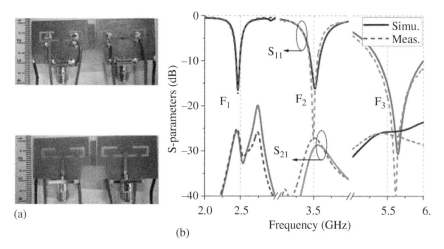

Figure 8.11 Sensing antenna in MIMO (a) topology and (b) S-parameter responses (*Source*: S. Pandit et al. [24]. © 2018, IEEE).

MIMO system can be used to detect the signal as well as the sensing. Further, adding the reconfigurability to the antennas gives an additional advantage of the using the same set of antennas over the multiple bands as the multifunctional antennas. A reconfigurable MIMO antenna for the wireless sensor application is shown in [24]. The antenna is capable of operating over multiple bands where the communication frequency bands have been converted to the antenna-based sensing band. The architecture of the MIMO antenna for wireless sensor applications is shown in Figure 8.11. The frequency reconfigurable sensing MIMO is designed on a 0.8 mm-thick low-cost FR-4 substrate and the space between two antennas are filled with an isolating structure to allow antennas to operate independently.

8.6 MIMO Antennas for 5G Systems

In 5G communication systems, due to the extended applications in automobile, medical, and space-borne systems, it is bound to increase the number of potential users. To efficiently utilize the 5G communication systems in these applications, the base stations need to increase the overall throughput of the system. In the 4G communication systems, the system throughput has been increased by using MIMO technique but that may not be sufficient and thus the massive MIMO antennas are being explored for the 5G communication systems [25–28]. Multiple-user multiple input-multiple output (MU-MIMO) is one of the better technique to serve at the base station of the wireless communication system [29]. This technique is characterized by the MU-MIMO sub-array or shared array architectures with good energy and spectral efficiencies. In this technique, the antenna is capable of generating multiple beams to provide connectivity to a mobile station at the base station which is obtained by digital beamforming [30]. In addition, MU-MIMO is also capable to offer flexible beam patterns and beam-steering capabilities. This type of MIMO offers a directional beam pattern to maximally radiate the energy in specific direction and thus reducing the adjacent channel interference. MU-MIMO antennas need to have good isolation among ports and low loss and high gain, apart from the low ECC. A linearly polarized series-fed 8×16 patch antenna arrays to serve at 28 GHz is proposed in [31]. The architecture of the antenna etched on a Rogers RT Duroid 5880 substrate with $\varepsilon_r = 2.2$ and thickness h = 20 mil is shown in Figure 8.12a. The antenna exhibits traveling wave characteristics which are expected from the series-fed microstrip patch antennas. The overall size of the antenna is 155 mm $\times$ 143 mm $\times$ 0.508 mm. The beam-steering capability of the antenna is shown in Figure 8.12b. The beam is steered by the progressive phase shift between two series of the antennas. The antenna is capable to steer the beam between $\pm 42°$.

To steer the beam, a beamforming circuit is required. The possible architecture of a digital beamforming circuit for a receiver system is shown in Figure 8.13 [31].

8.7 Massive MIMO Array

A high isolation cavity-backed dual slant polarization antenna for the massive MIMO configuration is reported in [32]. The operation is based on the principle of operation of the cavity-backed antenna. This cavity-backed antenna offers a 20.18% fractional bandwidth along with the low mutual coupling. The design detail of the single element antenna is shown in Figure 8.14 which consists of cross-printed diploe on either side of a 30 mil-thick Roger 5880 substrate and the structure is cavity backed. To feed the diploe antenna, a tapered balun transmission line has been used.

The effect of the change in the cavity height influences the impedance matching and the radiation performance of the antenna. For the smaller and larger cavity heights, the radiating elements are close to walls and it affects the radiation from the dipole. Thus, the height of the cavity walls is appropriately selected and the effect of the height is shown in Figure 8.15.

Using the cavity-backed cross dipole elements, a 4×4 sub-array for the massive MIMO applications is designed. Elements are placed at quarter wavelength distance from the cavity walls which gives in-phase reflection giving a directional radiation pattern. The structure of the antenna is shown in Figure 8.16. The distance between each element is about 0.6λ at 6 GHz. This distance is taken to meet the operational requirement of the massive MIMO. For the six sector-based base

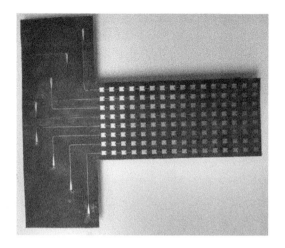

(a)

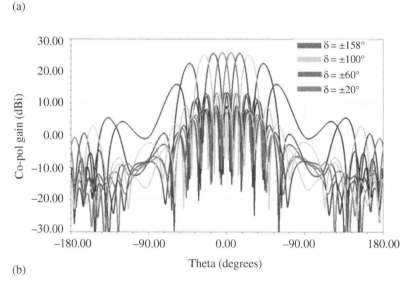

(b)

Figure 8.12 Series-fed (a) microstrip patch array and (b) its beamforming capability. *Source: S. Krishna*[31].

stations, the beam steering of at least ±30° is required to maintain the data link in the coverage sector. The inter-element spacing of 0.6λ ensures the beam steering up to ±45° without the grating lobes. For the massive MIMO implementation of these antennas, the information of coupling between elements and the total reflection coefficient (TARC) is analyzed using (8.5) [33, 34].

$$Active - S_m = \sum_{n=1}^{N} S_{mn} \frac{a_n}{a_m}; \ m = 1...N \ and \ a_m \neq 0. \tag{8.5}$$

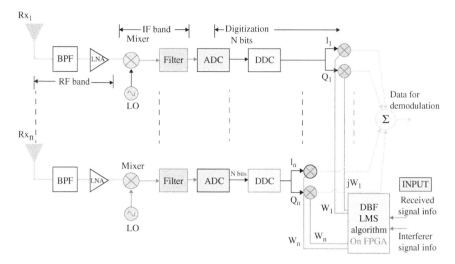

Figure 8.13 Digital beamforming architecture for a receiver system. *Source*: Krishna et al. [31].

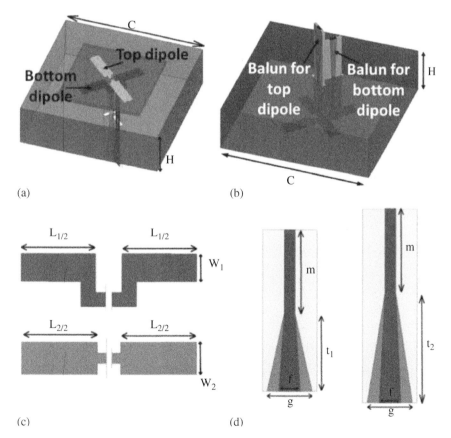

Figure 8.14 Cavity-backed (a) dipole, (b) cavity, (c) dipole, and (d) balun. *Source*: Komandla et al. [32].

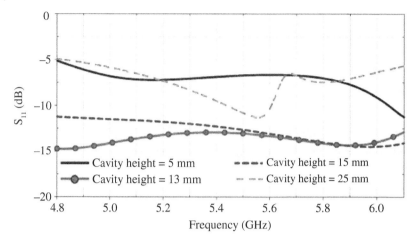

Figure 8.15 Impedance matching of cavity-backed cross dipole antenna. *Source:* Komandla et al. [32].

(a)   (b)

Figure 8.16 Cavity-backed dipole antenna: (a) top and (b) bottom view with SMA. *Source:* Komandla et al. [32].

TARC explains the behavior of the radiating elements and when it is zero, all the delivered power is radiated and when it is equal to one, all the power is reflected back or coupled to the other ports. The TARC calculation of a single element in Figure 8.17 shows that it degrades with the increasing beam scan angles due to the mutual coupling interactions between the array elements.

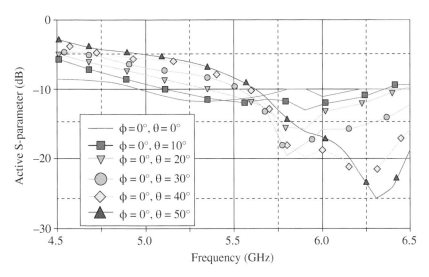

Figure 8.17 Active S-parameter of 4 × 4 antenna array. *Source:* Komandla et al. [32].

The massive MIMO should be an efficient radiator and thus the antenna efficiency and total radiated power (TRP) for the said massive MIMO has been measured using Eqs. (8.6–8.7a, 8.7b) [32].

$$\eta_{antenna} = \frac{TRP}{P_{in}}, \tag{8.6}$$

$$TRP = \frac{1}{4\pi} \int_0^{2\pi} \int_0^{\pi} \left(EIRP_\theta \left(\theta,\varphi\right) + EIRP_\varphi \left(\theta,\varphi\right) \right) \sin\theta \, d\theta \, d\varphi, \tag{8.7a}$$

$$TRP \approx \frac{\pi}{2MN} \sum_{n=0}^{N-1} \sum_{m=0}^{M-1} \left(EIRP_\theta \left(\theta_n,\varphi_m\right) + EIRP_\varphi \left(\theta_n,\varphi_m\right) \right) \sin\theta_n. \tag{8.7b}$$

In (8.6) and (8.7a, 8.7b), $\eta_{antenna}$, P_{in}, TRP, EIRP, N, and M are antenna efficiency, input power to the antenna, TRP, effective isotropic radiated power samples at N locations along the θ-axis, and M location along the Φ-axis for a total of $N \times M$ point measurements, respectively. The beamforming depends on its amplitude and phase weights.

The beam division multiple access (BDMA) is an expected technology of the future generation communication systems. The beamforming is analogous to the use of the spatial filter which is capable to steer the peak of the beam in the desired direction [35]. To steer the beam, antennas are added with the additional circuitries which operate using the analog or the digital techniques. With the

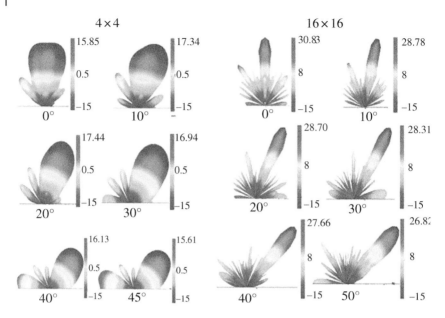

Figure 8.18 Beamforming of 16×16 dipole array at 5.3 GHz. *Source*: Komandla et al. [32].

advancement of the technology and the inherent limitation of component tolerance and signal drift, the digital beamforming technique is preferred. The digital beamforming technique consists of high-speed analog-to-digital converters (ADC) and digital down converters (DDC). In the multiuser environment, the beamforming techniques enable antenna arrays to serve with multiple techniques. It also eliminates the need of the phase shifters and attenuators. The weighing function needed to form the beam in the desired direction is generated at the baseband level and thus no RF phase shifters are needed. The sidelobe levels (SLL) are also controlled using the appropriate window function. The architecture of a digital beamforming system developed in the simulation environment in Keysight SystemVue to steer the beam is shown in Figure 8.18. Using this architecture, the antenna beams of a 4×4 and 16×16 have been steered in the desired direction and the beams have been generated using the Taylor window function. The simultaneous multi-beam generated by 16×16 array is shown in Figure 8.19 and directivity of these antennas is better than 20 dBi.

8.8 Dielectric Lens for Millimeter Wave MIMO

Multiple antennas when used in an array and operated with a specific diversity algorithm can offer a high degree of diversity. However, at any instant of time when using any of the diversity algorithms, it is possible that all antennas in the array are

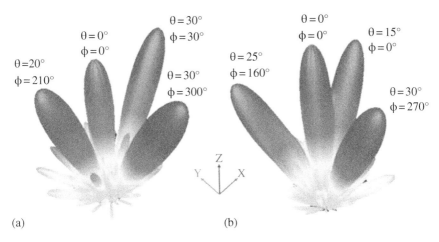

Figure 8.19 Multi-beam capability of 16 × 16 dipole array. *Source*: Komandla et al. [32].

not used simultaneously. Under this condition, the use of the array of antennas becomes impractical at the small portable platforms. As discussed in Chapter 7, the reconfigurable antennas which operate over a number of communication bands are the desirable solution. When such antennas support various applications, they can be treated as the multifunctional antennas. Basically, due to the reconfigurability, a single antenna can replace the multiple antennas. This feature makes it more attractive for the wireless communication systems where space is the constraint.

Due to the reconfigurability, the single antenna can accomplish various tasks and the radiation characteristics can be controlled to govern various applications such as frequency agility, pattern diversity, and beam steering. Depending upon the applications, these applications are becoming the integral part of the communication systems. Multiple radiating elements configured in the MIMO are being used to increase the throughput of the communication systems. In the future generation of the communication system starting from 5G, the role of the MIMO is further being extended as the massive MIMO with the capability of the beam steering to support the BDMA or the multiband antenna systems. This chapter discusses about the multi-functionality of the MIMO antenna for the newer applications.

It is expected that the millimeter-wave technology will be widely used in the 30–300 GHz range. This part of spectrum is capable to support multi-Gigabits/sec speed but suffers from the heavy path loss, channel sparsity, and hardware limitations. This limitation of the millimeter wave (mmWave) communication is partially overcome by the use of the high gain antenna and deployment of massive antenna arrays. Since the size of the antennas are small at high frequencies, deployment of a large number of antennas are possible where various beamforming techniques are used to govern the antenna beam pattern to make it directional. These beamforming techniques include (a) digital, (b) analog, and

(c) hybrid. Among these three techniques, the latter is different from others proposed for the MIMO applications below 6 GHz. Due to need of one radio frequency channel per antenna, it becomes an expensive technique in the massive MIMO design. The analog method uses phase shifters to shape output beam with only one RF chain. It does not address the shadowing and channel sparsity at mmWave. In the hybrid techniques, the benefits of digital and analog beamforming techniques are harnessed to optimize the performance. The digital technique deals with the shadowing and channel sparsity while the analog technique with the antenna directivity. In the hybrid technique, three types of connected networks are feasible. These are fully connected, sub-connected, and single connected antenna architectures. In the first method, each RF chain is connected to all antennas via phase-shifters. This type of the architecture generates highly directional beams but suffers from the complete beam selection. In the second technique, to reduce the complexity, each RF chain is connected to a subgroup of antennas. However, in the third method, several RF chains are connected to a lens antenna via switches. The transceiver antennas generate a few orthogonal beams to achieve multiplexing gain to better utilize the bandwidth. However, with the use of the reconfigurable antennas in the massive MIMO applications, the number of radiating elements required is significantly reduced with the similar performance. The advantage of using the reconfigurable antenna with hybrid beamforming technique further reduces the complexity and enhances the performance. The architecture of a reconfigurable dielectric lens antenna for the mmWave MIMO application is demonstrated in Figure 8.20 [36].

The hybrid architecture provides a balance between the analog and digital beamforming techniques. This architecture is based on the lens antenna array network. Unlike other two, this method uses a selection network comprising of switches instead of phase-shifters. The feeding structure for the antennas are placed beneath the lenses to offer significant directive gain and due to the use of the lens, the antenna system generates narrow beams. Due to the low scattering of the electromagnetic wave scattering at mmWave, the number of effective propagation path between transceiver antennas is lesser than the number of radiating beams, i.e. the number of antenna elements connected to the lens. Thus, the complexity of the network selection is significantly reduced. In addition, the multiplexing gain of the antennas in the hybrid architecture is equal to the number of the effective propagation path at the mmWave. In [37], it is stated that the lens array antenna system does not fall into the category of traditional MIMO antenna system where in the traditional MIMO, only the independent communication path links are increased. However, due to the sparsity in mmWave channels, in the lens array, the maximum number of independent paths is lesser than the number of radiating beams. This problem is solved by using multiple lens antenna arrays in the input–output fashion as shown in Figure 8.20. The architecture of the lens antennas shown in this figure has two fundamental properties: (a) each

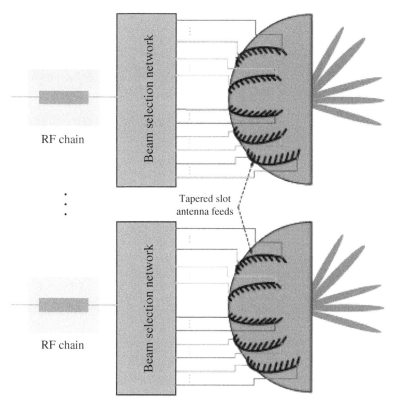

Figure 8.20 Dielectric lens for MIMO applications. *Source*: Almasi et al. [36]. © 2018, IEEE.

antenna connects to one RF chain which radiates multiple beams simultaneously and (b) transmitting and receiving antennas are equipped with reconfigurable antennas. The complete layout of the architecture with the 5G standard handheld device is shown in Figure 8.21. This architecture provides the degree of freedom to ensure the number of MIMO channel paths to be dictated by antennas rather than mmWave channel. Further, the hybrid beamforming technique in conjunction with STBCs offers full-diversity gain to combat the significant path loss and shadowing at mmWave frequencies.

8.9 Beamforming in Massive MIMO

In the digital beamforming architecture of a 5G communication system, the signal from a sender to a receiver can be sent in the line of sight (LOS) direction. However, for the non-line of-sight (NLOS) users, they reach through scattered

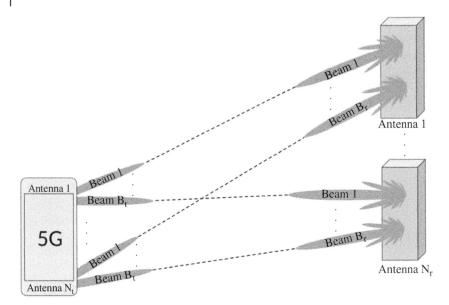

Figure 8.21 Uplink MIMO system with multibeam antenna. *Source*: Almasi et al. [36]. © 2018, IEEE.

beams by buildings, trees, and other environmental features. The NLOS beams are 20–30 dB lesser than the LOS beam and thus many of NLOS multipath signals at the receiver provide sufficient energy to correctly receive the communication. The low-profile scattered beams constructively interfere with each other at the user's location and nearby users experience the noise of relatively reduced magnitude. Antenna arrays with the constructive interference are capable to serve multiple users because the net superposition of the constructive interference at each antenna location is different. The array of antennas capable to explore this property is the massive MIMO antenna system. When antennas operate in array, their working depends on the progressive phase shift among elements which results in steering of the main beam of the antenna in desired direction, and finds its advantage in the 5G communication system. The progressive phase between elements can be applied using analog, digital, or the hybrid beamforming techniques [38, 39].

In analog beamforming technique, data stream is processed via a set of transceivers and data converters. In each branch, the signal is phase shifted and then transmitted by the antenna element. A typical method of beam steering using analog phase shifters is shown in Figure 8.22. The analog beamforming-based technique handles only one data stream at a time.

In the digital beamforming, each antenna is associated with dedicated transceiver and data converter. Thus, this technique can generate multiple antenna

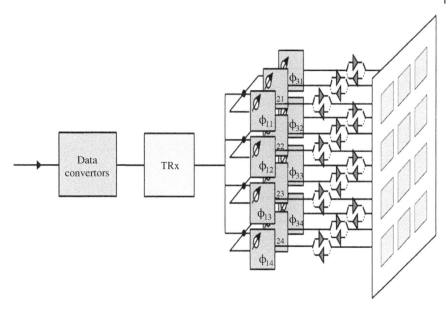

Figure 8.22 Analog beamforming technique. *Source*: Rik Jos,Dealing with power dissipation in 5G antenna design. https://www.eetindia.co.in/dealing-with-power-dissipation-in-5g-antenna-design/ © AspenCore, Inc. [38].

beams simultaneously. The required phase difference to steer the beam is created in the baseband. Due to the dedicated signal processing path, the technique helps in generating multiple beams to serve multiple users. This technique requires more hardware and puts a lot of burden on the digital signal processing chain. Typical digital beamforming architecture for the multiple beam antenna is shown in Figure 8.23.

8.10 MIMO in Imaging Systems

With the emphasis to automate different systems, the role of multifunction sensors has increased many folds and it is preferred over the single sensor-based approach [40–42]. The multifunctional systems are capable of collecting and merging information from different sources in a more compact way and transfer them on the unoccupied space on the main system. The integration of devices reduces the overall cost of the system by sharing the hardware among them. Due to its all-weather capabilities and ability to penetrate through the material and provide three-dimensional (3-D) sensing capabilities, radar is one of the most popular sensors. The radar sensor can be combined with other types of sensors

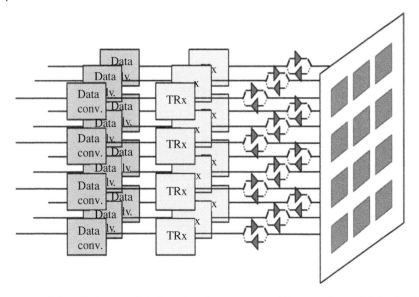

Figure 8.23 Typical multiple beamforming antenna architecture. *Source*: Rik Jos,Dealing with power dissipation in 5G antenna design. https://www.eetindia.co.in/dealing-with-power-dissipation-in-5g-antenna-design/ © AspenCore, Inc. [38].

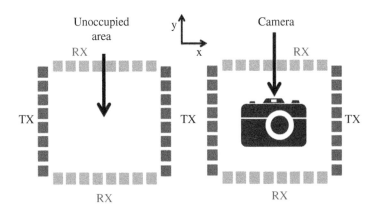

Figure 8.24 Architecture of 3DFMCW MIMO radar. *Source*: Miralles et al. [43]. © 2018, IEEE.

such as a light source, antennas for communication, a weapon, and the thermal camera. In addition, recently the MIMO radar has been integrated to increase the multi-functionality of the system [43]. The architecture of the frequency-modulated continuous wave (FMCW) radar with the camera is shown in Figure 8.24.

A 3-D FMCW MIMO radar with a camera and a two-axis gimbal is shown in Figure 8.25. For the given number of transmit (TX) and receive (RX) modules, a

Figure 8.25 FMCW MIMO radar.
Source: Miralles et al. [43].
© 2018, IEEE.

MIMO radar offers high angular resolution. This architecture uses the spatial separation of the antennas and forms the virtual array which demonstrates the angular resolution. The TX and RX antenna arrays are placed in a rectangular configuration. Since the center of the antenna aperture remains unoccupied, the camera and gimbal are placed to increase the functionality and the camera can be steered to detect the target. The radar system shown in Figure 8.25 contains 16 transmitting and 16 receiving antennas and the radar operates in 16–17 GHz band. The system finds its applications in ground-based surveillance including hazardous areas, energy transmission infrastructure with the easily deployable platform.

In the FMCW, chirp is created by running the algorithm in the laptop. This chirp is transmitted to the antenna board and it is distributed into two channels. Each chain can choose between eight TX antennas or a matched load to enable the time domain multiplexing (TDM) configuration and thus the orthogonal signals are obtained. Here, a MIMO cycle is composed of up-ramps where each ramp is sent via a different TX antenna. The TDM scheme helps transmitting antennas to obtain orthogonal signals and avoid interferences. With the TDM, it is possible to assign TX antennas to RX signals as per the timing scheme. At the beginning, the system waits for the start command from the human–machine interface (HMI) and after that the first TX antenna sends the FMCW chirp signal. RX signals are acquired by the ADC and stored in FPGA and processed in the digital signal processing unit. This process is repeated equal to the number of transmitting antenna (16 in this case) and then beamforming algorithm is run in the matrix laboratory (MATLAB) environment and the total data are transferred to the workstation and radar signal processing algorithm is launched. The angular resolution of the FMCW radar using MIMO technique depends on the number of transmitting and receiving antennas. For the case of 16 Tx (NTx) and 16 RX (NRX) antennas, there are total NTx + NRX antennas which lead to the design of a NTx × NRx virtual element arrays. The virtual array is calculated as the discrete convolution of all RX and TX antenna element positions. The range resolution of this type of the radar is obtained by the fast Fourier transform (FFT) processing and the FMCW ramp setting. For a case of radar cross section (RCS), area of $150\,m^2$ placed at a distance of $23.1\,m$ from the radar has been measured and the measured azimuthal and elevation half power angular resolutions are $4.7°$ and $3.6°$, respectively. Next, the dependency of the received power from the target at a certain distance in the azimuth and elevation planes have been estimated in terms of field of view (FOV) and it is $FOV_\theta = 100°$ and $FOV_\phi = 60°$, respectively. The camera integrated to FMCW radar can be used for the zonal surveillance applications. The FOV of the camera used in the system has also been calculated and it is $FOV_h = 31.3°$ and $FOV_v = 23.5°$, where h and v represents the horizontal and vertical planes, respectively. Since the FOV of radar is greater than that of the camera, the gimbals can be used to point the camera in the right direction. For the target detection, the camera image is superimposed on the radar data and to still see the scene, the amplitude of the reflection defines the transparency level and the color of the radar cell. In this method, there is a strong reflection from the opaque area in the image and it helps in identifying the closeness of the target with respect to the camera. Further zoom facility of the camera can be used to magnify the previously detected radar target. Due to the use of the zoom future of the camera in conjunction with the radar detection ability, the overall detection probability of the target by the FMCW radar can be enhanced.

8.11 MIMO Antenna in Medical Applications

8.11.1 Ex-VIVO Applications

Antennas are also indispensible in the life sciences. A number of implementable antennas comprising of a single radiating element in the form of single-input-single-output (SISO) [44, 45] have been successfully used in the biomedical applications. To increase the rate of transmission and to increase the immunity to noise and interferences, it becomes interesting to explore the additional antenna characteristics [46] and the multiple antennas in the form of MIMO are capable to solve this problem to a great extent without increasing the transmission power which is an important consideration for the implementable devices in the body. In the success of the implementable MIMO antennas, the two main parameters (a) reflection coefficient and (b) mutual coupling play the pivot role because the operating environment is completely different with respect to the free space propagation. These characteristics are governed by the phantom of the body parts. To enhance the data throughput, a 4×4 MIMO antenna designed on 18.5 mm × 18.5 mm substrate area and 1.27 mm thickness is shown in Figure 8.26. The miniaturized antenna has rounded corner to reduce the sharp edge damage to the living tissue in which it is implemented. The implementation scenario of the MIMO antenna is shown in Figure 8.27.

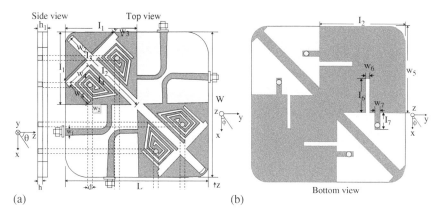

Figure 8.26 MIMO antenna for medical application. *Source*: Fan et al. [47]. © 2018, IEEE.

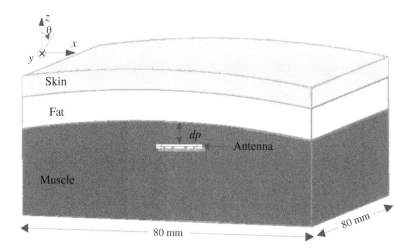

Figure 8.27 Implanted MIMO antenna in the tissue. *Source*: Fan et al. [47]. © 2018, IEEE.

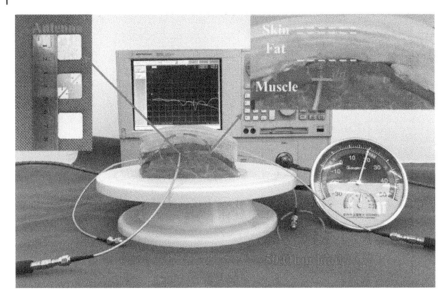

Figure 8.28 Measurement setup for indigestible MIMO antenna. *Source*: Fan et al. [47]. © 2018, IEEE.

The antenna shown in Figure 8.26 operates at the 2.4 GHz ISM band where S_{ii} and S_{ij} are less than −10 and −15.99 dB. The maximum gain of the antenna is −15.18 dBi. The low-gain antenna has the better specific absorption ratio (SAR) and the exposure of the surrounding tissue to the antenna is minimal. The measurement setup for the scattering parameters and SAR analysis is shown in Figure 8.28.

To look at the behavior of the antenna in the medical applications, it was placed inside a pork belly at the depth of 7.5 mm from the top of the muscle. The specimen contains skin, fat, and muscle similar to the diagram shown in Figure 8.27. The ex-vivo measured and simulated results are shown in Figure 8.29.

While designing implementable MIMO antenna, the health safety regulation is to be adhered and the SAR needs to be less than 2 W/kg when tested on the 10 g tissue. The SAR can be calculated from the received power by the antenna when the transmitted power and gain of the reference antenna is known. In the present case, for the 1 W of transmitted power, the antenna-received power is less than 27.3 mW and thus the expected SAR is quite low. For the implementable antennas where the gain is significantly low, the link margin analysis is used to evaluate the wireless communication performance between MIMO and the base antenna and it is greater than 0 dB in up to the distance of 15 mm from the site of implantation and thus the communication with the outside transmitter placed at this distance is possible.

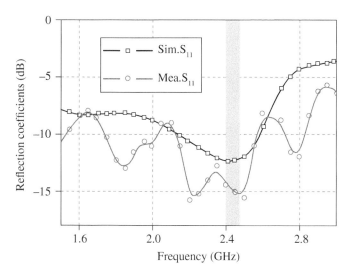

Figure 8.29 Ex-vivo measurement of implementable antenna. *Source*: Fan et al. [47]. © 2018, IEEE.

8.11.2 MIMO Antenna for Medical Imaging

In another application, the use of the MIMO antenna in medical applications is demonstrated in [48]. It can be used in localization of the cancerous cells by comparing the malignant and normal body cells as they show two different dielectric properties at microwave frequencies. The method is safer and inexpensive. The use of multiple antennas increases cross-range resolution of the imaging systems. However, the mutual coupling of the UWB band antennas restricts its applications in the medical electronics. Further, it also needs the isolation from the LAN. A low-profile UWB antenna with compact size and low mutual coupling is shown in Figure 8.30.

8.11.3 Wearable MIMO Antenna

In recent years, the use of the wearable gadgets has increased many folds and the antenna is the critical part of these devices. The design of wearable antenna is challenging which needs to be compact, low profile, and lightweight and to be easily integrated to the other devices. Due to the proximity to the human body, the SAR is also needed to be within the prescribed limit. In general, single antenna is used in the body area networking which does not meet the present-day data transmission capacity. Thus, in place, a MIMO wearable antenna is more appropriate. A wearable antenna in the MIMO configuration using characteristic mode

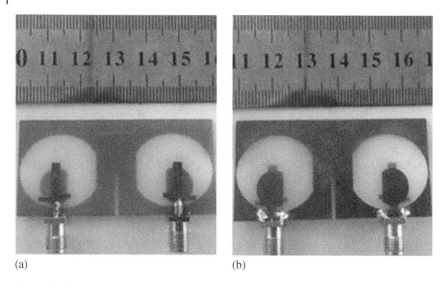

(a)   (b)

Figure 8.30 UWB MIMO for medical applications. *Source*: Latif et al. [48].

analysis (CMA) is presented in [49]. The CMA analysis begins with the analysis of the Eigen value of the perfectly electric conducting surface. The CMA is used to find the resonating mode and the number of radiating elements are added in such a way that mode of interest remains unchanged. While maintaining the same resonance, multiple conductors can be added to the perfectly electric conductor (PEC) to obtain the MIMO operation. Such antenna designed on a 38.1 mm × 38.1 mm substrate is shown in Figure 8.31.

The wearable antennas are conformal to the human body and cloths which also causes some deformation due to the movement of body parts. While designing such an antenna, the deformation aspect must be considered and the sensitivity analysis against the deformation may be used to overcome this problem while designing such type of antennas.

8.11.4 MIMO Indigestion Capsule

In recent years, wireless ingestible capsules are widely being used for the gastro-intestinal (GI) tract endoscopy. The procedure is painless and noninvasive in nature. One of the most limitations of these capsules is the limited data handling capacity in the real-time applications [50]. For the better resolution of the image captured by the capsule, the high data rate transmission is required which can be obtained by the use of the larger bandwidth of the transmitting antenna. However, this approach is not a rational solution as Shannon's law imposes an upper limit

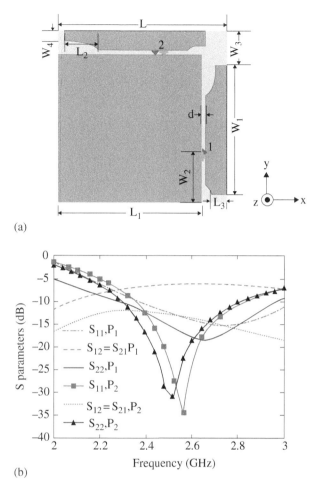

(a)

(b)

Figure 8.31 Wearable MIMO antenna: (a) structure and (b) S-parameter response. *Source*: Li et al. [49]. © 2018, IEEE.

on the channel bandwidth in the single antenna system. This problem can be solved by using MIMO antenna in ingestible capsules [51]. A MIMO antenna to be used in the endoscopy capsule is proposed in [52]. The ingestible antenna is designed on a conformal surface were two loop elements are placed orthogonally. Due to the orthogonal positions, these antennas have good isolation to decouple themselves. The layout of the antenna is shown in Figure 8.32.

The capsule is of dimension 10 mm × 14 mm (diameter × height), where two loop elements are attached. The capsule has been designed by wrapping on a Tactonic TLX-6 substrate ($\varepsilon_r = 2.65$, tan δ = 0.0019) of 0.254 mm thickness.

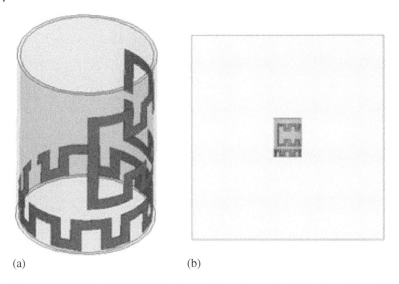

(a)   (b)

Figure 8.32 Indigestible MIMO antenna: (a) conformed and (b) planar views. *Source*: Xu et al. [52]. © 2017, The Institution of Engineering and Technology.

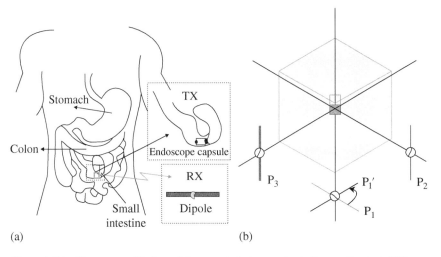

(a)   (b)

Figure 8.33 Placement of indigestible capsule in human body. *Source*: Xu et al. [52]. © 2017, The Institution of Engineering and Technology.

During the GI tract investigation, the capsule is placed deep inside the body as shown in Figure 8.33. The electromagnetic wave emitting from the capsule needs to pass through the human tissue and there is the significant loss of the transmitted power and as a result, antenna gain is low in comparison to the free space-radiating antennas. However, the communication link budget is calculated using

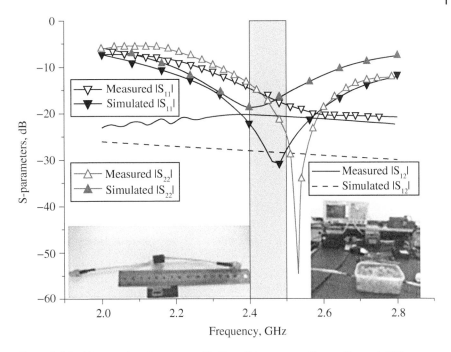

Figure 8.34 Measured performance of indigestible antenna. *Source*: Xu et al. [52]. © 2017, The Institution of Engineering and Technology.

the far-field method. To calculate the link budget, the communication link is established between an ingestible capsule and a half wave dipole attached to the arm of the human which operates in the same frequency band. For the MIMO operation, the reflection coefficient magnitude and isolation are important parameters and for the ingestible antennas, these are shown in Figure 8.34.

8.11.5 Reconfigurable Antennas in Bio-Medical Engineering

Due to the adaptability to the operating environment, the reconfigurable antennas also find numerous applications in the biomedical engineering. A polarization reconfigurable antenna consisting of four dipole elements for the body-centric wireless communication is reported in [53]. The multi-dipole antenna is switchable to 0°, +45°, 90°, and −45° linear polarization which is able to overcome the polarization mismatching and the multipath distortion in the complex wireless channels. Since, it comprises of simple dipoles, the design can be extended for the MIMO implementation. The application scenario of such an antenna in the biomedical applications is shown in Figure 8.35.

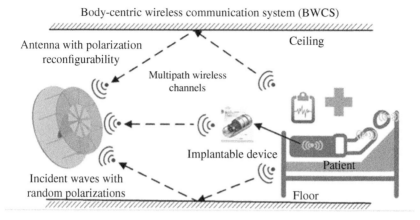

Figure 8.35 Application of reconfigurable antenna in body-centric wireless communication. *Source*: Wong et al. [53]. © 2017, IEEE.

The dipole antennas along with their biasing networks are etched on a 0.79 mm-thick RT Duroid substrate (ε_r = 2.2). The antennas are excited with the help of the ground tapered balun which is vertically placed. The polarization reconfiguration is controlled by switching ON/OFF the p-i-n diodes associated with the biasing lines. For four dipoles, there are four DC biasing lines to generate the polarization states along 0°, +45°, 90°, and −45°. The measurement setup of the antenna in the human tissue is shown in Figure 8.36.

The antenna has a wide 34% fractional bandwidth extending from 2.2 to 3.1 GHz. For the various polarization states, transmission coefficient S_{21} (dB) between human body implantable and the polarization reconfigurable antennas have been measured which shows the mitigation of the polarization mismatching problem. At the same time, the S_{21} measurement is also helpful in predicting the orientations of the implantable antenna inside human body in practical applications.

A combined frequency and polarization reconfigurable antenna designed on a wearable textile is shown in Figure 8.37. The agility of the antenna is controlled by four p-i-n diodes. Further, the circular polarization states have also been realized in global systems for mobile (GSM) and industrial, scientific, and medical (ISM) bands. Since, it is a simple disc-type monopole antenna designed on a flexible substrate and thus, can also be replicated in the MIMO implementation which may be useful in high data rate communication in the biomedical engineering.

In the last couple of years, the use of the flexible antennas in various sections including medical applications has increased many folds. Further, adding the reconfigurability in the MIMO implementation increases its usability many times. To meet this expectation, a reconfigurable flexible antenna in the MIMO applications has been demonstrated in [55]. The simple CPW-fed wearable

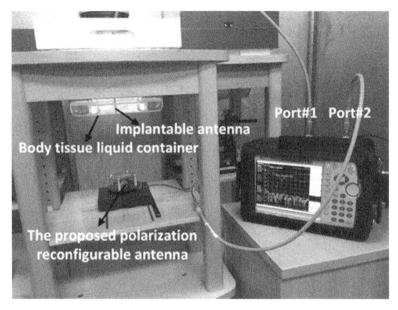

Figure 8.36 Polarization reconfigurable antenna (a) in human tissue and (b) measurement setup. *Source*: Wong et al. [53].

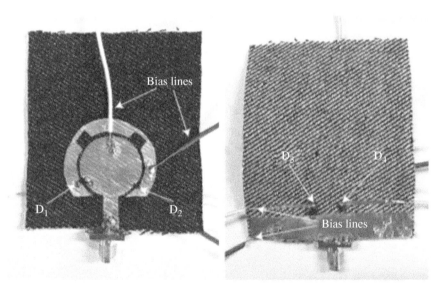

Figure 8.37 Frequency and polarization reconfigurable wearable antenna. *Source*: Saha et al. [54].

 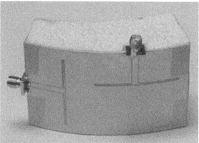

Figure 8.38 Flexible reconfigurable MIMO antenna. *Source*: Saeed et al. [55].

reconfigurable MIMO antenna to operate over two reconfigured communication bands 2.4 and 5.2 GHz is shown in Figure 8.38.

The antenna (Figure 8.38) consists of the slot and stub which have the orthogonal polarization to each other to enhance the isolation between them. The size of the MIMO antenna is 86 mm × 56 mm which is made on a 1.52 mm-thick flexible substrate Rogers RO3003 ($\varepsilon_r = 3$, tan $\delta = 0.0013$). The reconfigurability is controlled by the p-i-n diode. Since, the antenna is capable to operate in the ISM 2.45 GHz band, it is a potential candidate for the medical applications.

With the advancement in the wireless communication technology, high data rate communication is also being expected in the medical applications where these antennas in MIMO implementations can enhance the throughput tremendously. Also, by using the frequency-agile antennas, a device can transfer the data over different communication standards which may be the backbone of the future biomedical applications.

8.12 Conclusion

In this chapter, the potential application of the MIMO antennas has been discussed. In the beginning, various classifications of MIMO antennas are discussed. Due to the multiple resonances, this type of antenna can operate as the multifunctional antenna which helps in reducing the antenna footprint on the overall communication equipment. With the advancement in the technology, MIMO antennas are finding their applications in the radar systems where a number of antennas in MIMO configurations are being used to detect the target which offers a high-resolution image of the detected objects. The massive MIMO antenna is advancement in the existing MIMO systems where a large number of antenna elements are used to obtain the beamforming to meet the next-generation communication requirements. The MIMO antennas also find their application in imaging and can be integrated around the camera. The medical science is also experiencing the need of the high data rate communication systems. The recent trend shows the use of the reconfigurable MIMO antenna in the medical science.

References

1 Hajjar, M. and Hanzo, L. (2010). Multifunctional MIMO systems: a combined diversity and multiplexing design perspective. *IEEE Wirel. Commun.* 17 (2): 73–79.

2 Tarokh, V., Naguib, A., Seshadri, N., and Calderbank, A.R. (1999). Combined array processing and space-time coding. *IEEE Trans. Inf. Theory* 45 (4): 1121–1128.

3 Onggosanusi, E.N., Dabak, A.G., and Schmidl, T.A. (2002). High rate space-time block coded scheme: performance and improvement in correlated fading channels. *IEEE Wireless Communication and Networking Conference*, Orlando, 194–99 (17–21 March 2002).

4 Jongren, G., Skoglund, M., and Ottersten, B. (2002). Combining beamforming and orthogonal space-time block coding. *IEEE Trans. Inf. Theory* 48 (3): 611–627.

5 Liu, J. and Gunawan, E. (2003). Combining ideal beamforming and Alamouti space-time block codes. *Electron. Lett.* 39 (17): 1258–1259.

6 Lei, G.P. and Hao Li, S. (2018). Triple-band, dual-mode and dual-polarization antenna. *Progr. Electromagn. Res. Lett.* 73: 45–51.

7 Papapolymerou, J. and Bernhard, J.T. (2006). Special issue on multifunction antennas and antenna systems. *IEEE Trans. Antennas Propag.* 54 (2): 314–316.

8 Guan, D.F., Qian, Z.P., Cao, W.Q. et al. (2015). Compact SIW annular ring slot antenna with multi-band multi-mode characteristics. *IEEE Trans. Antennas Propag.* 63 (12): 5918–5922.

9 Cao, W.Q., Zhang, B.N., Liu, A.J. et al. (2012). Multi-frequency and dual-mode patch antenna based on electromagnetic band-gap (EBG) structure. *IEEE Trans. Antennas Propag.* 60 (12): 6007–6012.

10 Cao, W.Q., Liu, A.J., Zhang, B.N. et al. (2013). Dual-band spiral patch-slot antenna with omnidirectional CP and unidirectional CP properties. *IEEE Trans. Antennas Propag.* 61 (4): 2286–2289.

11 Morant, M. and Llorente, R. (2018). Performance analysis of carrier-aggregated multiantenna 4×4 MIMO LTE-A fronthaul by spatial multiplexing on multicore Fiber. *J. Lightwave Technol.* 36 (2): 594–600.

12 Chen, Y., Ye, L., Zhuo, J. et al. (2017). Frequency reconfigurable circular patch antenna with an arc-shaped slot ground controlled by PIN diodes. *Int. J. Antennas Propag.* 2017, Article ID: 7081978: 7.

13 Liu, Y.Z. and Xu, X.J. (2015). Azimuth sidelobe suppression technique for near field MIMO radar imaging. *Proc. SPIE Image Signal Process. Remote Sens. XXI* 9643: 96431E-1-8.

14 Zhuge, X., Yarovoy, A.G., Savelyev, T., and Ligthart, L. (2010). Modified Kirchhoff migration for UWB MIMO array-based radar imaging. *IEEE Trans. Geosci. Remote Sens.* 48 (6): 2692–2703.

15 Charvat, G.L., Kempel, L.C., Rothwell, E.J., Coleman, C.M., and Mokole, E.L. (2010). An ultrawideband (UWB) switched-antenna-array radar imaging system. *Proc. IEEE Int. Symp. Phased Array Syst. Technol.*, Waltham, MA, 543–550 (12–15 October 2010).

16 Zhuge, X. and Yarovoy, A.G. (2011). A sparse aperture MIMO-SAR-based UWB imaging system for concealed weapon detection. *IEEE Trans. Geosci. Remote Sens.* 49 (1): 509–518.

17 Zhuge, X.D. and Yarovoy, A.G. (2012). Three-dimensional near-field MIMO array imaging using range migration techniques. *IEEE Trans. Image Process.* 21 (6): 3026–3033.

18 Liu, Y., Xu, X., and Xu, G. (2018). MIMO radar calibration and imagery for near-field scattering diagnosis. *IEEE Trans. Aero. Electron. Syst.* 54 (1): 442–452.

19 Butt, F.A., Naqvi, I.H., and Riaz, U. (2018). Hybrid phased-MIMO radar: a novel approach with optimal performance under electronic countermeasures. *IEEE Commun. Lett.* 22 (6): 1184–1187.

20 Butt, F.A., Naqvi, I.H., and Najam, A.I. (2015). Radar deception jamming prevention using bi-static and mono-static radars. US Patent 9,069,066, filed 30 June 2015.

21 Butt, F.A. and Jalil, M. (2013). An overview of electronic warfare in radar systems. *Int. Conf. Technol. Adv. Elect., Electron. Comput. Eng.*, Konya, 213–217 (9–11 May 2013).

22 Mitola, J. (2009). Cognitive radio architecture evolution. *Proc. IEEE* 97 (4): 626–641.

23 Hussain, R., Khan, M.U., and Sharawi, M.S. (2018). An integrated dual MIMO antenna system with dual-function GND-plane frequency-agile antenna. *IEEE Antennas Wirel. Propag. Lett.* 17 (1): 142–145.

24 Pandit, S., Mohan, A., and Ray, P. (2017). Compact frequency-reconfigurable MIMO antenna for microwave sensing applications in WLAN and WiMAX frequency bands. *IEEE Sensor Letters* 2 (3) Article id: 3500804: 1–4.

25 Marzetta, T. (2015). Massive MIMO: an introduction. *Bell Labs Tech. J.* 20: 11–22.

26 Larsson, E., Edfors, O., Tufvesson, F., and Marzetta, T. (2014). Massive MIMO for next generation wireless systems. *IEEE Commun. Mag.* 52 (2): 186–195.

27 Lu, L., Li, G., Swindlehurst, A. et al. (2014). An overview of massive MIMO: benefits and challenges. *IEEE J. Sel. Topics Signal Process.* 8 (5): 742–758.

28 Björnson, E., Larsson, E.G., and Marzetta, T.L. (2016). Massive MIMO: ten myths and one critical question. *IEEE Commun. Mag.* 54 (2): 114–123.

29 [Online]. https://code.facebook.com/posts/1072680049445290/introducing facebook's new terrestrial connectivity systems terragraph and project-aries.

30 Langston, J.L. and Hinman, K. (1990). A digital beamforming processor for multiple beam antennas. *Int. Antennas and Propagation Society International Symposium, Merging Technologies for the 90's*, Dallas, TX, 388–391 (7–11 May 1990).

31 Krishna, S., Mishra, G., and Sharma, S.K. (2018). A series fed planar microstrip patch array antenna with 1D beam steering for 5G spectrum massive MIMO applications. *IEEE Radio and Wireless Symposium*, Anaheim, CA, 209–212 (15–18 January 2018).

32 Komandla, M.V., Mishra, G., and Sharma, S.K. (2017). Investigations on dual slant polarized cavity backed massive MIMO antenna panel with beamforming. *IEEE Trans. Antennas Propag.* 65 (12): 6794–6799.

33 Lechtreck, L. (1965). Cumulative coupling in antenna arrays. *Antennas and Propagation Society International Symposium*, Washington, DC, 144–149 (30 August–1 September 1965).

34 Kahn, W. (1969). Active reflection coefficient and element efficiency in arbitrary antenna arrays. *IEEE Trans. Antennas Propag.* 17 (5): 653–654.

35 S. Montebugnoli, G. Bianchi, A. Cattani et al. Some Notes on Beamforming. The Medicina IRA-SKA Engineering Group. *Tech. Rep.*

36 Almasi, M.A., Mehrpouyan, H., Vakilian, V., Behdad, N., and Jafarkhani, H. (2018). A new reconfigurable antenna MIMO architecture for mmWavecommunication. *IEEE International Conference on Communications*, Kansas, 1938–1883 (20–24 May 2018) 1–7.

37 Brady, J., Behdad, N., and Sayeed, A.M. (2013). Beamspace MIMO for millimeter-wave communications: system architecture, modelling, analysis,and measurements. *IEEE Trans. Antennas Propag.* 61 (7): 3814–3827.

38 https://www.eetindia.co.in/dealing-with-power-dissipation-in-5g-antenna-design/

39 https://www.eenewseurope.com/search/node/Managing%20power%20 dissipation%20in%205G%20antenna%20

40 Fasano, G., Accardo, D., Moccia, A. et al. (2008). Multi-sensor-based fully autonomous non-cooperative collision avoidance system for unmanned air vehicles. *J. Aerosp. Comput. Inf. Commun.* 5 (10): 338–360.

41 Dittrichand, J.S. and Johnson, E.N. (2002). Multi-sensor navigation system for an autonomous helicopter. *Proc. 21st Digit. Avion. Syst. Conf.* 2: 8C1-1–8C1-19.

42 Hall, D.L. and Linas, J. (1997). An introduction to multisensor data fusion. *Proc. IEEE* 85 (1): 6–23.

43 Miralles, E., Multerer, T., Ganis, A. et al. (2018). Multifunctional and compact 3D FMCW MIMO radar system with rectangular array for medium-range applications. *IEEE Aerosp. Electron. Syst. Mag.* 33 (4): 46–54.

44 Hall, P.S. and Hao, Y. (2012). *Antennas and Propagation for Body-Centric Wireless Communications*, 2e. Artech House: Norwood, MA.

45 Kiourti, A. and Nikita, K.S. (2012). A review of implantable patch antennas for biomedical telemetry: challenges and solutions. *IEEE Antennas Propag. Mag.* 4 (6): 340–349.

46 Felício, J.M., Fernandes, C.A., and Costa, J.R. (2016). Wideband implantable antenna for body-area high data rate impulse radio communication. *IEEE Trans. Antennas Propag.* 64 (5): 1932–1940.

47 Fan, Y., Huang, J.H., Chang, T.H., and Liu, X.Y. (2018). A miniaturized four-element MIMO antenna with EBG for implantable medical devices. *IEEE J. Electromag. RF Microwaves Med. Biol.* 21 (4): 226–233.

48 Latif, F., Tahir, F.A., Khan, M.U., and Sharawi, M.S. (2017). An ultra-wide band diversity antenna with band-rejection capability for imaging applications. *Microwave Opt. Technol. Lett.* 59 (7): 1661–1668.

49 Li, H., Sun, S., Wang, B., and Wu, F. (2018). Design of compact single-layer textile MIMO Antenna for wearable applications. *IEEE Trans. Antennas Propag.* 66 (6): 3136–3141.

50 Duan, Z., Guo, Y.-X., Je, M., and Kwong, D.-L. (2014). Design and in vitro test of a differentially fed dual-band implantable antenna operating at MICS and ISM bands. *IEEE Trans. Antennas Propag.* 62 (5): 2430–2439.

51 Izdebski, P.M., Rajagopalan, H., and Rahmat-Samii, Y. (2009). Conformal ingestible capsule antenna: a novel chandelier meandered design. *IEEE Trans. Antennas Propag.* 57 (4): 900–909.

52 Xu, L.-J., Li, B., Zhang, M., and Bo, Y. (2017). Conformal MIMO loop antenna for ingestible capsule applications. *Electron. Lett.* 53 (23): 1506–1508.

53 Wong, H., Lin, W., Huitema, L., and Arnaud, E. (2017). Multi-polarization reconfigurable antenna for wireless biomedical system. *IEEE Trans. Biomed. Circuits Syst.* 11 (3): 652–660.

54 Saha, P., Mitra, D., and Parui, S.K. (2020). A frequency and polarization agile disc monopole wearable antenna for medical applications. *Radioengineering* 29 (1): 74–80.

55 Saeed, S.M., Balanis, C.A., and Birtcher, C.R. (2017). Flexible reconfigurable antenna with MIMO configuration. *2017 IEEE International Symposium on Antennas and Propagation*, San Diego, 1643–1644 (09–14 July 2017).

9

Metamaterials in Reconfigurable Antennas

Saeed I. Latif and Satish K. Sharma

9.1 Introduction

Metamaterials were introduced decades ago to boost the performance of antennas and reduce their size. Recently, reconfigurable metamaterials have attracted a lot of attention as they provide extra degrees of freedom for synthesizing innovative adaptive communication systems for modern wireless, satellite, and biomedical applications. The unique electromagnetic field manipulation properties of 2-D/3-D periodic objects arbitrarily and in real time have enabled reconfigurability in metamaterials [1]. Field properties are manipulated by exploiting the change in negative refractive index, negative phase velocity, and negative permittivity/ permeability to develop multifunctional and reconfigurable antenna systems. This has triggered the development of devices at microwaves, terahertz, and optical frequencies with enhanced performances, flexibility, efficiency, and robustness. Along with varying the electromagnetic behavior of metamaterial-inspired structures by changing its physical or geometrical properties, reconfigurable metamaterials still utilize other common tuning mechanisms, such as the use of micromechanical systems, various semiconductor components, e.g. diodes, varactors, etc., microfluids, liquid crystals, and graphene. This chapter discusses various reconfigurabilities, namely frequency, pattern, and polarization, from metamaterial-inspired antenna structures.

9.2 Metamaterials in Antenna Reconfigurability

Metamaterials are artificial structures composed of periodic subwavelength metal or dielectric materials. They exhibit unique properties upon the incident of electromagnetic fields that are not available in nature. Recent developments in

Multifunctional Antennas and Arrays for Wireless Communication Systems, First Edition.
Edited by Satish K. Sharma and Jia-Chi S. Chieh.
© 2021 John Wiley & Sons, Inc. Published 2021 by John Wiley & Sons, Inc.

metamaterial engineering have provided an increased flexibility to alter the material properties in unprecedented ways. 3-D metamaterials composed of ordered collection of subwavelength resonators, such as metallic rods and split ring resonators, have high losses and strong dispersion associated with resonance. These structures are difficult to fabricate in the micro- and nanoscales as well. An alternate to 3-D metamaterials is planar metamaterials with subwavelength thickness, popularly known as metasurfaces. They are made of single-layer or multilayer stacks of thin planar structures that can be easily fabricated using nanoprinting and photolithography techniques. These ultrathin metasurfaces are less lossy compared to typical 3-D metamaterials due to the high suppression of surface waves along the direction of the wave propagation [2].

Metamaterials and metasurfaces have found applications in innovative electromagnetic devices, such as antennas, waveguides, resonators, and filters. Reconfigurability in metamaterials achieved through the alteration of their electromagnetic properties in the presence of external electromagnetic fields appears to be a novel capability that has enabled the development of high-performance reconfigurable antennas [3]. This alteration or control of the electromagnetic properties has mostly been implemented by using some sort of switches: electrical, optical, or micro-electro-mechanical-system (MEMS), or by introducing mechanical changes in the entire interconnected panel or region of metamaterial resonators or metasurfaces, or by introducing changes in the material properties by using tunable material. Along with the complexities in regular metamaterial fabrication, incorporating these tuning mechanisms make the overall device development really challenging. All electronically controlled systems, MEMS-based devices, and chemically, thermally, or static electric field-controlled liquid crystal-based devices that are used for tuning require bias signals. External circuits producing these bias signals are to be integrated into the resonant unit-cell geometry with minimal impact to the desired electromagnetic response necessary to realize artificial materials, i.e. metamaterials. To this end, altered material properties from metamaterials provide additional degrees of freedom in realizing reconfigurability in antennas with respect to static structures at the expense of fabrication simplicity. Some of these realizations of frequency, pattern, and polarization reconfigurability in metamaterial-inspired antennas will be discussed in Section 9.3. Metasurface-based reconfigurable antennas will be addressed in Section 9.4.

9.3 Metamaterial-Inspired Reconfigurable Antennas

All types of reconfigurability – frequency, pattern, and polarization – have been implemented in metamaterial-inspired antennas using typical tuning mechanisms, such as electronic or MEMS-based switches, or utilizing tuning materials [4–19].

Both frequency tuning and beamscanning capabilities are reported using reconfigurable metamaterials [20–22]. In the case of metamaterials, the basic idea for reconfigurability is to dynamically change the functionality of an antenna by deforming the geometrical structure of the unit cell of a metamaterial to alter the material properties or electromagnetic response when exposed to external fields.

9.3.1 Metamaterial-Based Frequency Reconfigurability

Frequency reconfigurable antennas are useful when a single wireless device is required to access multiple communication services. Thus, frequency reconfigurability has become a dominant feature in wireless systems to support more than one wireless standard to make the wireless devices compact in size, cost effective, and simple. It is advantageous to design an antenna system that has a small size, more bandwidth, and can operate at more than one frequency band. Metamaterial-inspired frequency reconfigurable antennas are studied and developed extensively in recent time [4–12].

A compact composite right-/left-handed transmission line (CRLH-TL) reconfigurable antenna is discussed in [12]. The propagation constant of a left-handed (LH) material is negative, representing a phase advance, whereas that of a right-handed (RH) material is positive, representing a phase lag. When LH and RH structures are combined, a composite RH–LH structure is achieved. In recent years, CRLH-TL has attracted much attention because of its many unique properties, especially the zeroth-order resonance. This resonance can make the antenna miniaturized since it is independent of the physical length of the antenna structure. The zeroth-order resonance is excited by properly terminating one end of the CRLH-TL. For an unbalanced CRLH-TL, there are two zeroth-order resonances: the series resonance which depends on the series inductance and series capacitance, and shunt resonance which depends on the shunt capacitance and shunt inductance. While for the open-ended resonator, the shunt resonance depends on shunt elements, the series resonance for the short-ended resonator depends on the series elements. A PIN diode can be used to realize the open-ended or short-ended termination to excite two resonances to obtain frequency reconfigurability.

The geometry of the proposed reconfigurable antenna is shown in Figure 9.1, which can be considered as a CRLH-TL cell. A PIN diode is used as a switch to reconfigure the antenna to operate in two different frequency bands by changing the CRLH-TL termination. The antenna was fabricated on RO4350B Rogers substrate having a thickness of 0.762 mm, relative permittivity of 3.48, and loss tangent of 0.004. The overall size of the antenna is 14 mm × 19 mm. The antenna is excited by a 50-Ω asymmetric coplanar waveguide which is connected to a rectangular matching patch. The matching patch is coupled to the rectangular radiating patch with a 0.4 mm gap. A PIN diode is placed in the gap between the radiating patch and the ground plane to change the termination states of the CRLH-TL from ON (short-ended) to OFF (open-ended). A simple DC bias network without any block capacitor is used.

The photograph of the fabricated antenna is shown in Figure 9.1b. The reflection coefficient of the fabricated antenna is compared with simulated data in Figure 9.2 for ON and OFF states. A low-frequency band of operation from 1.84 to

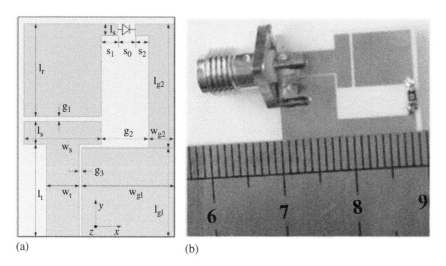

(a)   (b)

Figure 9.1 (a) Geometry of the proposed reconfigurable composite right/left-handed transmission line (CRLH-TL) antenna and (b) the photograph of the fabricated prototype. *Source*: Yang et al. [12]. Licensed under CC BY 4.0.

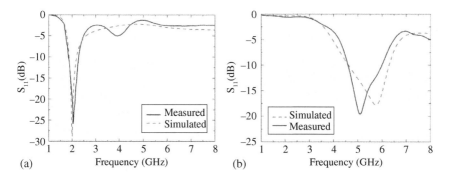

Figure 9.2 Measured and simulated S_{11} of the reconfigurable composite right/left-handed transmission line (CRLH-TL) antenna for the (a) ON state and (b) OFF state. *Source*: Yang et al. [12]. Licensed under CC BY 4.0.

2.28 GHz was obtained in the ON state, and a high-frequency band of operation (4.6–6 GHz) was obtained in the OFF state. The antenna demonstrates omnidirectional patterns in the E-plane and H-plane.

9.3.2 Metamaterial-Based Pattern Reconfigurability

Pattern reconfigurable antennas are essential for multiple-input-multiple-output (MIMO) systems, as they can increase the system capacity and potentially avoid the interference from noise sources by redirecting the null position. Metamaterial-inspired pattern reconfigurable antennas have attracted researchers significantly [13–15]. A common approach to improve the system capacity in MIMO systems is to employ pattern diversity using reconfigurable antennas to combine the monopole-like pattern (omnidirectional) and the patch-like pattern (broadside). Such a pattern reconfigurable metamaterial antenna made of CRLH-TL is discussed in [15], where both positive and negative phase velocities are exploited to generate positive, zero, and negative modes. Thus, pattern diversity is achieved from this CRLH-TL antenna with both broadside and omnidirectional patterns by exciting appropriate modes. RF switches are used to alter the dispersion properties of the CRLH-TL antenna to obtain reconfigurability. The reconfigurable metamaterial structure in Figure 9.3 has a rectangular microstrip antenna partially filled with composite right-/left-handed cells implemented with mushroom topologies. The LH inductance of the cells is realized by the shorting pins to the ground and the LH capacitance is developed for the gaps between the cells. RF switches are used to connect the partially filled CRLH-TL structure and the microstrip patch. When all the switches are OFF (State 2), the structure generates all three resonant modes. When all the switches are ON (State 1), the gaps within the structure are shorted that eliminate LH capacitances, exciting only zero and positive modes.

The detailed design with all dimensions of this antenna is shown in Figure 9.4. It has two 1.524 mm-thick RO4003 Rogers substrate layers (permittivity 3.5 and loss tangent 0.0027). The patch and the CRLH-TL cells are on the top of the upper substrate, and the coupling structure is on the bottom of this substrate. The ground plane is on the top of the lower substrate, and the impedance matching network is on its side. Skyworks DSM8100-000 beam-lead PIN switches were considered mounted on the top layer to switch the structure for reconfigurability.

Figure 9.5 shows the fabricated prototype of the antenna and the measured and simulated reflection coefficients are compared in Figure 9.6. In both bands, it is evident that the antenna demonstrates dual-band operation around 2.5 and 3.5 GHz. The limited bandwidth in the lower band is because of the inherently

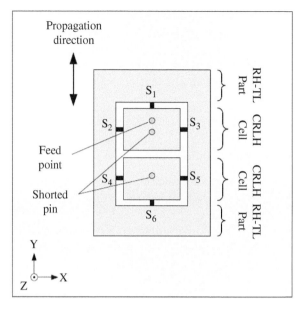

Figure 9.3 A microstrip patch antenna partially filled with a CRLH-TL structure. S1–S6: RF switches. *Source*: Zhang et al. [15]. © 2018, IEEE.

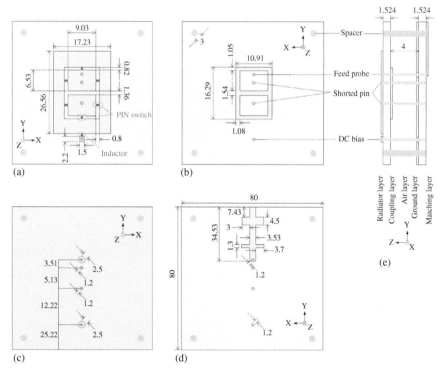

Figure 9.4 Topology of the microstrip patch antenna partially filled with a CRLH-TL structure: (a) metal layer 1: radiator layer, (b) metal layer 2: coupling layer, (c) metal layer 3: ground layer, (d) metal layer 4: matching layer, and (e) side view showing all layers (Unit in mm). *Source*: Zhang et al. [15]. © 2018, IEEE.

Figure 9.5 Fabricated prototype of the microstrip patch antenna partially filled with CRLH-TL cells. *Source*: Zhang et al. [15]. © 2018, IEEE.

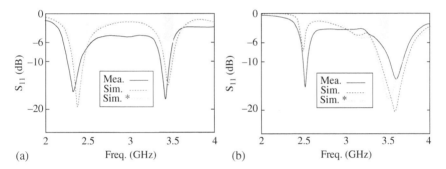

Figure 9.6 Reflection coefficient plots of the microstrip patch antenna partially filled with CRLH-TL: (a) State 1 and (b) State 2. Yellow rectangles indicate the ISM and the WiMAX bands: 2.4835–2.5 and 3.4–3.5 GHz, respectively. The green dashed lines are for the case with the ground plane size of $70 \times 70 \, mm^2$. *Source*: Zhang et al. [15]. © 2018, IEEE.

narrowband nature of the negative mode. The normalized radiation patterns are shown in Figure 9.7. In State 1, the antenna exhibits a typical omnidirectional pattern at 2.49 GHz and a broadside pattern at 3.45 GHz. In State 2, the pattern type switches with a broadside pattern at 2.49 GHz and an omnidirectional pattern at 3.45 GHz.

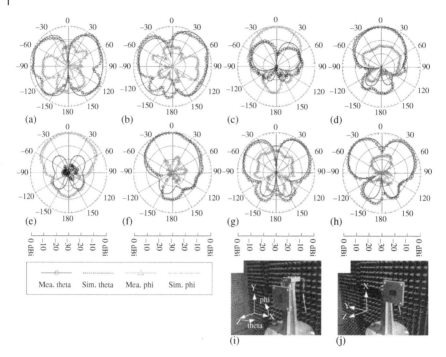

Figure 9.7 Normalized radiation patterns of the microstrip patch antenna partially filled with CRLH-TL structure in State 1 at 2.49 GHz in the (a) xz plane and (b) yz plane, at 3.45 GHz in the (c) xz plane and (d) yz plane, and in State 2 at 2.49 GHz in the (e) xz plane and (f) yz plane, at 3.45 GHz in the (g) xz plane and (h) yz plane. Radiation pattern measurement setup in the (i) xz plane and (j) yz plane. *Source*: Zhang et al. [15]. © 2018, IEEE.

9.3.3 Metamaterial-Based Polarization Reconfigurability

Polarization reconfigurability features from antennas in wireless and satellite communication systems help maximize signal strength to avoid signal fading in multipath environment and have been implemented using reconfigurable metamaterials [16–19]. A metamaterial-inspired polarization reconfigurable antenna is presented, which is based on a polarization-rotation artificial magnetic conductor (PRAMC) structure [18]. It combines a PRAMC-based multi-polarized dipole antenna with a RFIC switch-based network to obtain three polarization states, namely right-handed circular polarization (RHCP), left-handed circular polarization (LHCP), and +45° linear polarization (LP) by controlling the DC bias of the switches accordingly.

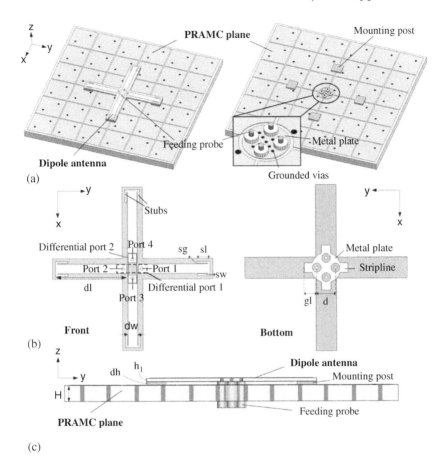

Figure 9.8 Topology of the polarization reconfigurable antenna using the polarization-rotation artificial magnetic conductor (PRAMC) structure: (a) 3-D view with and without dipole, (b) top view of the dipole antenna, and (c) side view. *Source*: Yang et al. [18]. © 2015, IEEE.

The PRAMC structure, shown in Figure 9.8, consists of a square metal patch with 6 ×6 cells, two sets of grounded vias along the diagonal direction, and a ground plane. The metal patch is etched on a FR4 dielectric substrate having the thickness, H = 2.4 mm, relative permittivity, ε_r =4.4, and loss tangent, tan δ =0.02. The PRAMC structure has two TM transmission zeros, one at f_1 =4.6 GHz and other at f_2 = 5.65 GHz, with the TM reflected wave solely suppressed and the orthogonal TE component significantly excited. Therefore, the polarization state of the incident wave is rotated 90° at two frequency points, which are called the polarization-rotation (PR) frequencies. The PR frequency bandwidth between the PR points f_1 and f_2 for $|\Gamma_{TM/TM}| < -10\,dB$ is about 29.1%. Additionally, the phase at the center frequency (5.2 GHz) is 90°, which indicates that the incident TM wave lags the TE-reflected wave by 90°. This guarantees that a RHCP wave can be readily generated from this structure at 5.2 GHz. A differentially driven multi-polarized dipole antenna is designed accordingly, which consists of a pair of linearly polarized horizontal dipole antennas placed crosswise with the PRAMC structure as the ground plane. For the crossed dipole antenna, Rogers Duroid 5880 ($h_1 = 0.5\,mm$, $\varepsilon_r = 2.2$, $\tan\delta = 0.0009$) is used. Two stubs are introduced at the end of each arm of the crossed dipole and a circular metal plate with four striplines is etched on the back of the dipole to achieve good impedance matching. Two pairs of probes, each pair providing a differential excitation, are inserted from the bottom of the PRAMC structure into the circular metal plate on the back of the dipole for feeding the antenna. Differential port 1 is formed by differential signals from probes 1 and 2, and similarly differential port 2 from probes 3 and 4. To maintain the same potential on the outer wall of the probes, another circular metal plate is etched on the PRAMC surface with several grounded vias between the ground and this circular metal plate. The optimized antenna parameters of the antenna are $h_1 = 0.5\,mm$, $d = 3\,mm$, $dw = 1.5\,mm$, $dl = 12\,mm$, $dh = 0.5\,mm$, $gl = 2\,mm$, $sw = 0.3\,mm$, $sg = 0.1\,mm$, and $sl = 2\,mm$.

When differential signals are applied to differential port 1, the arms of the cross dipole along the y-direction are excited, which results in a RHCP radiation. The differential port 2 excites the arms along x-direction for LHCP radiation. When both differential ports are excited simultaneously, two orthogonal polarizations are superimposed to produce the +45° LP mode. Switching to various polarizations is achieved employing one single-pole three-throw (SP3T) and two single-pole double-throw (SPDT) RF switches. While the SP3T RFIC switch (SKY13373-460LF from Skyworks Corporation) provides three channels corresponding to three polarization states, two SPDT RFIC switches (PE42424 from Peregrine Semiconductor Corporation) provide isolation between the selected channel and other channels.

The prototype of the reconfigurable dipole antenna on the PRAMC structure was fabricated and tested. The measured impedance bandwidths for three cases are about 19.6% (4.6–5.6 GHz), and are compared with the simulated data in Figure 9.9. Gain and axial ratio (AR) performance of this antenna for various polarization states, obtained by controlling the DC bias signals, are compared in Figure 9.10. In the RHCP state, the measured 3-dB AR bandwidth is 15.3%

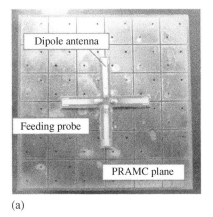 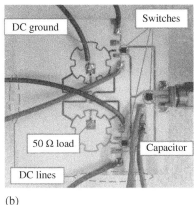

(a)   (b)

Figure 9.9 Photographs of the fabricated polarization-reconfigurable PRAMC structure-based dipole antenna: (a) top view and (b) bottom view with the switch network. *Source*: Yang et al. [18]. © 2015, IEEE.

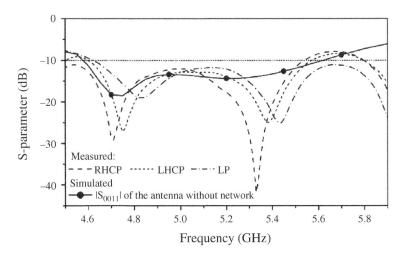

Figure 9.10 Measured reflection coefficients of the polarization-reconfigurable PRAMC structure-based dipole antenna for three polarization states. *Source*: Yang et al. [18]. © 2015, IEEE.

(4.8–5.6 GHz), which is slightly less than the simulated data. The measured RHCP gain is stable with a peak value of 4.25 dBi at 5.2 GHz, but is lower compared to the simulated gain of the antenna without the switch network. The reduction is due to the loss of the switches, the insertion loss of the hybrid ring used in the

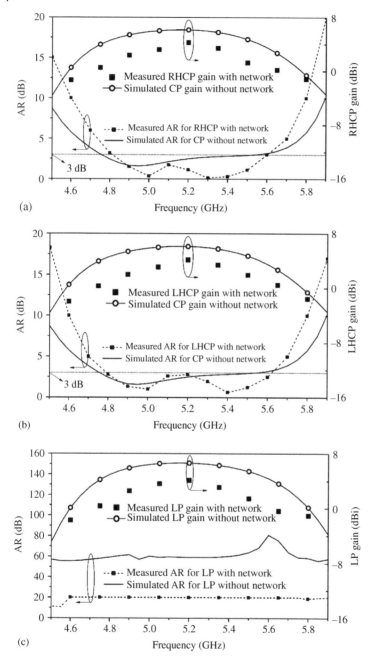

Figure 9.11 Measured ARs and gains of the proposed reconfigurable antenna for three polarization states: (a) RHCP, (b) LHCP, and (c) LP. *Source*: Yang et al. [18]. © 2015, IEEE.

switching network, and the transition loss of the SMA connector. In the LHCP state, the 3-dB AR bandwidth of the antenna is about 15.8% with the peak gain of 4.15 dBi. In the LP state, the peak gain is 4.1 dBi at 5.2 GHz with an average AR value of 20 dB in the entire band indicating the LP operation of the antenna. These results are shown in Figure 9.11.

9.4 Metasurface-Inspired Reconfigurable Antennas

3-D metamaterials made of metallic structures suffer from high losses and strong dispersion associated with the resonant responses, as well as difficulty in fabricating in the micro- and nanoscale levels. A low loss and fabrication-friendly solution is planar metamaterials, popularly known as metasurfaces, consisting of single-layer or few-layer stacks of planar structures that can be readily fabricated using well-known lithography and nanoprinting methods, and the ultrathin thickness in the wave propagation direction can greatly suppress the undesirable losses. A huge amount of research effort has been devoted to reconfigurable metasurfaces to obtain all forms of reconfigurability [23–42].

An electronically controlled broadband orbital angular momentum (OAM) mode reconfigurable metasurface antenna, a new type of reconfigurability, is presented in [35]. The OAM is a subset of the MIMO system, which does not increase the system capacity, but offers an additional degree of freedom in spatial diversity that essentially enhances the channel capacity and spectrum efficiency in wireless communication applications. The layered perspective view of the OAM metasurface array antenna and its element structure are illustrated in Figure 9.12. The element structure consists of the metasurface patch layer and the driven patch layer both on Rogers 4350B substrates ($\varepsilon_r = 3.48$, $\tan \delta = 0.0037$) having a thickness of h_1, and the feeding network layer on a Rogers 5880 substrate (i.e., $\varepsilon = 2.2$, $\tan \delta = 0.0009$) with a thickness of h_5. The separation between the metasurface layer and the driven patch layer is h_2. The metasurface is a 2-D structure which has square patches with periodicity M_p, and an interval M_g between two contiguous metal strips. The metasurface is placed as parasitic elements over the driven patch to achieve wideband operation. The size of the entire antenna structure is $125 \times 125 \times 3.624 \, \text{mm}^3$ ($2.3\lambda_o \times 2.3\lambda_o \times 0.069\lambda_o$, where λ_o is the wavelength at 5.7 GHz).

By using antenna arrays, multiple OAM modes can be generated by a single antenna. One of the array structures proposed to generate OAM beams in the microwave and RF bands is the uniform circular array. The maximum mode generated by the OAM antenna is determined by the number of elements in this

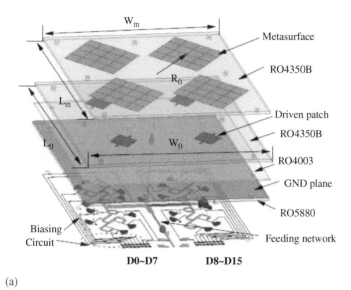

(a)

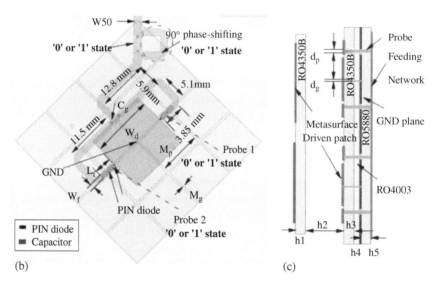

(b)   (c)

Figure 9.12 Geometry of the OAM reconfigurable metasurface array antenna: (a) layered perspective view, (b) element upward 2-D view, and (c) side view. *Source*: Wu et al. [35]. © 2019, IEEE.

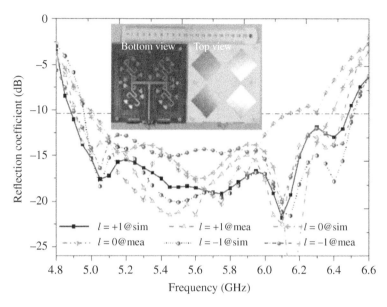

Figure 9.13 Simulated and measured reflection coefficients of the OAM reconfigurable metasurface array antenna for the three OAM modes. *Source*: Wu et al. [35]. © 2019, IEEE.

circular array. In order to generate mode $|l| = 1$ with $N = 4$ circular array, the antenna elements should be divided into four uniform amplitudes with a sequential $2\pi l/N = \pm\pi/2$ phase difference between the consecutive elements. The OAM mode $l = +1$ for a continuous phase difference with clockwise direction is 90° and for the mode $l = -1$, it is −90°. A reconfigurable feed network (RFN) is used for this antenna having 0° and 90° phase shifting, in which each path is composed of two PIN diodes (MADP-000907-14020P from MACOM). Three phase-shifting schemes are implemented to generate three modes by changing the coding sequences of the PIN diodes' DC bias voltage. These reconfigurable OAM mode states are achieved in a wide frequency band.

The fabricated prototype of the metasurface OAM antenna, developed with the printed circuit board technology, is shown in Figure 9.13, along with the reflection coefficients of the triple-mode OAM states. It is evident that −10 dB reflection coefficient bandwidths of triple-mode states ($l = \pm1$ and $l = 0$) are 28.1% from 4.9 to 6.5 GHz and 24.6% from 5.0 to 6.3 GHz, respectively. The measured and

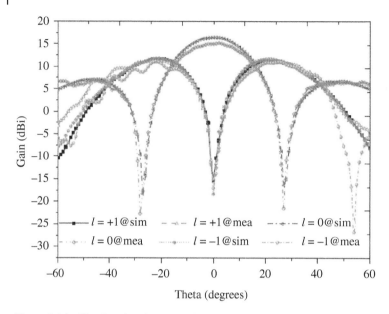

Figure 9.14 Simulated and measured gains of the OAM reconfigurable metasurface array antenna in three different modes at 5.7 GHz. *Source*: Wu et al. [35]. © 2019, IEEE.

simulated gains of three different modes at 5.7 GHz are shown in Figure 9.14, and the peak gains of l = +1, l = −1, and l = 0 modes are 11.17, 11.05, and 15.19 dBi, respectively, for the azimuth angle of 0°.

9.5 Conclusion

In this chapter, reconfigurable metamaterial antennas are discussed that can provide frequency, pattern, and polarization reconfigurability. Recent developments in the metamaterial technology will enable the use of these new class of reconfigurable antennas in emerging wireless applications. Modern communication systems and satellite technologies will have to tackle challenges as new methods and experiments are attempted to meet future demands. Reconfigurable metamaterials will play a vital role in providing solutions to those challenges in the near future with adaptivity, flexibility, and multi-functionality.

References

1 Oliveri, G., Werner, D.H., and Massa, A. (2015). Reconfigurable electromagnetics through metamaterials – a review. *Proceedings of the IEEE* 103 (7): 1034–1056.

2 Guo, Y.J. and Qin, P.Y. (2016). Reconfigurable antennas for wireless communications. In: *Handbook of Antenna Technologies* (eds. Z.N. Chen, D. Liu, H. Nakano, et al.). Singapore: Springer.

3 Turpin, J.P., Werner, D.H., and Wolfe, D.E. (2015). Design considerations for spatially reconfigurable metamaterials. *IEEE Transactions on Antennas and Propagation* 63 (8): 3513–3521.

4 Jang, Y., Choi, J., and Lim, S. (2012). Frequency reconfigurable zeroth-order resonant antenna using RF MEMS switch. *Microwave and Optical Technology Letters* 54 (5): 1266–1269.

5 Cheribi, H., Ghanem, F., and Kimouche, H. (2013). Metamaterial-based frequency reconfigurable antenna. *Electronics Letters* 49 (5): 315–316.

6 Li, L., Wu, Z., Li, K. et al. (2014). Frequency-reconfigurable quasi-Sierpinski antenna integrating with dual-band high-impedance surface. *IEEE Transactions on Antennas and Propagation* 62 (9): 4459–4467.

7 Liang, B., Sanz-Izquierdo, B., Parker, E.A., and Batchelor, J.C. (2015). A frequency and polarization reconfigurable circularly polarized antenna using active EBG structure for satellite navigation. *IEEE Transactions on Antennas and Propagation* 63 (1): 33–40.

8 V, R. and S, R. (2015). A compact metamaterial inspired triple band antenna for reconfigurable WLAN/WiMAX applications. *AEU – International Journal of Electronics and Communications* 69 (1): 274–280.

9 Saeed, S.M., Balanis, C.A., Birtcher, C.R. et al. (2017). Wearable flexible reconfigurable antenna integrated with artificial magnetic conductor. *IEEE Antennas and Wireless Propagation Letters* 16: 2396–2399.

10 Nasir, U., Afzal, A.S., Ijaz, B. et al. (2017). A compact frequency reconfigurable CPS-like metamaterial-inspired antenna. *Microwave and Optical Technology Letters* 59 (3): 596–601.

11 Ijaz, B., Khan, M.S., Asif, S.M. et al. (2017). Metamaterial-inspired series-fed frequency reconfigurable array with zero-phase CRLH interconnects. *Microwave and Optical Technology Letters* 60 (1): 140–146.

12 Yang, X., Xiao, J., Wang, J., and Sheng, L. (2019). Compact frequency reconfigurable antennas based on composite right/left-handed transmission line. *IEEE Access* 7: 131663–131671.

13 Zhang, Y., Wei, K., Zhang, Z. et al. (2015). A compact dual-mode metamaterial-based loop antenna for pattern diversity. *IEEE Antennas and Wireless Propagation Letters* 14: 394–397.

14 Yan, S. and Vandenbosch, G.A.E. (2016). Radiation pattern-reconfigurable wearable antenna based on metamaterial structure. *IEEE Antennas and Wireless Propagation Letters* 15: 1715–1718.

15 Zhang, J., Yan, S., and Vandenbosch, G.A.E. (2018). Realization of dual-band pattern diversity with a CRLH-TL-inspired reconfigurable metamaterial. *IEEE Transactions on Antennas and Propagation* 66 (10): 5130–5138.

16 Chen, D., Yang, W., Che, W. et al. (2019). Polarization-reconfigurable and frequency-tunable dipole antenna using active AMC structures. *IEEE Access* 7: 77792–77803.

17 Wu, Z., Tang, M., Li, M., and Ziolkowski, R.W. (2020). Ultralow-profile, electrically small, pattern-reconfigurable metamaterial-inspired Huygens dipole antenna. *IEEE Transactions on Antennas and Propagation* 68 (3): 1238–1248.

18 Yang, W., Che, W., Jin, H. et al. (2015). A polarization-reconfigurable dipole antenna using polarization rotation AMC structure. *IEEE Transactions on Antennas and Propagation* 63 (12): 5305–5315.

19 Feng, B., Li, L., Zeng, Q., and Sim, C. (2018). A low-profile metamaterial loaded antenna array with anti-interference and polarization reconfigurable characteristics. *IEEE Access* 6: 35578–35589.

20 Mirzaei, H. and Eleftheriades, G.V. (2011). A compact frequency-reconfigurable metamaterial-inspired antenna. *IEEE Antennas and Wireless Propagation Letters* 10: 1154–1157.

21 Sleasman, T., Imani, M.F., Xu, W. et al. (2016). Waveguide-fed tunable metamaterial element for dynamic apertures. *IEEE Antennas and Wireless Propagation Letters* 15: 606–609.

22 Lago, H., Soh, P.J., Jamlos, M.F., and Zakaria, Z. (2018). Beam-reconfigurable crescent array antenna with AMC plane. *International Journal of RF and Microwave Computer-Aided Engineering* 28 (7).

23 Zhu, H.L., Liu, X.H., Cheung, S.W., and Yuk, T.I. (2014). Frequency reconfigurable antenna using metasurface. *IEEE Transactions on Antennas and Propagation* 62 (1): 80–85.

24 Kandasamy, K., Majumder, B., Mukherjee, J., and Ray, K.P. (2015). Low-RCS and polarization-reconfigurable antenna using cross-slot-based metasurface. *IEEE Antennas and Wireless Propagation Letters* 14: 1638–1641.

25 Zhu, H.L., Cheung, S.W., and Yuk, T.I. (2015). Mechanically pattern reconfigurable antenna using metasurface. *IET Microwaves, Antennas & Propagation* 9 (12): 1331–1336.

26 Johnson, M.C., Brunton, S.L., Kundtz, N.B., and Kutz, J.N. (2015). Sidelobe canceling for reconfigurable holographic metamaterial antenna. *IEEE Transactions on Antennas and Propagation* 63 (4): 1881–1886.

27 Yurduseven, O., Marks, D.L., Gollub, J.N., and Smith, D.R. (2017). Design and analysis of a reconfigurable holographic metasurface aperture for dynamic focusing in the Fresnel zone. *IEEE Access* 5: 15055–15065.

28 Araújo, F.F., D'Assunção, A.G., Costa, L.F.V.T., and Alves, W.S. (2017). Design of rotatable metasurface microstrip antenna with reconfigurable polarization. *Proc. Int. Appl. Comput. Electromagn. Soc. Symp. (ACES)*, Florence (26–30 March 2017), 1–2.

29 Cai, Y.-M., Yin, Y., and Li, K. (2017). A low-profile frequency reconfigurable metasurface patch antenna. *Proc. IEEE Int. Symp. Antennas Propag. & USNC/URSI Nat. Radio Sci. Meeting*, San Diego, CA, 1375–1376.

30 Imani, M.F., Sleasman, T., and Smith, D.R. (2018). Two-dimensional dynamic metasurface apertures for computational microwave imaging. *IEEE Antennas and Wireless Propagation Letters* 17 (12): 2299–2303.

31 Ni, C., Chen, M.S., Zhang, Z.X., and Wu, X.L. (2018). Design of frequency- and polarization-reconfigurable antenna based on the polarization conversion metasurface. *IEEE Antennas and Wireless Propagation Letters* 17 (1): 78–81.

32 Hu, J., Luo, G.Q., and Hao, Z. (2018). A wideband quad-polarization reconfigurable metasurface antenna. *IEEE Access* 6: 6130–6137.

33 Sun, S., Jiang, W., Gong, S., and Hong, T. (2018). Reconfigurable linear-to-linear polarization conversion metasurface based on PIN diodes. *IEEE Antennas and Wireless Propagation Letters* 17 (9): 1722–1726.

34 Gao, X., Yang, W.L., Ma, H.F. et al. (2018). A reconfigurable broadband polarization converter based on an active metasurface. *IEEE Transactions on Antennas and Propagation* 66 (11): 6086–6095.

35 Wu, J., Zhang, Z., Ren, X. et al. (2019). A broadband electronically mode-reconfigurable orbital angular momentum metasurface antenna. *IEEE Antennas and Wireless Propagation Letters* 18 (7): 1482–1486.

36 Wu, Z., Liu, H., and Li, L. (2019). Metasurface-inspired low profile polarization reconfigurable antenna with simple DC controlling circuit. *IEEE Access* 7: 45073–45079.

37 Wang, H., Qu, S., Yan, M. et al. (2019). Design and analysis of multi-band polarisation selective metasurface. *IET Microwaves, Antennas & Propagation* 13 (10): 1602–1609.

38 Yu, J., Jiang, W., and Gong, S. (2019). Low-RCS beam-steering antenna based on reconfigurable phase gradient metasurface. *IEEE Antennas and Wireless Propagation Letters* 18 (10): 2016–2020.

39 Li, H., Man, X., and Qi, J. (2019). Accurate and robust characterization of metasurface-enabled frequency reconfigurable antennas by radially homogeneous model. *IEEE Access* 7: 122605–122612.

40 Hosseininejad, S.E., Rouhi, K., Neshat, M. et al. (2019). Digital metasurface based on graphene: an application to beam steering in terahertz plasmonic antennas. *IEEE Transactions on Nanotechnology* 18: 734–746.

41 Lin, M., Huang, X., Deng, B. et al. (2020). A high-efficiency reconfigurable element for dynamic metasurface antenna. *IEEE Access* 8: 87446–87455.

42 Vasić, B., Isić, G., Beccherelli, R., and Zografopoulos, D.C. (2020). Tunable beam steering at terahertz frequencies using reconfigurable metasurfaces coupled with liquid crystals. *IEEE Journal of Selected Topics in Quantum Electronics* 26 (5): 1–9.

10

Multifunctional Antennas for User Equipments (UEs)

Satish K. Sharma and Sonika P. Biswal

10.1   Introduction

In modern wireless communication, the quest to fulfill the demand of the users to get access of multiple communication features from an antenna system has encouraged the antenna designers to innovate, design, and implement multifunctional antenna (MFA) on user terminal (UE) host ground plane. Multifunctional antenna system (MFAS) can be categorized into three types. In the first category, MFAs are multiband antennas having the ability to operate at all necessary operating bands at a time [1]. In the second category, they are reconfigurable/tunable antennas having the feasibility to switch the operating band with suitable polarization based on user's demand [2]. In the third category, they are a combination of multiple antennas in a single PCB, where all antennas can operate independently at a particular band or different band with suitable polarization [3]. In this category, antennas can operate from lower cellular band to high *mm*-wave band. Antenna research under this category is in high demand after the emergence of the 5G spectrum.

It is well known that the lower band (LB) spectrum is very crowded with narrow bandwidth. This has enforced the researchers to explore the *mm*-wave band which is less crowded and expected to deliver Giga bits per second (Gbps) data rate over a wider bandwidth [4]. During the propagation of *mm*-wave, UEs have larger path loss as compared to antennas operating below 6 GHz. Therefore, a high-gain directional antenna is preferred to overcome these challenges. The lower cellular band/sub-6 GHz band antennas in UEs use omnidirectional radiation behavior with moderate gain, which can achieve an approximate 3D spherical coverage [5, 6]. The coverage issue of a *mm*-wave array can be overcome by enabling the beam

---

Multifunctional Antennas and Arrays for Wireless Communication Systems, First Edition.
Edited by Satish K. Sharma and Jia-Chi S. Chieh.
© 2021 John Wiley & Sons, Inc. Published 2021 by John Wiley & Sons, Inc.

steering ability. Integration of multiple beam steerable arrays in a single UE helps to achieve a coverage range like LB omnidirectional antennas [7]. Researchers are also trying to integrate the exiting LB antennas with 5G antenna array in a common ground plane to endure the requirements of multiple wireless communication band traffic. Many challenges and requirements can be resolved during the integration. The small form factor of UEs brings out one such design challenge considering that the antennas acquire more space especially at the LB frequencies below 2 GHz. If other antennas operate on nearby frequency spectrum, it can impact the performance of the LB antennas. The overall area assigned for the antenna elements on a small form factor UE is really restricted which imposes a major concern due to high coupling between the adjacent antenna elements. Therefore, selection of the proper radiating elements for both the LB and *mm*-wave band antennas should be done very carefully so that minimal interference between each other is present.

First, we present some lower cellular band and sub-6 GHz band MIMO antennas. Next, we discuss high-gain beam steerable 5G array antennas. We also discuss collocated antennas covering both lower and *mm*-wave band antennas in a single platform. At the end, we discuss some important parameters on radio frequency/ electromagnetic field (RF/EMF) exposure limits for UEs.

10.2 Lower/Sub-6 GHz 5G Band Antennas

In this section, we discuss MFAS for lower operating band (less than 6 GHz) that is suitable for UEs, e.g. mobile phones. Multiple lower cellular band and sub-6 GHz antennas are mounted on a single ground plane to provide high data rate. Massive multi-input multi-output (MIMO) system based on multiple antennas is one of the core technologies for 5G operation for modern UE, as it provides high data rate, enhanced channel capacity, and good reliability of the link. The minimum number of antenna elements in a massive MIMO array for UEs is approximately six to eight which can work in the lower cellular bands and sub-6 GHz bands.

The first example is a dual polarized antenna-based eight-element MIMO antenna which is reported for 5G MIMO smartphone application, as shown in Figure 10.1 [8]. Here, all eight elements operate at the desired 2.6 GHz band (2.55–2.65 GHz). Two different antenna types are utilized here. Four L-shaped monopole slots (Ant-1 to Ant-4) are placed at four corners of the ground plane that radiate linearly polarized (LP) wave along the narrow edge of the ground plane. The radiation is along the x-axis and defined as horizontal polarization. The purpose of employing an L-shaped monopole slot is to improve the isolation between the adjacent ports (e.g. between Ant-1 and 2 and between Ant-3 and 4). Figure 10.2 provides the simulated surface current distribution and E-field

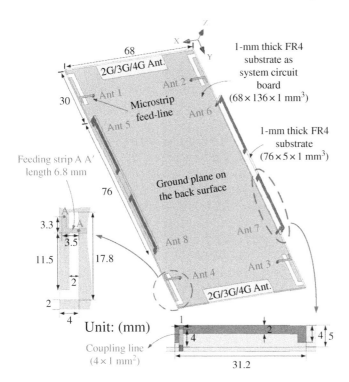

Figure 10.1 Eight-element array for 5G MIMO operation in a smartphone, including the dimensions of the L-shaped monopole slot antenna and C-shaped coupled-fed antenna. *Source*: Li et al. [8]. © 2016, IEEE.

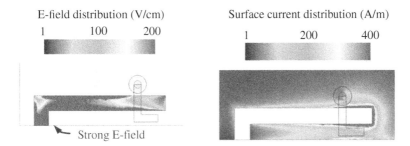

Figure 10.2 E-field and surface current distribution of the L-shaped monopole slot antenna. *Source*: Li et al. [8]. © 2016, IEEE.

distribution of the single L-shaped monopole slot at 2.6 GHz. It can be observed that strong E-fields are present at the open end of the slot which gradually decrease toward the closed end. The surface current flow along the edges of monopole slot shows a $\lambda_g/4$ resonant path which is provided by the slot. Another four C-shaped coupled-fed antennas (Ant-5 to Ant-8) are placed in the middle section of the ground plane, which radiate LP waves along the long edges of the ground. The antenna unit having length $\lambda_g/2$ at 2.6 GHz is coupled fed by a 50 Ω feed line and is mounted perpendicular to the ground plane. The radiation is along the y-axis and defined as the vertical polarization. Figure 10.3 provides the simulated surface current distribution of this antenna. The concentration of surface current gradually decreases from the center toward both ends. Bending at both ends in two vertical sections toward the ground does not influence the polarization. The isolation between any two adjacent antenna elements is better than 12 dB.

The simulated and measured results of both the horizontally and vertically polarized antennas show an acceptable 10-dB impedance matching performance at 2.6 GHz band (2.55–2.65 GHz) (Figure 10.4a). The isolation between any antenna pair is better than 12 dB (Figure 10.4b).

It is also observed that Ant-1 and -4 mainly radiate toward +x direction and Ant-2 and -3 mainly radiate toward −x direction. The antenna efficiencies including mismatching losses are shown in Figure 10.5. The measured antenna efficiencies of the L-shaped monopole slot antennas (Ant 1–Ant 4) are approximately 48–55% and the simulated efficiencies are 53–60% over the operating band. For

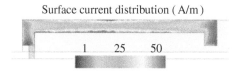

Surface current distribution (A/m)

Figure 10.3 Surface current distribution in C-shaped coupled-fed antenna. *Source*: Li et al. [8].© 2016, IEEE.

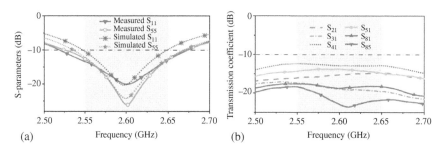

Figure 10.4 Simulated S-parameters of the dual polarized eight-element MIMO antenna: (a) reflection coefficients (dB) (b) transmission coefficient (dB). *Source*: Li et al. [8]. © 2016, IEEE.

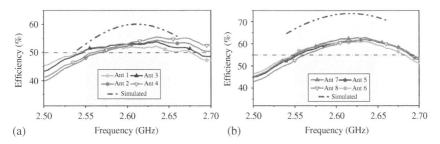

Figure 10.5 Efficiencies of the antennas: (a) C-shaped coupled-fed antennas and (b) L-shaped monopole slot antennas. *Source*: Li et al. [8]. © 2016, IEEE.

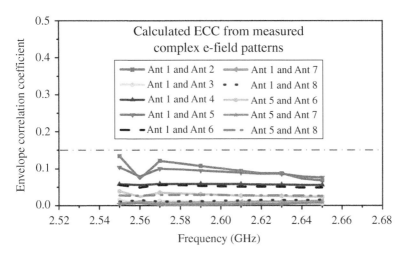

Figure 10.6 ECC of C-shaped coupled-fed antennas (Ant 5–Ant 8) and L-shaped monopole slot antennas (Ant 1–Ant 4) based on 3D field patterns. *Source*: Li et al. [8]. © 2016, IEEE.

C-shaped coupled-fed antenna elements (Ant 5–Ant 8), the measured antenna efficiencies are 55–63% while the simulated efficiencies are 65–68% over the operating band.

The envelope correlation coefficient (ECC) is calculated using the measured 3-D electric field patterns and shown in Figure 10.6. The calculated ECC values are less than 0.15 over the entire operating band, which fulfills the acceptable criterion of ECC for MIMO operation (<0.5). The mean effective gain (MEG) for the L-shaped monopole slot antennas (Ant 1–Ant 4) varies from −5.90 to −5.45 dBi, whereas for the C-shaped coupled-fed antennas (Ant 5–Ant 8), it varies from −5.76 to −4.79 dBi over the operating band. This indicates the differences in the MEG are less than 1 dB.

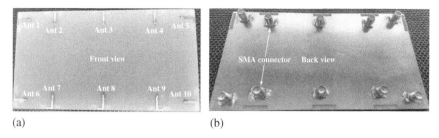

Figure 10.7 Fabricated 10-port MIMO antenna for LTE bands 42/43/46 in a sub-6 GHz mobile handset: (a) top view and (b) bottom view. *Source*: Li et al. [9].

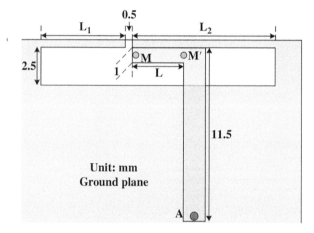

Figure 10.8 Sub-6 GHz antenna element. *Source*: Li et al. [9]. © 2018, IEEE.

The second example is about a 10-element MIMO antenna for LTE 42/43/46 reported in [9] and is shown in Figure 10.7. Four horizontal antennas (Ants 1, 5, 6, and 10) are placed at four corners of the PCB and the rest six vertical antennas (Ants 2, 3, 4, 7, 8, and 9) are placed at mid-section of the PCB. Perpendicular alignment of the radiators mounted at the four corners of the substrate helps to attain polarization diversity and provides good isolation performance.

Each element is a dual mode/dual band T-shaped slot antenna, as shown in Figure 10.8. A rectangular-shaped slot having a size of 16 mm × 2.5 mm is initially etched horizontally from the ground plane. A narrow strip of size 0.5 mm × 0.5 mm is further etched vertically from the ground plane forming a T-shaped slot. Here the slot is coupled fed by L-shaped strip which is placed at the opposite side of the ground plane. The antenna element occupies the 3.4–3.8 GHz band and 5.15–5.925 GHz band, which are the combination of the uplink and downlink

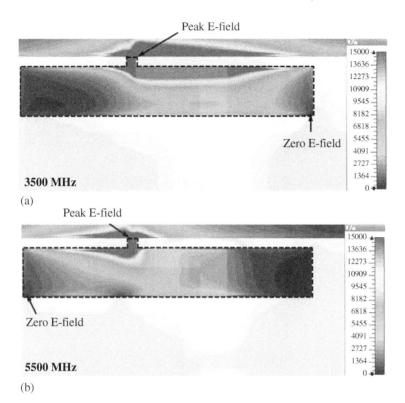

Figure 10.9 (a) 3500 MHz and (b) 5500 MHz. Surface current distribution and (E-field distribution) of the sub-6 GHz antenna element. *Source*: Li et al. [9]. © 2018, IEEE.

frequency ranges of the LTE bands 42, 43, and 46. Figure 10.9 provides the E-field distribution within the slot at 3.5 and 5.5 GHz, respectively. It can be observed that the surface E-field exhibits its maximum value at the opening end and minimum value is observed at the closed end, which identifies it as a $\lambda_g/4$ of the fundamental mode [9]. At 3.5 GHz, the length of E-field path is $\lambda_g/4$ along the longer open slot and at 5.5 GHz, the length of E-field path is $\lambda_g/4$ along the shorter open slot.

The S-parameters of the antennas (Ants 1, 2, 3, 4, and 5) are provided in Figure 10.10. The 10 dB matching bandwidth covers LTE bands 42/43/46 with isolation greater than 12 dB. The measured total efficiencies of the antennas provided in Figure 10.11a are better than 40 and 60%, respectively, in both bands which is acceptable for achieving lower capacity loss. The calculated ECCs are provided in Figure 10.11b, which are less than 0.15 and 0.05, respectively, over the low and high bands, satisfying a good diversity performance.

A third example is another eight-element-based MIMO for 5G thin terminal application [10]. The antenna element is a printed planar inverted F antenna

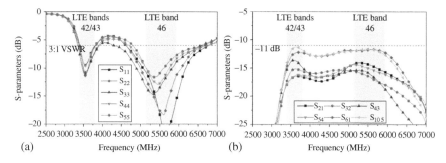

Figure 10.10 Measured S-parameters. (a) Reflection coefficients and (b) transmission coefficients. *Source*: Li et al. [9]. © 2018, IEEE.

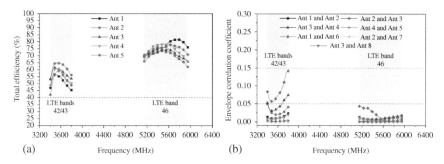

Figure 10.11 (a) Measured total efficiency and (b) calculated ECC of the antennas. *Source*: Li et al. [9]. © 2018, IEEE.

(PIFA), which occupies GSM1900 (1880–1920 MHz), LTE2300 (2300–2400 MHz), and LTE2500 (2540–2620 MHz). The detailed view of the antenna element is shown in Figure 10.12, where A is the grounding point and **B** is the feeding point. Branch 1 and 2 are one quarter wavelength at 1.9 and 2.4 GHz, respectively. It can be observed from the current distribution in Figure 10.13 that the LB (1.9 GHz) is controlled by both branches and the higher band (2.4 GHz) is controlled by branch 2. The antenna system consists of eight symmetric PIFAs. Here, Ant-1, 2, 3, and 4 have strong radiation along the x-axis and Ant-5, 6, 7, and 8 have strong radiation along the y-axis.

The measured S-parameters of the antenna elements are provided in Figure 10.14. It can be observed the 10 dB impedance matching bandwidth covers the desired operation with mutual couplings below −10 dB. The total efficiencies of the antennas are provided in Figure 10.15a. The measured ECCs based on uniform 3D channel model and Verizon LTE test plan channel are less than 0.1, as shown in Figure 10.15b, which satisfies good diversity performance in the MIMO system.

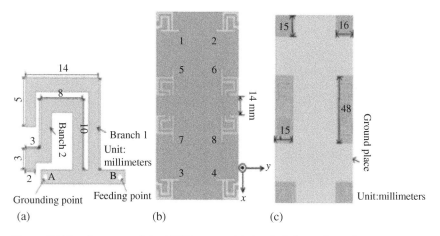

Figure 10.12 Geometry of the MIMO system: (a) detailed dimensions of antenna element, (b) configuration of eight-element MIMO system, and (c) ground of PCB. *Source:* Qin et al. [10]. © 2016, IET.

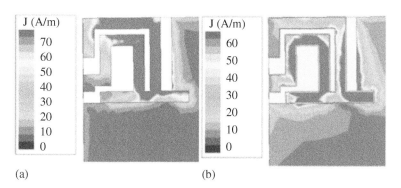

Figure 10.13 Simulated surface current distribution of the antenna element at (a) 1.9 GHz and (b) 2.4 GHz. *Source:* Qin et al. [10]. © 2016, IET.

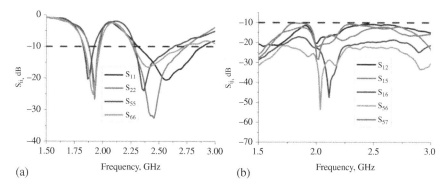

Figure 10.14 Measured S-parameters of the antenna: (a) reflection coefficient (dB) and (b) isolation (dB). *Source:* Qin et al. [10]. © 2016, IET.

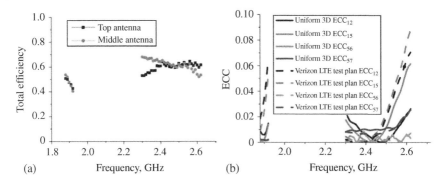

(a)   Frequency, GHz   (b)   Frequency, GHz

Figure 10.15 (a) Measured total efficiency and (b) measured ECCs of the prototype. *Source*: Qin et al. [10]. © 2016, IET.

The antenna elements mounted on the PCB of all the examples discussed above are operating in the same lower frequency band. In the next few examples, we will discuss about the antenna system where multiple antennas are operating in different lower frequency bands and are mounted in a single PCB. This enables the user to access multiple bands with increased data rate.

A 10-element-based antenna is proposed in [11] for future 4G/5G MIMO applications. The antenna model is provided in Figure 10.16. Ant-1 and Ant-2 are designed to cover the 4G band GSM850/900/1800/1900, UMTS2100, and

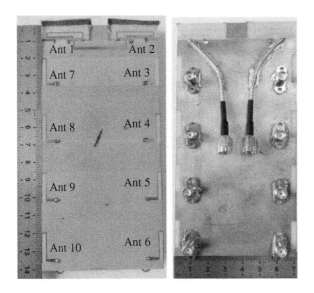

Figure 10.16 10-element-based antenna for 4G/5G application. *Source*: Ban et al. [11].

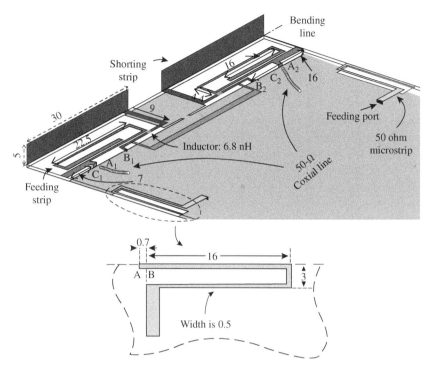

Figure 10.17 Detailed view of the 4G and 5G antenna elements (Unit in mm). *Source*: Ban et al. [11]. © 2016, IEEE.

LTE2300/2500 operating bands. Ant 3–Ant 10 are designed to cover a sub-6 GHz 5G band (3.4–3.6 GHz).

The detailed view of the 4G and 5G antenna elements are provided in Figure 10.17. The 4G antenna elements mounted at the top edge of the PCB consist of two feeding strips fed by 50 Ω mini coaxial feed line at point A_1 and A_2. Two 6.8 nH chip inductors are added at the bent shorting strips, where two points B_1 and B_2 are shorted to the ground plane. The protruded ground plane offers decoupling effect in the upper bands and improves matching at the LB. However, it also deteriorates the isolation level in the LBs. Therefore, a neutralization line (NL) is added between the two 4G antenna elements. Here, the feeding strip and shorting strip excite a fundamental resonant mode at 950 and 850 MHz, respectively, occupying the LBs for GSM850/900 operations. The shorting strip provides a higher-order resonance at 1800 MHz and is added with the resonance at about 2600 MHz (introduced by the protruded ground) to cover the GSM1800/1900/UMTS operations. The 5G antennas (Ant 3–Ant 10) are mounted along the long side of the PCB. The antenna element is a bending monopole strip placed above the clearance region on the ground plane. Here, the monopole strip resonates at 3.5 GHz with the excitation of its higher order resonance mode, where the total length of the strip is approximately $0.41\lambda_0$. Although the fundamental resonance is observed at 1.75 GHz, its poor matching has insignificant influence on the 4G antenna module operating band.

The measured S-parameters are provided in Figure 10.18. The 4G antennas have desirable measured 6-dB impedance bandwidths from 823 to 968 MHz and from 1697 to 2706 MHz with the isolation better than 10 dB in both desired frequency bands. The 5G antennas also cover the 3.5 GHz operating band (3.4–3.6 GHz) because the isolations of nonadjacent 5G elements with isolation are better than 10 dB.

The antenna efficiencies and gain measured in the SATIMO chamber are shown in Figure 10.19. The measured efficiencies at the lower and upper bands were more than 40 and 60%, respectively, and the corresponding antenna gain varied from −0.5 to 0.43 dBi and 0.99 to 3.7 dBi. The measured efficiencies of the 5G antennas were about 62–78%, and the antenna gain varied from 1.9 to 3.2 dBi over

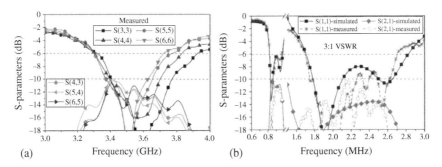

Figure 10.18 Measured S-parameters of antenna in Figure 10.17: (a) reflection coefficient and (b) transmission coefficient. *Source*: Ban et al. [11]. © 2016, IEEE.

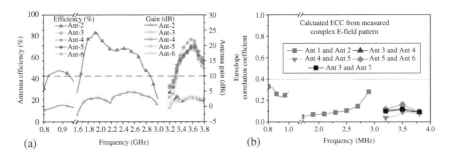

Figure 10.19 (a) Measured efficiencies and gain, and (b) calculated ECC of the antenna elements. *Source*: Ban et al. [11]. © 2016, IEEE.

the desired 3.5 GHz bands (3400–3600 MHz). The calculated values of ECC are less than 0.5 satisfying a good diversity performance.

Another example is a 12-port MIMO antenna reported in [12] for sub-6 GHz band applications. The antenna model is provided in Figure 10.20 which consists of three different types of antenna elements, i.e. (i) inverted π-shaped antenna (Ant-1 and Ant-2), (ii) longer inverted L-shaped open slot antenna (Ant-3, 5, 7, 8, 10, and 12), (iii) shorter inverted L-shaped antenna (Ant-4, 6, 9, and 11).

The inverted π-shaped antenna provides a dual resonant mode to cover both LTE bands 42/43 and LTE band 46. This is composed of two branch strips, namely, strip 1 (ABCEFG) that provides the fundamental λ/4 mode at around 3.5 GHz for LTE bands 42/43, and strip 2 (ABDHIJ) that provides its high-order mode at approximately 5.5 GHz to fully cover LTE band 46. The long L-shaped open slot antenna is excited at its fundamental resonant mode, when the length of the slot is 0.126λ at 3.6 GHz. Similarly, the short L-shaped open slot antenna is excited at its fundamental resonant mode, when the length of the slot is 0.117λ at 5.5 GHz.

Figure 10.21 shows the measured S-parameters of the antennas. The inverted π-shaped antenna (Ants 1 and 2) provides 6 dB matching bandwidths from 3040 to 3840 MHz) which occupies the LTE bands 42/43 and LTE band 46, with isolation (S_{21}) of better than 20 dB. The longer inverted L-shaped open slot antennas (Ants 3, 5, 7, 8, 10, and 12) provide 6 dB bandwidth from 3400 to 3838 MHz, which can occupy LTE bands 42/43 with the isolation between the adjacent ports better than 12 dB. The shorter inverted L-shaped open slot antennas (Ants 4, 6, 9, and 11) provide 6 dB bandwidth from 3400 to 3838 MHz, which can occupy LTE band 46 with the isolation between the adjacent ports better than 12.5 dB.

The measured total efficiencies of the antennas for LTE bands 42/43 and for LTE band 46 vary approximately between 41 and 82% and between 47 and 79%, as shown in Figure 10.22a. The calculated ECC values were lower than 0.15 and 0.1 in the LTE bands 42/43 and LTE band 46, respectively, as shown in Figure 10.22b. Some other lower cellular band and sub-6 GHz band antenna-based MFA systems are reported in [13–17].

10.3 5G *mm*-Wave Antenna Arrays

The fifth generation (5G) wireless access is the big evolution for modern wireless communication systems which aims to provide all possible connectivity for various devices in user's environment. Compared with 4G cellular systems, higher attenuation of the signal and coverage issue is one of the major challenges in 5G cellular systems. The best solution is to utilize high gain directional phased array antenna at both the UEs and base station, which is one of the key technologies in 5G cellular systems. This provides a motivation to develop *mm*-wave phased array

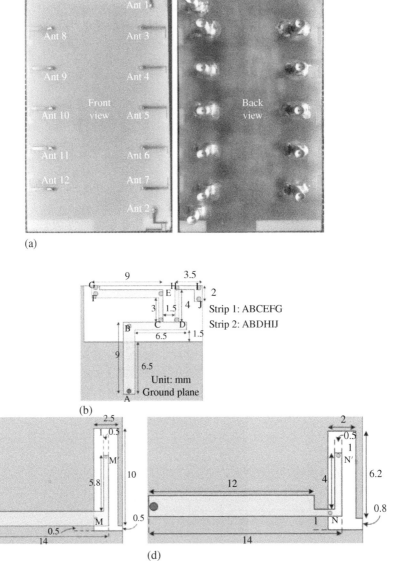

Figure 10.20 (a) 12-port antenna for LTE 42/43/46 application, (b) inverted π-shaped antenna with dimension, (c) shorter inverted L-shaped slot antenna with dimension, and (d) longer inverted L-shaped slot antenna with dimension (unit in mm). *Source*: Li et al. [12]. © 2017, IEEE.

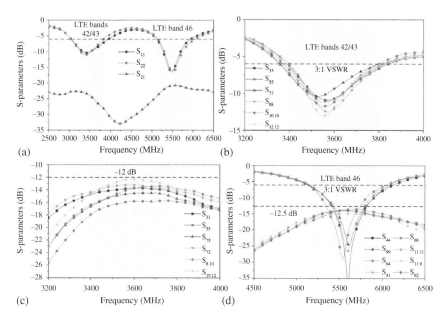

Figure 10.21 Measured S-parameters. (a) Reflection coefficients and isolation of the Ants 1 and 2, (b) reflection coefficients of Ants 3, 5, 7, 8, 10, and 12, (c) isolations between the adjacent antennas, and (d) reflection coefficients of Ants 4, 6, 9, and 11 and isolations between the adjacent antennas. *Source*: Li et al. [12]. © 2017, IEEE.

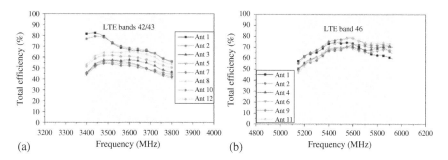

Figure 10.22 Total efficiencies of the antennas operating for (a) LTE bands 42/43 and (b) LTE band 46. *Source*: Li et al. [12]. © 2017, IEEE.

in UEs with their corresponding radio frequency (RF) transceiver, beam steering topologies, and algorithms to meet the stringent requirements of 5G cellular systems. The end-fire fan-beam radiation pattern-based phased arrays are also recommended for *mm*-wave 5G mobile devices as an alternative to this broadside planar phased array. In this section, we will discuss some *mm*-wave phased array reported for UE application.

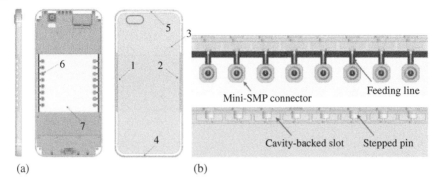

Figure 10.23 (a) Antenna model and (b) design of one of the beam steering arrays. *Source*: Yu et al. [18]. © 2017, IEEE.

The first example is based on the two eight-element *mm*-wave arrays consisting of cavity-backed slot antennas with beam steering capability which are built on the left- and right-side edges of the device and are shown in Figure 10.23 [18]. The eight-element phased array is based on the eight identical cavity-backed slot antenna elements [18]. Two symmetric phased arrays are mounted at left and right edges to achieve a fan beam-like radiation pattern. The center to center distance is 6.5 mm of the two slot elements, which is close to half-guided wavelength at 28 GHz. The slot length is 5.8 mm, and its width is 1.5 mm. A directional radiation pattern is achieved from a single element because of the reflection from the cavity. Here, a stepped pin acts as a feeding structure for the antenna element. It must be noted that the stepped pin structure has wider impedance bandwidth as compared with a straight pin or screw. All eight cavity-backed slot antennas are fed by a small stepped pin structure. The narrow open-end of the pin is soldered to a feeding microstrip line which is linked directly to a 28 GHz front-end RF integrated circuit (RFIC) chip. The microstrip feed lines are printed on a 0.254 mm-thick Rogers 5880 substrate ($\varepsilon_r = 2.2$ and tan $\delta = 0.0009$), which is integrated with the mini-SMP connectors for antenna measurement purposes.

Figure 10.24 shows the front-end module of the proposed beam steering array. The *mm*-wave antenna of mobile device should be regarded as part of the front-end module and placed within extremely close proximity to the 5G RF transceiver chip (using compact, low-loss interface technologies). The assembly of 28 GHz RFIC chip is performed by using the standard flip chip attachment process with standard solder reflow and underfill.

The measured S-parameters are provided in Figure 10.25a. The 10 dB bandwidth occupies the 28 GHz band (27.5–28.35 GHz) with an adjacent port isolation greater than 17 dB. The beam steering characteristic of one set of the eight-element phased array in different scanning angles from 0° to 60° at 28 GHz is

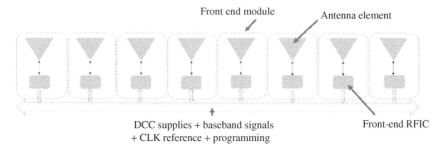

Figure 10.24 Front-end module of the eight-element beam steering array. *Source*: Yu et al. [18]. © 2017, IEEE.

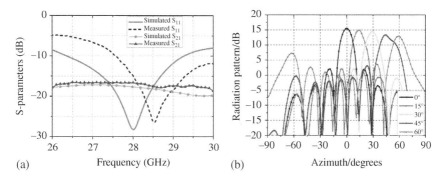

Figure 10.25 (a) Measured S-parameters and (b) beam steering performances of the array. *Source*: Yu et al. [18]. © 2017, IEEE.

provided in Figure 10.25b, which shows that the simulated gain at boresight has a maximum gain of 15.6 dBi.

The second example is based on a capacitive coupled patch-based antenna array provided in Figure 10.26 [19]. The proposed antenna is a pair of compact capacitive coupled symmetric patches as shown in Figure 10.26b. The overall size of the proposed design is 3.7 mm × 3.25 mm. Rogers RT5880 of a $\varepsilon_r = 2.2$ and loss tangent 0.0009 is considered as the substrate for printed circuit board (PCB). The antenna element covers 24–28 GHz for future 5G smartphone applications. The capacitive feed helps to reduce the effective patch size. The parasitic patch improves the bandwidth for 10 dB compared with a single patch antenna. The operating principle of the proposed antenna is studied using the equivalent circuit model and theory of characteristic modes [19]. Four sub-arrays each consisting of 12-element linear array are mounted in a 3D-shaped PCB. The antenna elements are placed at a distance of 6.3 mm which is close to half wavelength at 24 GHz. The four

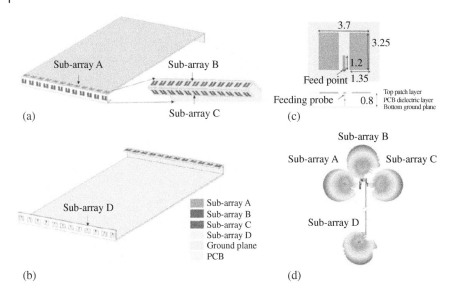

Figure 10.26 Proposed antenna design and array configuration: (a) perspective view of mobile phone PCB, (b) flipped image of PCB, (c) single antenna element, and (d) beam steering performance. *Source*: Stanley et al. [19]. © 2018, IEEE.

sub-arrays provide a 360° coverage whereas each sub-array provides a 90° coverage. The antenna array provides a 10-dB bandwidth 24–28 GHz with isolation better than 16 dB. The simulated beam steering ability of a single array is ±60° along the array axis with a peak gain of 16.5 dBi at the boresight direction. The proposed antenna exhibited a stable gain and uniform radiation pattern over the desired frequency band.

Another example is a compact beam-steerable antenna array for 5G mobile terminals at 28 GHz such as shown in Figure 10.27. The proposed array consists of one active element and two passive parasitic elements [20]. Two switches are utilized in the design instead of phase shifters. Each parasitic element is terminated with short-circuited transmission lines of different lengths via one switch. By controlling the two switches, different reactive impedance is loaded on two parasitic elements.

The radiation pattern of the active element can be placed into different directions by two parasitic elements. The length of the whole array is less than 0.81 wavelength. The designed array covers the 28–29 GHz band with the scan angle ±90°. By placing two arrays on each long chassis edge, 360° beam steering can be realized. Some other examples of *mm*-wave array enabled UEs with other essential constraints for *mm*-wave array design are provided in [21–24].

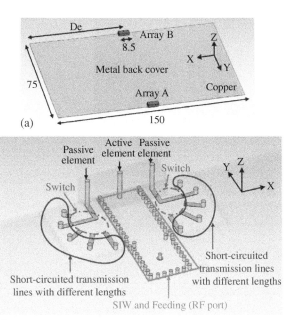

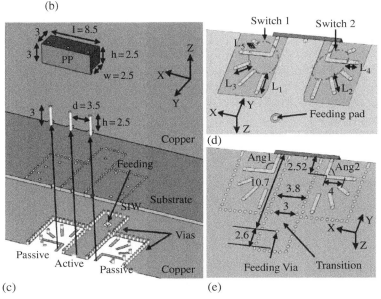

Figure 10.27 Schematics of the beam-steerable array: (a) 3-D perspective view, (b) general view, (c) 3-D exploded view, (d) back view, and (e) back view with surface copper hidden (unit: mm). *Source*: Zhang et al. [20]. © 2018, IEEE.

10.4   Collocated Sub-6 GHz and *mm*-Wave 5G Array Antennas

In this section, some collocated LB antennas and *mm*-wave antennas over a single host ground plane are discussed, which helps the user to get access from LBs to *mm*-wave band. Despite the shared volume, design strategy and operating principles of all antennas can be considered separately, at least in the beginning, with a goal to lower the coupling and de-correlate the individual patterns. However, before finalization, all these antennas should be studied together to notice any interactions between them. Limited works have been reported under this category in the literature so far.

The first example is a codesigned *mm*-wave and long-term evolution (LTE) antennas in a metal-rimmed handset [25]. LTE antennas are specified to operate at LTE lower band (LTE LB) frequencies 700–960 MHz and LTE higher band (LTE HB) 1710–2690 MHz with 6 dB matching level. The *mm*-wave band antennas are designed to operate from 25 to 30 GHz with 10 dB matching level. Figure 10.28 illustrates the antenna model of the codesigned *mm*-wave and LTE antennas. Here, Ports 1 and 2 refer to the feed for LTE LB and LTE HB, whereas Ports 4 and 5 are used for aperture matching of the LTE bands. Necessary equivalent circuit model for aperture matching circuit is provided in [25]. Aperture matching helps to increase the bandwidth and improves antenna efficiency. Port 3 is the feed point for the *mm*-wave array. The LTE antennas are integrated at the side edge/rim for efficient use of the restricted volume. The substrate used for LTE antennas is a 0.8 mm-thick FR-4 (ε_r = 4.3, loss tangent = 0.025), and copper plating models the display of the phone. The ground clearance for the LTE antennas is taken as 10 mm. The designed LTE antenna is integrated into the metal rim. The metallic

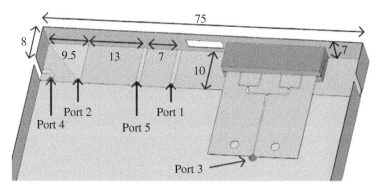

Figure 10.28   Schematic of codesigned LTE and *mm*-wave antennas. *Source*: Kurvinen et al. [25] (Unit: mm). Licensed under CC-BY-4.0.

rim behaves as a capacitive coupling element (CCE). It covers LTE LB at
700–960 MHz and LTE HB at 1710–2690 MHz. CCE is selected for its simple and
compact design and its capability to achieve wide bandwidth and due to these
characteristics it is also used in metal rim-based smart phones. The simulated
measured matching performances for the LTE antennas are provided in
Figure 10.29. It can be seen that the array satisfies 6 dB matching bandwidth from
0.7 to 0.96 GHz for LTE LB and 1.71–2.69 GHz for LTE HB. The mutual coupling
level is below −15 dB. The total efficiency is around 60% for both LTE LB and HB
antennas in simulation as shown in Figure 10.30. The measured efficiency for the
LTE LB varies from 50 to 90% and for LTE HB varies from 60 to 80%. The radiation
behavior of these antennas is also nearly omnidirectional.

The *mm*-wave antenna is a conventional Vivaldi antenna with microstrip to
slot-line transition, which is further used to implement a four-element linear

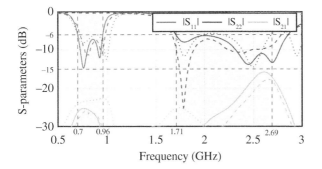

Figure 10.29 Simulated and measured S-parameters of the LTE LB and HB antennas.
Source: Kurvinen et al. [25]. Licensed under CC-BY-4.0.

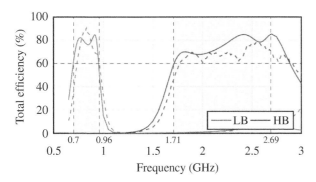

Figure 10.30 Simulated and measured total efficiencies of the LTE LB and HB antennas.
Source: Kurvinen et al. [25]. Licensed under CC-BY-4.0.

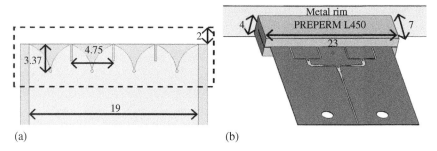

(a)   (b)

Figure 10.31 (a) Structure of the mobile Vivaldi antenna and (b) Vivaldi structure is enclosed with PREPERM L450 plastic (Unit: mm). *Source*: Kurvinen et al. [25]. Licensed under CC-BY-4.0.

array. The antenna model is shown in Figure 10.31. For appropriate radiation performance, an aperture is cut into the metal rim. The height of the aperture should be at least half of the effective wavelength and favors for filling the window with high permittivity material to keep the window practically small. The array is mounted on a separate PCB, a 0.101 mm-thick Rogers RO4350B substrate ($\varepsilon_r = 3.48$, loss tangent = 0.0037). Such thin substrate provides good rigidity and low losses. A plastic material (PREPERM L450 with $\varepsilon_r = 4.5$, and loss tangent = 0.0005) is used to enclose the aperture and allows the antenna to radiate through the metal rim. A suitable dielectric is chosen considering the matching performance and size of the hole created due to the array in the metallic rim. Essential factors for matching of the Vivaldi antenna are dimension of antennas, spacing between the elements, plastic to air interface, and size of the aperture. The tapering of the slot is designed based on the equations provided in [25, 26].

The ideal option to obtain beam steering is to feed independently the antenna elements through separate RF chain or phase shifter. Three different PCBs considering different progressive phase shifts are implemented between the elements to analyze beam steering. Three setups with phase shift of 0°, 50°, and 100° between the elements are considered with three different feed networks.

The simulated measured matching performances for three setups are provided in Figure 10.32. It can be observed that the array satisfies 10 dB matching bandwidth from 25 to 30 GHz for all the three setups. The beam steering performances of the array are also provided for 26 and 28 GHz in Figure 10.33a and b, respectively. It can be noted that at 26 GHz, the scanning ability of the array is ±30° along the array axis with peak realized gain variation around 5–7 dBi. At 28 GHz, the scanning ability of the array is almost ±40° along the array axis with peak realized gain of 5 dBi.

Another example of a collocated antenna system which is a combination of 4G MIMO antenna along with a *mm*-wave slot array for broadband high data rate

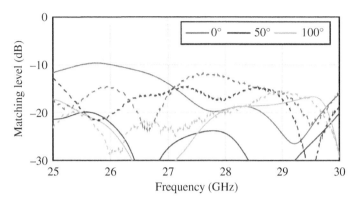

Figure 10.32 Matching performances of the antenna array for three setups with phase difference of 0°, 50°, and 100°. *Source*: Kurvinen et al. [25]. Licensed under CC-BY-4.0.

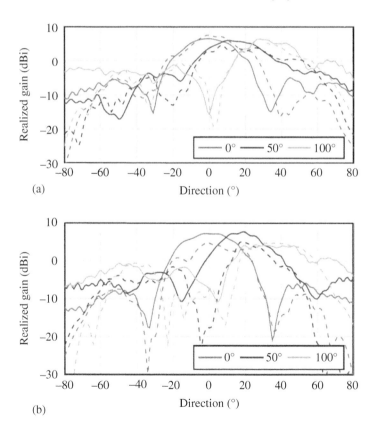

Figure 10.33 Simulated and realized gain of the antenna array with three setups at (a) 26 GHz and (b) 28 GHz. *Source*: Kurvinen et al. [25]. Licensed under CC-BY-4.0.

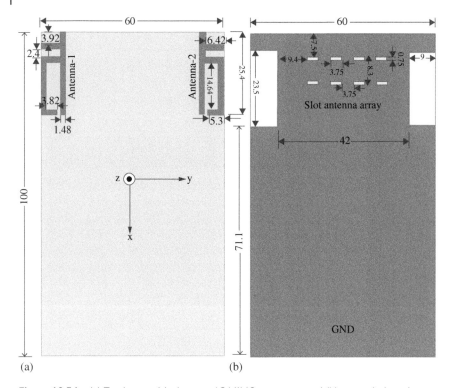

Figure 10.34 (a) Top layer with the two 4G MIMO antennas and (b) ground plane layer with the mm-wave slot array (Unit: mm). *Source*: Hussain et al. [27]. © 2017, IET.

applications is presented in [27]. The antenna model is shown in Figure 10.34. The MIMO antenna system consists of two reactive loaded monopoles that cover the frequency range 1.87–2.53 GHz for 4G wireless systems. The two elements are separated by a distance of 42 mm to be placed at the two top corners of the ground plane and mounted on the top layer of the board. The MIMO antennas are mounted on a 0.76 mm-thick Rogers RO3003 board with $\varepsilon_r = 3$ and loss tangent $= 0.0013$. The *mm*-wave array consists of a 2×4 planar slot array, which is etched on the ground plane of the 4G antennas. This is fed using a corporate feed network consisting of four parallel microstrip lines separated by 7.5 mm and each line is feeding two slot elements in series. The array is mounted over a RO3003 substrate with a thickness of 0.13 mm. Both 4G and *mm*-wave array arranged themselves in a four-layer configuration. The detailed arrangement of the antennas and dimensions are provided in [27].

Figure 10.35 shows the S-parameters (simulated and measured) of the two-element MIMO and *mm*-wave array. The two-element MIMO antenna has a −6 dB

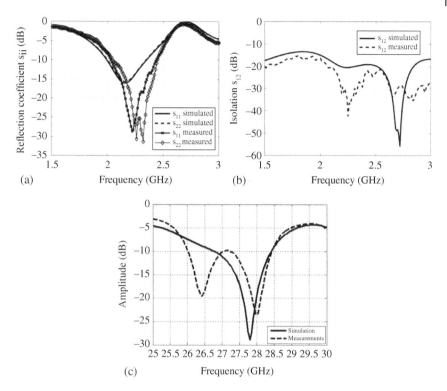

Figure 10.35 Simulated and measured S-parameters: (a) reflection coefficient (dB) of the two-element MIMO antenna, (b) isolation (dB) of the two-element MIMO antenna, and (c) reflection coefficient (dB) of the *mm*-wave array. *Source*: Hussain et al. [27]. © 2017, IET.

matching bandwidth from 1870 to 2530 MHz. The measured isolation was more than 15 dB between the MIMO antennas and more than 25 dB between the MIMO antennas and the *mm*-wave array for the entire band of 4G/5G operation. The 10 dB matching bandwidth of the *mm*-wave array is 26.8–28.4 GHz in simulation and 26.0–28.4 GHz in measurements. The maximum measured gain and the efficiency for the 4G two-element MIMO antenna system is 3.86 dBi and 83%, respectively. Good diversity performances of the MIMO system are also obtained. The simulated gain of the array is 9 dBi at 28 GHz.

Another example in [28] is based on three different types of MIMO antennas/array for lower cellular bands, sub-6 GHz band, and 28 GHz 5G band which are designed and developed on a common host ground plane. The simulated antenna model and fabricated antenna are shown in Figure 10.36. In the first type, one 2-port MIMO antenna (L_1 and L_2) is based on a new, compact unbalanced "T" shaped microstrip slot antenna for the operation in 1.6–2.2 GHz and 2.5–2.7 GHz

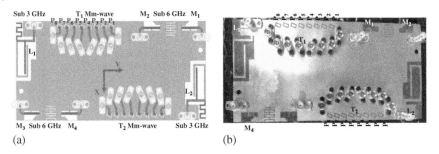

Figure 10.36 Antenna model consisting of lower cellular band antennas, sub-6 GHz band antennas, and *mm*-wave arrays: (a) simulated antenna model and (b) fabricated antenna [28].

that covers UMTS/DCS 1800/1900 PCS/1900 LTE bands and WiMAX applications. The first resonant mode is excited at 1.9 GHz, where the slot radiator behaves as a $\lambda_g/4$ magnetic monopole mode as the surface E-field propagates $\lambda_g/4$ from the top open end to the short-circuited end [12]. The coupling between monopole branch ($0.25\lambda_g$, at 1.9 GHz) and ground branch ($0.56\lambda_g$, at 1.9 GHz) plays a major role in the generation of mode-1. The second resonant mode is excited at 2.55 GHz, and is generated due to the coupling of the slot [12].

The S-parameters of the lower cellular band antennas are provided in Figure 10.37a. It can be observed that the Type-I MIMO antenna possesses a 6 dB matching bandwidth from 1.6 to 2.7 GHz, which covers the desired operating band with isolation greater than 20 dB between the antenna elements.

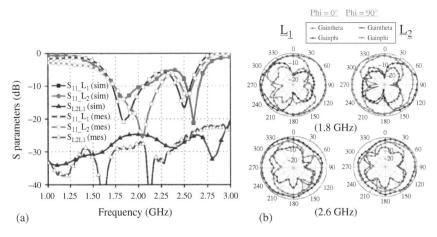

Figure 10.37 Antenna performances L_1 and L_2: (a) simulated and measured S parameters and (b) measured 2D radiation pattern at 1.8 and 2.6 GHz. *Source*: Biswal et al. [28].

Figure 10.37b presents the 2D radiation pattern of L_1 and L_2 at 1.8 and 2.6 GHz in two principal planes ($\phi = 0°$ and $90°$). The measured peak realized gain of L_1 and L_2 is nearly 3 and 4.1 dBi at 1.8 and 2.6 GHz, respectively. Computed ECC values are below 0.15 which are much less than 0.50 and satisfy the criteria for a good diversity performance.

In the second type, two 2-port MIMO antennas (M_1 and M_2) based on a new, compact "L" shaped open microstrip slot antenna is discussed for sub-6 GHz 5G band (3.2–4.8 GHz) application. A connected meander slot is also included as a decoupling slot. The S-parameters of the sub-6 GHz band antennas are provided in Figure 10.38a. The antennas possess a 10 dB bandwidth from 3.3 to 5 GHz, which occupies the sub-6 GHz band between 3.4 and 4.8 GHz with the measured isolation between the antennas being greater than 15 dB. Figure 10.38b presents the 2D radiation pattern of M_1 and M_2 at 3.5 and 4.6 GHz in two principal planes ($\phi = 0°$ and $90°$). The measured peak realized gain of M_1 and M_2 is nearly 3 and 4.1 dBi at 1.8 and 2.6 GHz, respectively. Computed ECC values are below 0.15 which are much less than 0.50 and satisfy the criteria for a good diversity performance.

In the third type, circular polarized (CP) linear phased array antennas (T_1 and T_2) based on a tilted rectangular microstrip slot loop radiator is designed to operate at 28 GHz 5G wireless communication band. The CP behavior is achieved due to the loading of a gap (capacitive reactance) along the length along Y-axis closer to the feed line. The length of the loop is $1.06\lambda_0$ at 28 GHz, where the gap is loaded at 0.25 mm from the central line along "$-Y$'" axis on one of the horizontal branches loaded by a feed line. To improve the CP purity of the antenna, the slot

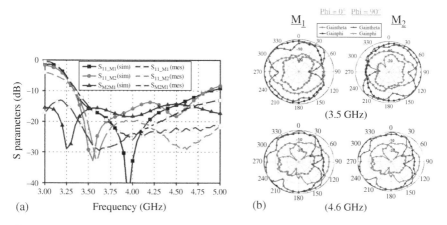

Figure 10.38 Antenna performances L_1 and L_2: (a) simulated and measured S parameters and (b) measured 2D radiation pattern at 3.5 and 4.6 GHz. *Source*: Biswal et al. [28].

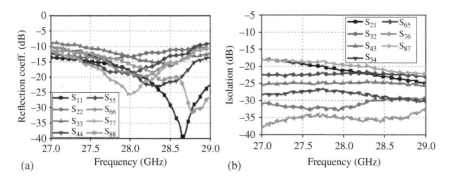

Figure 10.39 Measured S-parameters of the CP phased array: (a) reflection coefficient (dB) and (b) isolation level (dB). *Source*: Biswal et al. [28].

radiator is rotated clockwise with an angle of 20°. A rotated strip below the slot radiator is attached with the feed line to improve the matching performance. In this configuration, different length of feed lines is optimized for good impedance matching with 2.92 mm K-connectors. The S-parameters are provided in Figure 10.39. It can be observed that the 10 dB matching level occupies the desired operating 5G cellular band 27.5–28.35 GHz. Therefore, to obtain a maximum gain in broadside direction ($\theta_0 = 0°$), the input phase for each element is different than others. In order to achieve a beam scanning, all elements are excited by different phases to scan the main beam along the array axis ($\phi = 90°$).

The normalized patterns of the array at 28 GHz are computed using the measured individual active element patterns by using the following equations:

$$E_T\left(\theta,\phi\right) = \sum_{n=1}^{N} E_n\left(\theta,\phi\right) \cdot a_n \cdot e^{-j\left(kd\left(n-1\right)\sin(\theta)\sin(\phi)+\beta_{pi}\right)}, \tag{10.1}$$

$$\beta_{pi} = -\sum_{n=1}^{N} kd\left(n-1\right)\sin\left(\theta_0\right), \tag{10.2}$$

where $E_n(\theta, \phi)$ is the individual element pattern, N is the total number of radiating elements along the array axis (Y), k is the wave number, and d is the inter-element spacing. In order to scan the beam to a desired angle θ_0, the phase of the element is shifted progressively by factor $-kd \sin(\theta_0)$. The computed normalized patterns of the array for $\phi = 0°$ and 90° cut planes are shown in Figure 10.40a. A directive pattern is observed along the array axis ($\phi = 90°$) with a cross-polarization level of better than 20 dB and a sidelobe level (SLL) greater than 14 dB. The beam steering capability of the array is computed in MATLAB using the normalized measured individual radiation patterns and the array factor with different progressive phase shift values (β_p). The computed normalized RHCP patterns for

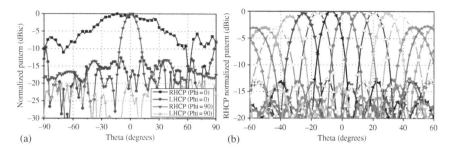

Figure 10.40 (a) Normalized 2D pattern and (b) beam steering performance of the array. *Source*: Biswal et al. [28].

different values of β_p along the array axis are plotted in Figure 10.40b. It can be noted that the array has the capability to steer the beams within the scanning range of $\pm 40°$ with a 1 dB gain reduction.

10.5 RF and EMF Exposure Limits

It is necessary to investigate the effect of interaction of user body on the antenna performance installed in user equipment. This provides a clear idea that how efficient are the antennas within the UE (e.g. smart phone, tablet, or any handheld terminals). For this, we will discuss some examples of user's hand effect on the handset antenna's performances. The hand tissue is a lossy material, which absorbs the radiated power from the antenna elements and affects the total efficiency of the antenna system. All the antenna performance of these examples is mentioned in Sections 10.2 and 10.3. Therefore, we will discuss their user hand effect only. Two modes of operation are considered here, i.e. (i) data mode and (ii) read mode. Data mode can be further divided into two types, i.e. (i) right-hand data mode (RHDM) and (ii) left-hand data mode (LHDM). As both hands are usually preferred in read mode, thus it can be termed as both hand read mode (BHRM). The MIMO antenna in [12] is studied in data and read mode. The antenna performance is provided in Section 10.2. The simulated antenna model is shown in Figure 10.41.

Next example is the antenna reported in [28] and their antenna performances are mentioned in Section 10.3. The simulated model of the prototype under RHDM, LHDM, and BHRM is shown in Figure 10.42. Figures 10.43a and c show the simulated total efficiencies for RHDM, LHDM, and BHRM of Type-I (L_1 and L_2), Type-II (M_1, M_2, M_3, and M_4), and Type-III antennas (T_1 and T_2), respectively. It can be observed that the simulated total efficiencies of Type-I antennas for

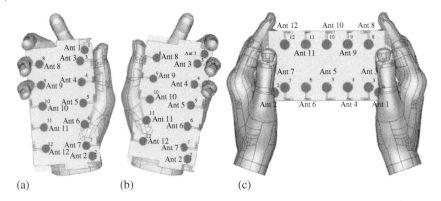

Figure 10.41 The proposed MIMO antenna array under (a) left-hand data mode (LHDM), (b) right-hand data mode (RHDM), and (c) both hand read mode (BHRM). *Source*: Li et al. [12]. © 2017, IEEE.

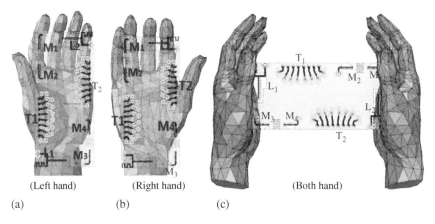

(a)   (b)   (c)

Figure 10.42 Simulated model for three different interactions of user hand with the antenna prototype: (a) left-hand data mode (LHDM), (b) right-hand data mode (RHDM), and (c) both hand read mode (BHRM). *Source*: Biswal et al. [28]. © 2020, IEEE.

RHDM and LHDM are better than 55% (Figure 10.43a). For BHRM, the total effi-ciencies have degraded as the antennas L_1 and L_2 are covered by the palms and fingers. The total efficiencies of Type-II antenna elements M_3 and M_4 for both RHDM and LHDM and M_2 and M_4 for BHRM are better than 60% (Figure 10.43b). Whereas, the total efficiencies of M_1 and M_2 for both RHDM and LHDM and M_1 and M_3 for BHRM have degraded by 10%. The total efficiencies of Type-III antenna arrays T_1 and T_2 for BHRM are better than 60%, as shown in Figure 10.43c. The total efficiencies of T_1 and T_2 for RHDM and LHDM have degraded by 10–12%. The power radiated from an open slot antenna is absorbed by the hand tissue,

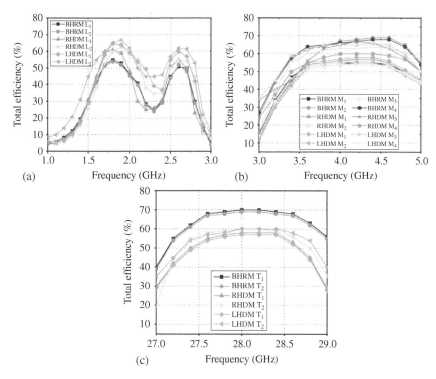

Figure 10.43 Total efficiency of three antennas at left-hand data mode (LHDM), right-hand data mode (RHDM), and both hand read mode (BHRM) for (a) Type-I antenna, (b) Type-II antenna, and (c) Type-III antenna array. *Source*: Biswal et al. [28]. © 2020, IEEE.

when the open section (having peak E-field) are covered by hands or fingers [29, 30].

The radio frequency (RF) electromagnetic fields (EMF) emitted from user equipment must be tested to satisfy the essential requirements and limits on human exposure to EMF [31–34]. The International Commission on Non-Ionizing Radiation (ICNIRP) regulates the guideline of widely adopted exposure limit worldwide [32]. The guideline for RF exposure in the United States is regulated by the Federal Communications Commission (FCC) [33]. The specific absorption rate (SAR) specifies the basic restrictions on RF EMF exposure for the frequencies used for 2G/3G/4G cellular systems. SAR indicates the rate at which electromagnetic radiation energy is absorbed per unit mass by a human body during the exposure of a RF wave. Proper guideline for SAR is prescribed for UEs considering the adverse health effects associated with excessive localized tissue heating and whole-body heat stress. The prescribed average SAR limit is 1.6 W/kg over a 1 g of tissue or of 2.0 W/kg over 10 g of tissue for frequency up to 6 GHz [35]. It is also

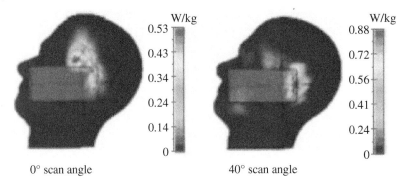

W/kg

0.53
0.43
0.34
0.24
0.14
0

0° scan angle

W/kg

0.88
0.72
0.56
0.41
0.24
0

40° scan angle

Figure 10.44 SAR distribution of the antenna at 0° and 40° scan angles. *Source*: Bang and Choi [36]. © 2018, IEEE.

essential to note that, by now, there are no such guidelines which recommend any prescribed SAR limit for RF wave exposure to the human body at *mm*-waves. Two examples are discussed here for *mm*-wave antennas. A 28 GHz beam-steering array antenna based on rotated slot radiators is reported in [36] for a metal-covered 5G cellular handset. The antenna model is provided in Figure 10.44. All necessary antenna performances are provided in [34]. For SAR analysis, the antenna proto-type is simulated in the presence of a head phantom available in CST Microwave Studio. The SAR performance of the antenna at 28 GHz is provided in Figure 10.44. The peak SAR values for the scan angles of 0° and 40° are 0.53 and 0.88 W/kg, respectively.

The second example is for the antenna prototype reported in [28]. Three different types of antennas are integrated on the same ground plane, i.e. Type-I for lower cellular band, Type-II for sub-6 GHz band, and Type-III for *mm*-wave array. The antenna performances are provided in Section 10.3. For SAR analysis, the antenna prototype is simulated in the presence of a head phantom available in Ansys HFSS. The input power of each antenna element is taken as 24 dBm. The simulated 1 g averaged SAR distributions are provided in Figure 10.45 for Type-I, -II, and -III antennas. The average SAR value over 1 g tissue is 0.12 W/kg at 1.8 GHz and 0.30 W/kg at 2.6 GHz for Type-I antenna. For Type-II antenna, the average SAR1g value is 0.16 W/kg at 3.5 GHz and 0.11 W/kg at 4.5 GHz. For Type-III array the SAR1g value 0.88 W/kg (for 0° scan) and 0.79 W/kg (for 40° scan) at 28 GHz. It can be noted that the SAR values of the Type-I and -II antennas lie within the prescribed limit. It is very tedious while simulating all three types of antennas considering the head model. It is important to note that the simulation time required for *mm*-wave is more (almost double) as compared with LB antennas.

Optimization of the maximum output power of 5G cellular handsets is also very essential as it directly provides impact in system capacity and coverage [37]. The

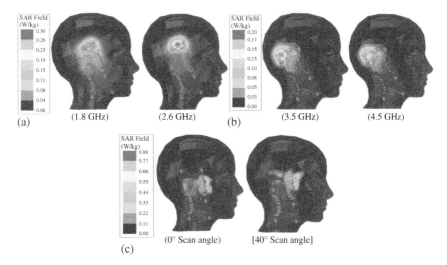

Figure 10.45 SAR distribution of the antenna at 0° and 40° scan angles for (a) Type-I, (b) Type-II, and (c) Type-III antennas. *Source*: Biswal et al. [28]. © 2020, IEEE.

RF EMF exposure regulations have some restriction on the maximum output power. Therefore, it is essential to study the EMF exposure property of the *mm*-wave phased array in mobile devices. There is another parameter, free space power density (PD, measured in W/m^2), that also evaluates the EMF exposure level for frequencies above 6 GHz based on the FCC regulations [38], above 10 GHz based on the ICNIRP regulations [32], and above 3 GHz-based IEEE limitations [39]. The free space PD evaluates the EMF penetration depth into the body that must be as much as smaller. According to the FCC regulations, the limited spatial peak value of PD is $10 W/m^2$.

The suggestion of the changing basic restriction from SAR to power density was investigated based on the maximum possible transmitted power (P_{max}) from a device (canonical dipole) used near the human body [40, 43]. It was observed that the existing exposure limits will result in a nonphysical discontinuity of several dB in P_{max} as the transition is made from SAR to power density-based basic restrictions. Therefore, to be compliant with applicable exposure limits at frequencies above 6 GHz, P_{max} might have to be several dB lower than the power levels used for current cellular technologies [40, 43]. It was also observed that the increase in skin temperature due to RF exposure from the same source, while transmitting at the maximum allowable power, should be compliant with these limits. The increase in maximum steady-state temperature was also found to display a similar discrepancy. IEEE C95.1-2005 provides the most consistent level of protection against thermal hazards of exposure over the frequency range 6–60 GHz [39]. In [37], the RF EMF exposure for phased arrays at 15 and 28 GHz was investigated based on the PD property. This was also selected by NTT Docomo [41] for their demo system of 5G communication. The maximum allowed output powers to satisfy the compliance with the peak PD of $10 W/m^2$ in different cases are also studied. It is observed from the investigation that an array with maximum number of antennas shows a lower PD in the reactive near-field and Fresnel region due to the larger form factor of its radiator. Generally, a larger array has a higher gain and needs less output power, which is beneficial for dropping the PD value in the near-field region. At higher frequency, larger number of elements is also required to reduce the peak PD value and compensate for the higher path loss. The study also considered the effects of a progressive phase shifts between the radiating elements but was restricted to the FCC exposure limits [38]. A study to investigate the maximum transmitted power and maximum EIRP to comply with all major RF EMF exposure standards has been also presented in [42] for array antennas intended for user equipment and low-power radio base stations in 5G mobile communication systems.

10.6 Conclusion

In this chapter, first we discussed the lower cellular band and sub-6 GHz 5G band-enabled user devices. Next, we discussed the 5G *mm*-wave array intended for UEs. Then we discussed few collocated *mm*-wave arrays and LB antennas for user equipment. Lastly, we discussed the effect of user hand for data and read mode and essential parameters to prescribe the RF/EMF exposure limits for user equipment.

References

1 Basavarajappa, V. and Vinoy, K.J. (2010). An integrated wideband multifunctional antenna using a microstrip patch with two U-slots. *Progress In Electromagnetics Research* 22: 221–235.

2 Manafi, S., Nikmehr, S., and Bemani, M. (2012). Planar reconfigurable multifunctional antenna for WLAN/wimax/UWB/pcsdcs/UMTS applications. *Progress In Electromagnetics Research* 26: 123–137.

3 Haidacher, F., Mathiae, S., Mierke, F. et al. (2006). *U.S. Patent No. 7,034,758*. Washington, DC: U.S. Patent and Trademark Office.

4 Andrews, J.G., Buzzi, S., Choi, W. et al. (2014). What will 5G be? *IEEE Journal on Selected Areas in Communications* 32 (6): 1065–1082.

5 Rappaport, T.S., Gutierrez, F., Ben-Dor, E. et al. (2012). Broadband millimeter-wave propagation measurements and models using adaptive-beam antennas for outdoor urban cellular communications. *IEEE Transactions on Antennas and Propagation* 61 (4): 1850–1859.

6 Rappaport, T.S., Sun, S., Mayzus, R. et al. (2013). Millimeter wave mobile communications for 5G cellular: it will work! *IEEE Access* 1: 335–349.

7 Zhao, K., Helander, J., Sjöberg, D. et al. (2016). User body effect on phased array in user equipment for the 5G mmWave communication system. *IEEE Antennas and Wireless Propagation Letters* 16: 864–867.

8 Li, M.Y., Ban, Y.L., Xu, Z.Q. et al. (2016). Eight-port orthogonally dual-polarized antenna array for 5G smartphone applications. *IEEE Transactions on Antennas and Propagation* 64 (9): 3820–3830.

9 Li, Y., Luo, Y., and Yang, G. (2018). Multiband 10-antenna array for sub-6 GHz MIMO applications in 5-G smartphones. *IEEE Access* 6: 28041–28053.

10 Qin, Z., Geyi, W., Zhang, M., and Wang, J. (2016). Printed eight-element MIMO system for compact and thin 5G mobile handest. *Electronics Letters* 52 (6): 416–418.

11 Ban, Y.L., Li, C., Wu, G., and Wong, K.L. (2016). 4G/5G multiple antennas for future multi-mode smartphone applications. *IEEE Access* 4: 2981–2988.

12 Li, Y., Luo, Y., and Yang, G. (2017). 12-port 5G massive MIMO antenna array in sub-6GHz mobile handset for LTE bands 42/43/46 applications. *IEEE Access* 6: 344–354.

13 Jha, K.R. and Sharma, S.K. (2018). Combination of MIMO antennas for handheld devices [wireless corner]. *IEEE Antennas and Propagation Magazine* 60 (1): 118–131.

14 Singh, H.S., Upadhyay, R., and Shubair, R.M. (2020). Performances study of compact printed diversity antenna in the presence of user's body for LTE mobile phone applications. *International Journal of RF and Microwave Computer-Aided Engineering* 30 (5): e21743.

15 Singh, A. and Saavedra, C.E. (2020). Wide-bandwidth inverted-F stub fed hybrid loop antenna for 5G sub-6 GHz massive MIMO enabled handsets. *IET Microwaves, Antennas & Propagation* 14 (7): 677–683.

16 Sun, L., Li, Y., Zhang, Z., and Wang, H. (2020). Self-decoupled MIMO antenna pair with shared radiator for 5G smartphones. *IEEE Transactions on Antennas and Propagation* 68 (5): 3423–3432.

17 Han, C.Z., Xiao, L., Chen, Z., and Yuan, T. (2020). Co-located self-neutralized handset antenna pairs with complementary radiation patterns for 5G MIMO applications. *IEEE Access* 8: 73151–73163.

18 Yu, B., Yang, K., and Yang, G. (2017). A novel 28 GHz beam steering array for 5G mobile device with metallic casing application. *IEEE Transactions on Antennas and Propagation* 66 (1): 462–466.

19 Stanley, M., Huang, Y., Wang, H. et al. (2018). A capacitive coupled patch antenna array with high gain and wide coverage for 5G smartphone applications. *IEEE Access* 6: 41942–41954.

20 Zhang, S., Syrytsin, I., and Pedersen, G.F. (2018). Compact beam-steerable antenna array with two passive parasitic elements for 5G mobile terminals at 28 GHz. *IEEE Transactions on Antennas and Propagation* 66 (10): 5193–5203.

21 Ojaroudiparchin, N., Shen, M., Zhang, S., and Pedersen, G.F. (2016). A switchable 3-D-coverage-phased array antenna package for 5G mobile terminals. *IEEE Antennas and Wireless Propagation Letters* 15: 1747–1750.

22 Hong, W., Baek, K.H., and Ko, S. (2017). Millimeter-wave 5G antennas for smartphones: overview and experimental demonstration. *IEEE Transactions on Antennas and Propagation* 65 (12): 6250–6261.

23 Hong, W., Baek, K.H., Lee, Y. et al. (2014). Study and prototyping of practically large-scale mmWave antenna systems for 5G cellular devices. *IEEE Communications Magazine* 52 (9): 63–69.

24 Hong, W. (2017). Solving the 5G mobile antenna puzzle: assessing future directions for the 5G mobile antenna paradigm shift. *IEEE Microwave Magazine* 18 (7): 86–102.

25 Kurvinen, J., Kähkönen, H., Lehtovuori, A. et al. (2018). Co-designed mm-wave and LTE handset antennas. *IEEE Transactions on Antennas and Propagation* 67 (3): 1545–1553.

26 Shin, J. and Schaubert, D.H. (1999). A parameter study of stripline-fed Vivaldi notch-antenna arrays. *IEEE Transactions on Antennas and Propagation* 47 (5): 879–886.

27 Hussain, R., Alreshaid, A.T., Podilchak, S.K., and Sharawi, M.S. (2017). Compact 4G MIMO antenna integrated with a 5G array for current and future mobile handsets. *IET Microwaves, Antennas & Propagation* 11 (2): 271–279.

28 Biswal, S., Sharma, S., and Das, S., (2020) "Collocated Microstrip Slot MIMO Antennas for Cellular Bands Along With 5G Phased Array Antenna for User Equipments (UEs)," *IEEE Access*, 8, 209138-209152.

29 Zhao, K., Zhang, S., Ying, Z. et al. (2013). SAR study of different MIMO antenna designs for LTE application in smart mobile handsets. *IEEE Transactions on Antennas and Propagation* 61 (6): 3270–3279.

30 Ban, Y.L., Qiang, Y.F., Chen, Z. et al. (2014). A dual-loop antenna design for hepta-band WWAN/LTE metal-rimmed smartphone applications. *IEEE Transactions on Antennas and Propagation* 63 (1): 48–58.

31 Council Recommendations of 12 (1999). *On the limitation of exposure of the general public to electromagnetic fields (0 Hz to 300 GHz) document 1999/519/ EC. Official Journal of European Communities* 199.

32 Ahlbom, A., Bergqvist, U., Bernhardt, J.H. et al. (1998). Guidelines for limiting exposure to time-varying electric, magnetic, and electromagnetic fields (up to 300 GHz). *Health Physics* 74 (4): 494–521.

33 FCC (1997). *Code of Federal Regulations CFR Title 47, Part 1.1310, Radio frequency Radiation Exposure Limits*. Washington, DC: Federal Commun. Commission.

34 Cleveland, R.F. Jr., Sylvar, D.M., and Ulcek, J.L. (Aug. 1997). *OET Bulletin 65, Evaluating Compliance with FCC Guidelines for Human Exposure to Radiofrequency Electromagnetic Elds*. Washington, DC: FCC.

35 Wu, T., Rappaport, T.S., and Collins, C.M. (2015). Safe for generations to come: considerations of safety for millimeter waves in wireless communications. *IEEE Microwave Magazine* 16 (2): 65–84.

36 Bang, J. and Choi, J. (2018). A SAR reduced mm-wave beam-steerable array antenna with dual-mode operation for fully metal-covered 5G cellular handsets. *IEEE Antennas and Wireless Propagation Letters* 17 (6): 1118–1122.

37 Zhao, K., Ying, Z., and He, S. (2015). EMF exposure study concerning mmWave phased array in mobile devices for 5G communication. *IEEE Antennas and Wireless Propagation Letters* 15: 1132–1135.

38 FCC (2010). FCC code of federal regulations. *CFR Title 47 Part 2.1093.*

39 Chou, C.K. (2006). *IEEE Standard for Safety Levels with Respect to Human Exposure to Radio Frequency Electromagnetic Fields, 3 kHz to 300 GHz (ieee std c95. 1-2005).* New York: The Institute of Electrical and Electronics Engineers Inc.

40 Foster, K. and Colombi, D. (2017). Thermal response of tissue to RF exposure from canonical dipoles at frequencies for future mobile communication systems. *Electronics Letters* 53 (5): 360–362.

41 NTT Docomo (2014). Docomo 5G White Paper [Online]. https://www.nttdocomo.co.jp/english/corporate/technology/whitepaper_5g/

42 Thors, B., Colombi, D., Ying, Z. et al. (2016). Exposure to RF EMF from array antennas in 5G mobile communication equipment. *IEEE Access* 4: 7469–7478.

43 Colombi, D., Thors, B., and Törnevik, C. (2015). Implications of EMF exposure limits on output power levels for 5G devices above 6 GHz. *IEEE Antennas and Wireless Propagation Letters* 14: 1247–1249.

11

DoD Reconfigurable Antennas

Jia-Chi S. Chieh and Satish K. Sharma

11.1 Introduction

The DoD utilizes many reconfigurable antennas for various purposes. The most prevalent reason is when pattern scanning is necessary. Multifunction RF systems are also becoming more prevalent, systems that can simultaneously support radar, communications, and electronic warfare/information operations. The ability to support various mission sets that have disparate operational requirements requires a certain level of reconfigurability. This chapter will discuss various DoD systems that utilize reconfigurability.

11.2 TACAN

The tactical air navigation system (TACAN) is a navigation system used by military aircraft. It is used to provide aerial users with bearing and distance to a demarcated ground station. It has many similarities to the VHF omnidirectional range (VOR) distance measuring equipment (DME) that is used in civil aviation. Although space-based navigational systems, such as GPS, are preferred, TACAN remains a critical capability.

TACAN operates between 960 and 1215 MHz and has 126 two-way operating channels each of 1 MHz spacing. Frequencies from 962 to 1024 and 1151–1213 MHz are for ground-to-air, while 1025–1150 MHz is for air-to-ground [1]. The distance measuring aspect of TACAN is similar to radar principles, where the distance is measured by the round-trip travel time of a pulsed RF signal. Since TACAN is a

Multifunctional Antennas and Arrays for Wireless Communication Systems, First Edition.
Edited by Satish K. Sharma and Jia-Chi S. Chieh.
© 2021 John Wiley & Sons, Inc. Published 2021 by John Wiley & Sons, Inc.

beacon transponder, the echo is synthetic rather than from natural reflections. The identification signal sent by each beacon allows the pilot to identify which beacon is transmitting, therefore allowing geographic location to be plotted [2]. The typical pulse pair has a width of $3.5 \pm 0.5\,\mu s$ with a rise/fall time of $2.5 \pm 0.5\,\mu s$.

11.2.1 TACAN Antenna

There have been several iterations on the TACAN antenna; however, the principle functionality remains the same. We will start by describing the first-generation TACAN antenna and move to modern versions.

The most basic TACAN antenna is composed of a cylindrical nested array. The central antenna element is a vertical monopole element which is the driven element. This center antenna is stationary and omnidirectional. This is shown in Figure 11.1. Around the central element is an inner diameter of 5 in. and contains a single vertical conductive wire. This wire acts as a parasitic reflector, allowing the TACAN antenna to have a semi-directional cardioid pattern. This inner cylinder rotates at a 15 Hz rate.

Figure 11.2 illustrates the cardioid pattern resulting from the single parasitic reflector element. This pattern is rotated at a rate of 15 Hz. Around the central element is an outer diameter of 33 in. with nine vertical conductive wires that act as reflectors. This outer diameter is rotated at a rate of 135 Hz and is used for fine bearing acquisition. The resulting pattern is that of a cardioid with nine minor ripples as shown in Figure 11.3. The pattern contains the basic 15 Hz envelope as well as a ninth harmonic 135 Hz envelope. The two 15 and 135 Hz signals can be distinguished by using filter elements.

Early implementations of the TACAN antenna required dual mechanical rotation of the 15 Hz parasitic and 135 Hz parasitic elements. For reliability, an electronically scanned version of the TACAN antenna was proposed and developed [2, 3]. Figures 11.4 and 11.5 show the electronically modulated TACAN antenna from [2]. It is composed of an array which is 12-elements high

Figure 11.1 Taken from figure 4 in [1]. *Source*: Adapted from Colin and Dodington [1].

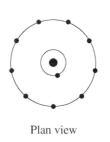

Plan view

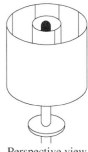

Perspective view

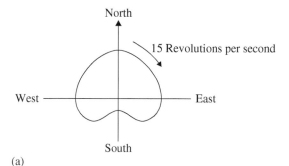

(a)

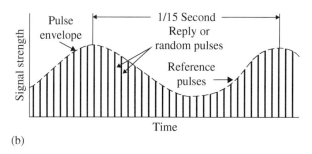

(b)

Figure 11.2 Single parasitic element pattern from figure 5 in [1]. *Source*: Adapted from Colin and Dodington [1].

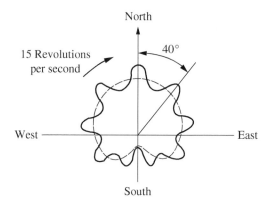

(a)

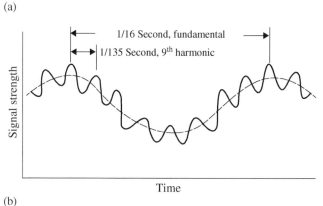

(b)

Figure 11.3 9-Parasitic element pattern from figure 6 in [1]. *Source*: Adapted from Colin and Dodington [1].

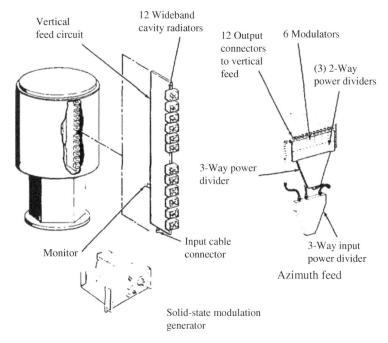

Figure 11.4 Electronically modulated TACAN antenna from figure 3 in [2]. *Source*: Shestag [2]. © 1974, IEEE.

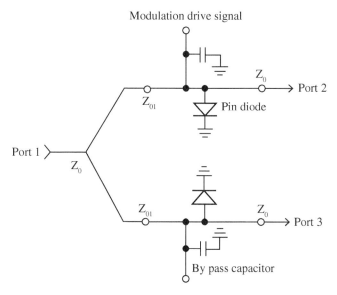

Figure 11.5 Modulator circuit from figure 6 in [2]. *Source*: Shestag [2]. © 1974, IEEE.

and 36-elements in circumference. The element used is a cubical cavity resonator, radiating through a slot on the cylindrical surface. A simple ON–OFF modulator circuit is then used to rotate the pattern synthetically, as shown in Figure 11.5. A square wave signal is used at 15 and 135 Hz to provide a drive input to the modulator circuit, which either shunts the RF signal to ground, or lets it pass through.

11.2.2 Course Bearing

Since the timing of the RF pulses is used to measure the distance, another medium must be used to convey the bearing. Amplitude modulation is used to convey the bearing by modulating the strength of the pulse through making a pseudo-directional pattern rotated around the vertical axis. This amplitude-modulated signal is then used to identify the aircraft direction with respect to the TACAN ground station. This pseudo-directional antenna pattern is rotated at 15 Hz and resembles a cardioid pattern.

Figure 11.6 Illustrates how coarse bearing acquisition is obtained. A 15 Hz reference burst signal is transmitted from the ground station along with the 15 Hz modulated echo signal. The reference signals are transmitted when the peak of the radiation pattern is pointing to east. The reference signals can be distinguished from echo signals as they contain exactly 12 pulse pairs spaced exactly 30 µs apart. Since the phasing between the 15 Hz modulated echo signal and the 15 Hz reference signal is dependent on the location of the aircraft, its bearing can be determined.

11.2.3 Fine Bearing

In order to improve bearing acquisition resolution, a fine bearing feature is also built into the TACAN system. The fine bearing measurement system relies on the secondary nine-parasitic radiating elements that rotate at a rate of 135 Hz. These secondary parasitic elements modify the cardiod radiation pattern such that there are local maxima and minima in each directional quadrant, which are spaced 40° apart; this is shown in Figure. As a result, for each single degree of bearing change, the resulting measured phase change is 9°, giving a magnifying effect in detecting changes in the bearing. It is evident that in any of the nine directions that are 40° apart, the aircraft will receive a 135-cycle signal with identical phase. This would result in location accuracy within some 40° bearing sector, but it would not know which sector. To solve this ambiguity, the coarse-bearing 15-cycle phase measurement is used. When used together, fine bearing acquisition is obtained.

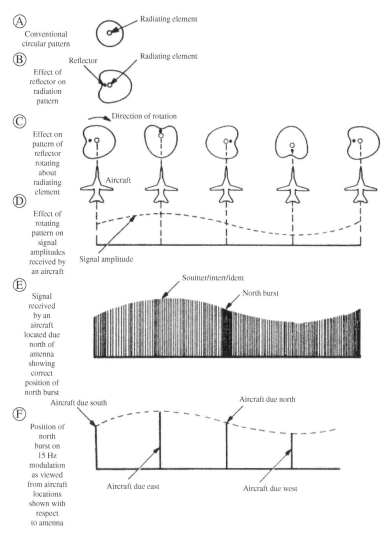

Figure 11.6 Coarse bearing acquisition, figures 8–20 from [4]. *Source*: Shipboard Electronics Materials Officer. [4]. © 1982, United States Naval Communications.

11.3 Sea-Based X-Band Radar 1 (SBX-1)

Ballistic missile detection utilizing phased array radars is one of the earliest applications of pattern reconfigurable antennas. Several ground-based phased array radars are employed today for missile detection, where the antenna beam needs to be scanned quickly. The Sea-Based X-band radar is unique in that it is

Figure 11.7 SBX-1 in Hawaii from [5]. *Source*: A brief history of the sea-based X-Band Radar-1 [5].

semiportable and the largest and most sophisticated X-band phased array radar in the world. The radar is capable of steering both electronically and mechanically to cover 360° in azimuth and 90° in elevation from horizon to zenith [5]. The array has 45 000 transmit/receive (T/R) modules and the radar can see an object the size of a baseball at a distance of 2500 miles. Figures 11.7 and 11.8 show the SBX-1 radar platform and the phased array antenna inside the radome.

11.4 The Advanced Multifunction RF Concept (AMRFC)

In the late 1990s, the Navy funded research in shared antenna apertures that could be reconfigured for various use cases including radar, communications, and electronic warfare. The AMRFC program was started in response to the growing number of antennas on the topside of U.S. Navy ships, which doubled from the 1980s to the 1990s. This is shown in Figure 11.9, which includes ship classes LPD (landing platform/dock), LCC (amphibious command ship), CVN (aircraft carrier), and DDGs (destroyer class). The proliferation of antennas on the ship's topside presents a number of problems including increased antenna blockage, electromagnetic interference (EMI), increased radar cross section (RCS), as well as maintenance issues resulting from the plurality of stove-piped systems. The goal of AMRFC was to develop one system that could simultaneously support multiple

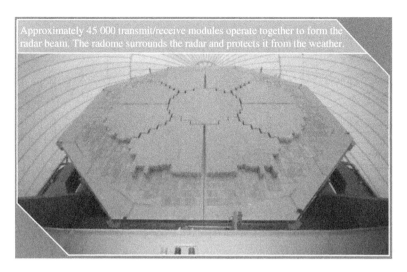

Figure 11.8   SBX-1 phased array from [5]. *Source*: A brief history of the sea-based X-Band Radar-1 [5].

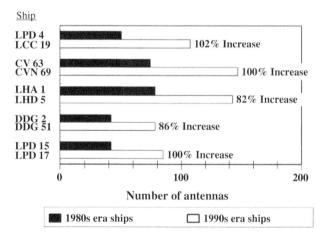

Figure 11.9   Topside antenna growth on U.S. Navy ships from [6]. *Source*: Tavik et al. [6]. © 2005, IEEE.

functions through a set of shared assets including the antenna aperture [6]. Figure 11.10 shows a conceptual diagram of the goals of the AMRFC program. This program was funded through the Office of Naval Research (ONR) and led by the Naval Research Laboratory (NRL). Three major contractors (Lockheed-Martine, Northrup Grumman, and Raytheon) provided the receive array, transmit array, and digital receiver hardware for the AMFRC test bed. In order to

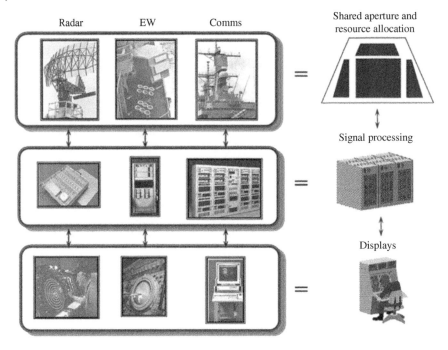

Radar   EW   Comms   Shared aperture and
                                          resource allocation

Signal processing

Displays

Figure 11.10 AMRFC concept from [6]. *Source*: Tavik et al. [6]. © 2005, IEEE.

demonstrate feasibility, the AMRFC test bed was developed at the Chesapeake
Bay Detachment Test Facility at the NRL. The AMRFC concept included separate
receive and transmit arrays for isolation needed for simultaneous operation, was
designed to operate between 6 and 18 GHz, provide 36 simultaneous receive
beams, and four simultaneous transmit beams. For communications, the AMRFC
array targeted Common Data Link (CDL) and MILSATCOM in the X-Band, and
Tactical Common Data Link (TCDL) and commercial SATCOM in the Ku-Band.
The AMFRC array demonstrated CDL links that could support up to 274 Mb/s
downlinks from an airborne platform and uplink data rates up to 10.7 Mb/s. One
of the primary functions for the AMFRC array is EW, which is divided into receive
and transmit categories of electronic surveillance (ES) for passive surveillance
and electronic attack (EA) for active countermeasures. Multiple embedded
antenna elements from the AMRFC receive array-formed vertical and horizontal
interferometers to provide precision direction finding in both azimuth and elevation. The EA function within the AMFRC was to provide countermeasures against
surveillance and targeting radars and missile seeker radars. The EA technique
that was demonstrated used coherent digital RF memory (DRFM) components as
part of the AMFRC signal generation subsystem. Finally, the AMFRC array also
demonstrated radar functionality. In low-power mode, the navigation radar has

an instrumented range to 25 nmi and scans a quadrant (±50° azimuth and 0° elevation) at a five-seconds update rate. The radar also demonstrated several radar modes including moving target indicator (MTI) and clutter cancellation.

The transmit array includes 1024 pairs of active radiating elements, which are segmented into four quadrants of 256 sites each, which each quadrant further divided into four sub-arrays. Each of the four quadrants is capable of independent simultaneous beamforming in any combiner of quarter, half, or full array. This is shown in Figure 11.11. The 1024 array radiators are dual-polarized wideband elements and mounted on a square grid with 0.413″ spacing to provide high-quality dual-polarized grating lobe-free beams. The orthogonal polarizations enable the production of all possible polarizations. This is shown in Figure 11.12.

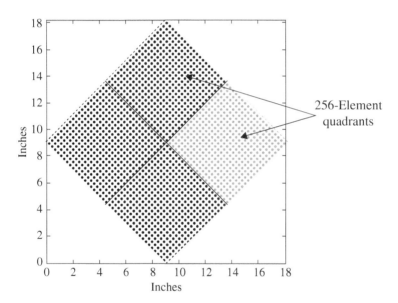

Figure 11.11 Transmit array from [6].*Source*: Tavik et al. [6]. © 2005, IEEE.

Figure 11.12 Receive array from [6].
Source: Tavik et al. [6]. © 2005, IEEE.

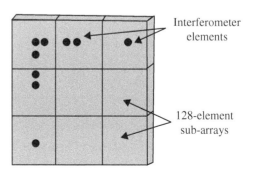

The receive array operates also from 6 to 18 GHz and consists of 1152 radiating elements. The array elements are grouped into 3 × 3 arrangements of sub-arrays each with 128 dual-polarized antenna elements. Figure 11.13 shows the three types of beamformers that are available in receive mode including analog, narrow-band digital, and wide-band digital.

The actual developed test bed is shown in Figure 11.14 at the Chesapeake Bay Detachment Facility NRL. The array is shown including all of the peripheral

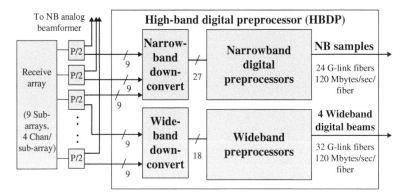

Figure 11.13 Receive beamformer from [6]. *Source*: Tavik et al. [6]. © 2005, IEEE.

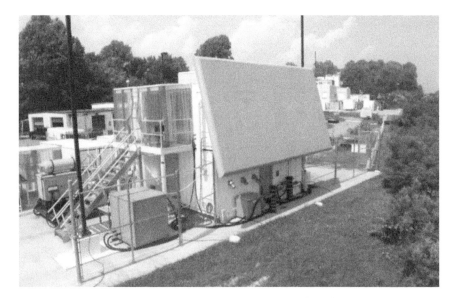

Figure 11.14 AMRFC test bed at Chesapeake Bay Detachment Facility NRL *Source*: De Long [7].

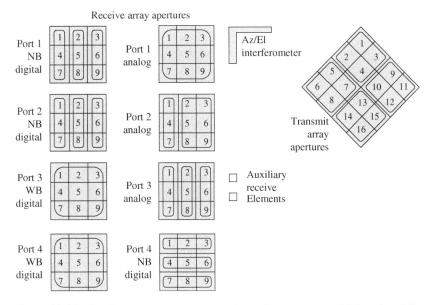

Figure 11.15 Receive and transmit array configurations to support RF functions [6]. *Source*: Tavik et al. [6]. © 2005, IEEE.

equipment necessary in stacked trailers including transmit and receive electronics, communications/radar/EW electronics, as well as power plant, dry air, and chiller unit.

Figure 11.15 (right) shows the transmit array with 16 sub-arrays, and the out-line of the four quadrants that consists of four sub-arrays each. A quadrant is the smallest aperture size that can be allocated in the transmit mode. Figure 11.15 (left) shows each eight replicas of the receive array, each with nine sub-arrays. The four rows represent the four output ports of each sub-array, which corre-spond to four channels in the receive module. The four ports are independent from each other. The two columns represent the two-way power splitter at the output of each sub-array port (from Figure 11.13). The outline indicates the smallest aperture size that can be allocated for an RF function on a given port. The type of sub-array beamforming is indicated in Figure 11.15. Figure 11.16 shows a chosen scenario that was used for live demonstration and utilizes the RF multifunction nature of this reconfigurable array. In this setting, the trans-mit array was segmented to form three beams, two used for communications (Ku-Band TCDL and X-Band MILSATCOM) and a single beam for navigational radar. On receive, the receive array can be configured for three communications beams (Ku-Band SATCOM, TCDL, and X-Band MILSATCOM) and a single radar beam.

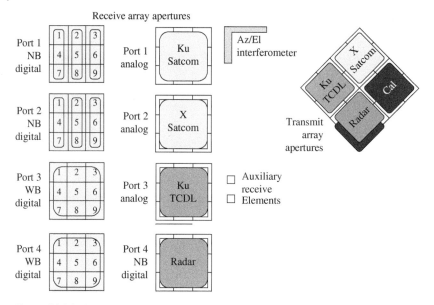

Figure 11.16 Normal scenario configuration of AMRFC [6]. *Source*: Tavik et al. [6]. © 2005, IEEE.

11.5 Integrated Topside (InTop)

While the AMFRC was successful in demonstrating the multifunction RF concept and the ability to build wideband RF systems capable of multiple simultaneous beams with simultaneous and independent functions, the cost was unaffordable for transitioning into Navy ships [7]. The size of the arrays were driven by receive requirements for SATCOM, namely, gain to temperature (G/T) which resulted in the need for large arrays spaced at $\lambda/2$ for grating-lobe free scanning, which resulted in large arrays with a large number of elements. The ability to reduce the topside antenna clutter on Navy ships through the use of a shared reconfigurable aperture for multifunction RF was still a priority. To address the challenge of low-cost wideband arrays, many research programs were initiated by the Navy and several key breakthroughs were made.

11.5.1 Wavelength Scaled Arrays

One approach that was adopted, to reduce the number of elements, was the wavelength scaled approach [8]. The core concept behind the wavelength scaled array is to create a wideband array with significantly fewer elements by utilizing antennas of various sizes that work as an aggregate to cover the entire frequency band. This approach assumes that the overall size of the array is dictated by the lowest frequency of operation and as a result there are more elements than required at the higher frequency of operation. The core operating principle is that several sized antenna elements are used to create a "nested" array of antenna elements. In their original study, flared notch radiators were used and Figure 11.17 illustrates the core concept of wavelength scaled arrays using a 2:1 scaling concept.

In the example shown in Figure 11.17, the array is required to operate from 2.5 to 20 GHz (8:1 bandwidth) and assumes a requirement for constant beamwidth. There are three different sized flared notch radiators that are used, shown in Figure 11.18. The first element operates from 2.5 to 20 GHz, the second element operates from 2.5 to 10 GHz, and the third from 2.5 to 5 GHz. For 2.5–5 GHz

Figure 11.17 Example of wavelength scaled array with three different flared notch radiators from [8]. *Source*: Kindt et al. [8]. © 2009, IEEE.

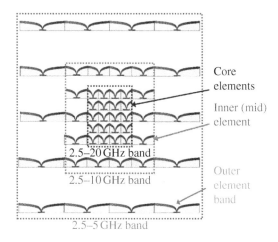

Core elements

Inner (mid) element

2.5–20 GHz band

2.5–10 GHz band

Outer element band

2.5–5 GHz band

Figure 11.18 Three different sized flared notch radiators used in figure 12.17 from [8]. *Source*: Kindt et al. [8]. © 2009, IEEE.

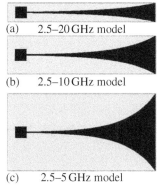

(a) 2.5–20 GHz model

(b) 2.5–10 GHz model

(c) 2.5–5 GHz model

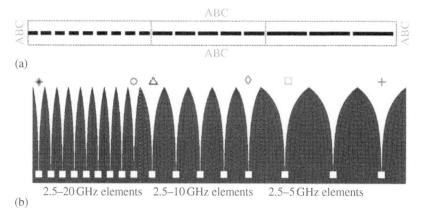

(a)

(b)

2.5–20 GHz elements   2.5–10 GHz elements   2.5–5 GHz elements

Figure 11.19 Simulations taking into account transitions between element types from [8]. *Source*: Kindt et al. [8]. © 2009, IEEE.

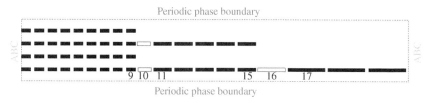

Periodic phase boundary

9 10 11   15 16 17

Periodic phase boundary

Figure 11.20 Wavelength scaled array configuration for examining end effects from [8]. *Source*: Kindt et al. [8]. © 2009, IEEE.

operation, all the elements are active, from 5 to 10 GHz, only the first and second elements are active, and for 10–20 GHz operation only the first type is active. The beam will have similar beamwidths across frequency because the size of the aperture changes. Discontinuity of the elements is a challenge as mutual coupling between elements plays a large role. Figure 11.19 shows FEM simulations that were run to analyze the effects of transitions between element types across the array. In order to understand the impact on phased array applications, the wavelength scaled array was analyzed under various steering conditions while observing the active VSWR. Figure 11.20 shows a boundary box of the finite array that was analyzed and Figure 11.21 shows the simulated active VSWR from 0 to 60° in the E-plane from 2 to 20 GHz. As can be seen, the active VSWR is below 2 for the entire band out to 40°. A prototype of this kind of wavelength scaled array was eventually fabricated and is shown in Figure 11.22. This is one of the breakthroughs to enable cost-effective phased arrays, one of the key challenges to the AMFRC program.

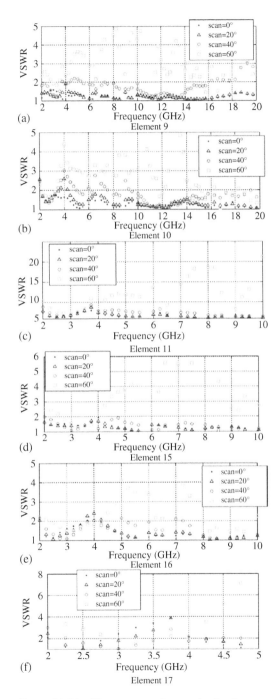

(a)
Element 9

(b)
Element 10

(c)
Element 11

(d)
Element 15

(e)
Element 16

(f)
Element 17

Figure 11.21 Element scanning results for the array configuration in Figure 11.20 from [8]. *Source*: Kindt et al. [8]. © 2009, IEEE.

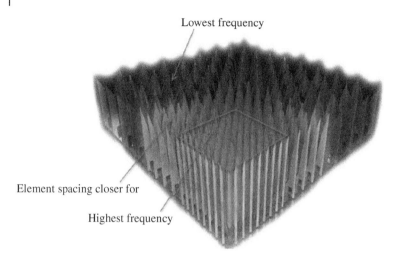

Lowest frequency

Element spacing closer for

Highest frequency

Figure 11.22 Designed prototype from [7]. *Source*: De Long [7].

11.5.2 Low-Cost Multichannel Microwave Frequency Phased Array Chipsets on Si and SiGe

The second breakthrough was reducing the cost of RF electronics and their associated complexity through higher levels of integration. Traditional T/R modules are designed as microwave integrated circuits (MICs) where several GaAs chips (PA, LNA, Switch, and Phase Shifter) are integrated on a ceramic substrate. Developing key T/R chips on GaAs is costly and limits the level of integration since multichannel phased array chipsets are not feasible in compound semiconductor technology (III-V). The key developmental challenge that was overcome was the ability to develop multichannel phased array chipsets in commercial silicon (Si) and silicon germanium (SiGe) technology. The ability to utilize Si and SiGe meant the ability to integrate many T/R channels on a single small RFIC, eliminating the need for multiple III-V chips and its associated integration/assembly. Furthermore, with Si and SiGe, chips can be mounted via flip-chip, removing the need for wire-bonding assembly. Figure 11.23 shows a T/R module developed out of the INTOP program utilizing SiGe and GaAs components side by side [7]. The T/R module utilizes a SiGe switch matrix and beamformer, while integrating gallium nitride (GaN) low-noise amplifiers (LNAs) and bulk acoustic wave (BAW) filters. This T/R chipset supports four simultaneous beams from X- through Ka-Band. Compared to the T/R modules used in the AMRFC program, the fully integrated SiGe solution consumes half the power and a third of the weight. Furthermore, since the SiGe design is utilizing commercial foundries, it is also more cost effective.

A single very influential breakthrough was the ability to design compact wideband phase shifters in Si and SiGe. Various phase shifter architectures had been

Figure 11.23 T/R module using SiGe and GaAs from [7]. *Source*: De Long [7].

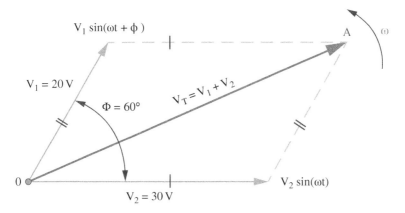

Figure 11.24 Phasor concept.

proposed for MMICs in the past; with the most prolific being switched line phase shifters and high-pass/low-pass phase shifters. The drawback to these common architectures is that it does not lend itself for compact designs which are necessary for high-density multichannel integration for phased array chipsets. These types of phase shifters require multistage switching of either delay lines or lumped element filters, which easily consumes chip space. The introduction of the vector modulator (phase interpolator) phase shifter solves many of these issues, paving the way to cost-effective phased array RF electronics. The principle of operation goes back to the phasor, shown in Figure 11.24.

Two signals can be defined as:

Signal 1 $\quad A_1 \cos(\omega t + \theta_1)$,

Signal 2 $\quad A_2 \cos(\omega t + \theta_2)$.

The sum of the two signals is defined as:

$$A_1 \cos(\omega t + \theta_1) + A_2 \cos(\omega t + \theta_2) = A_3 \cos(\omega t + \theta_3),$$

where

$$A_3^2 = (A_1 \cos\theta_1 + A_2 \cos\theta_2)^2 + (A_1 \sin\theta_1 + A_2 \sin\theta_2)^2,$$

$$\theta_3 = \tan^{-1}\left(\frac{A_1 \sin\theta_1 + A_2 \sin\theta_2}{A_1 \cos\theta_1 + A_2 \cos\theta_2}\right).$$

As can be seen, A_1 and A_2 are the magnitudes of the two vectors, and θ_1 and θ_2 are the corresponding phases of the two vectors. When the two vectors are combined, you form a third vector that has a magnitude and phase defined by the equation above. Figure 11.25 shows a top-level schematic of the first vector modulator-type phase shifter realized at microwave frequencies on SiGe [9]. The first step is to go from differential phase to quadrature phase, so that 360° phase shift is possible. The quadrature all-pass filter (QAF) does this operation, and then goes into a sign selection and a variable gain amplifier. The variable gain amplifier essentially gives the ability to change the magnitude of each of the four vectors. Those vectors are combined using a current summation circuit, and thus allows for the ability to both shift the phase of the signal, as well as the amplitude. The ability to have a built-in gain control allows for further phased array capability such as the realization of amplitude tapering for sidelobe suppression. A digital-to-analog controller (DAC) controls the step size of the phase shifter, and a logic encoder is

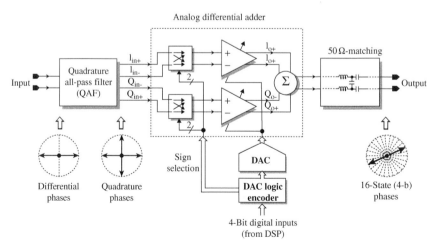

Figure 11.25 Vector modulator phase shifter from [9]. *Source*: Koh and Rebeiz [9].

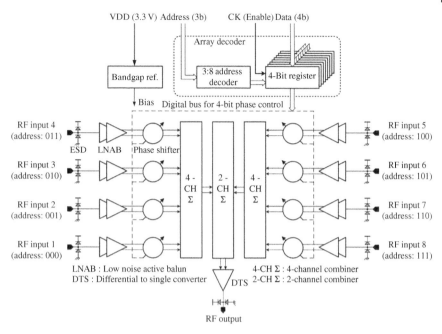

Figure 11.26 8-Element X/Ku-Band phased array chipset from [10]. *Source*: Koh and Rebeiz [10]. © 2008, IEEE.

built in such a way that a serial protocol interface (SPI) can be used to control the phase shifter. This type of phase shifter has paved the way for low-cost high-fidelity RF phase shifters, which coincidentally is also critical for 5G adoption. Figure 11.26 shows the RFIC floor plan for the eight-channel X/Ku-Band receive only phased array chipset. Figure 11.27 shows a micrograph of that chipset.

Figure 11.28 shows the measured performance of the phase shifter from [9] and as can be seen, the phase error is better than 4.5° from 6 to 16 GHz. From the S-parameter measurements, synthesized array factor radiation patterns were created and shown in Figure 11.29 and as can be seen, the patterns are close to ideal. This first reported phased array chipset designed in SiGe to operate in the microwave frequency regime demonstrated for the first time that a low-cost high-performance fully integrated multichannel phased array chipset could be realized. Finally, full T/R multichannel phased array chipsets have also been demonstrated in [10]. Figure 11.30 shows a high-level schematic of the T/R chain for Ka-band. Figure 11.31 shows the RFIC micrograph. Finally, Table 11.1 shows the performance of the designed chip.

These two breakthroughs and others in antenna array topology and RF components led to a refresh of the AMRFC program goals, which became the ONR

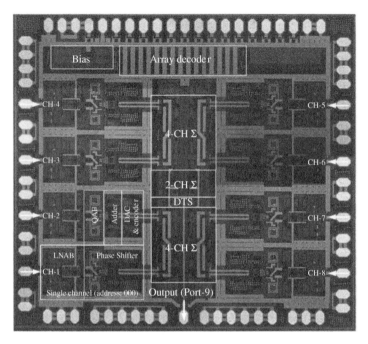

Figure 11.27 8-Element X/Ku-Band phased array micrograph. *Source*: Koh and Rebeiz [9].

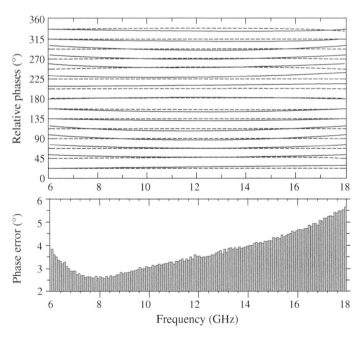

Figure 11.28 Vector modulator phase shifter performance from [10]. *Source*: Koh and M. Rebeiz [10].

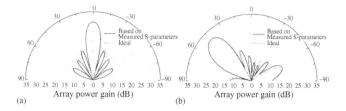

(a)   (b)

Figure 11.29 Simulated radiation patterns from S-parameter results from [10]. *Source*: Koh and Rebeiz [10]. © 2008, IEEE.

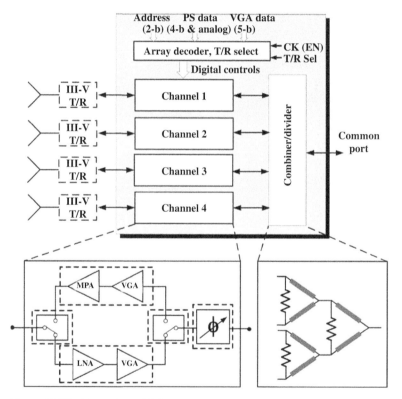

Figure 11.30 Four-channel T/R phased array chipset on Si from [11]. *Source*: Kang et al. [11] (C) 2009, IEEE.

Innovative Naval Prototype (INP) program Integrated Topside (InTop). The vision for InTop was to provide Navy platforms with adaptable RF capabilities at reduced cost by developing integrated sensors and communication solutions which are open, modular, and scalable. InTop covers radar, EW, communications, and information operations (I/O) functions utilizing shared RF apertures (Figure 11.32). In

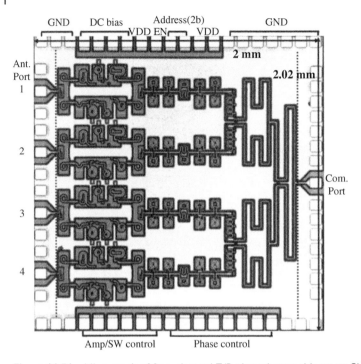

GND DC bias Address(2b) GND
VDD EN VDD

Ant. Port 1

2 mm
2.02 mm

2

Com. Port

3

4

Amp/SW control Phase control

Figure 11.31 Micrograph of four-channel T/R phased array chipset on Si from [11]. *Source*: Kang et al. [11] (C) 2009, IEEE.

2010, the EW/IO/Comms prototype design was started with goals to enable multiple simultaneous communication links, while supporting multiple simultaneous EW engagements. This program was eventually transitioned into a real Navy system, Surface Electronic Warfare Improvement Program (SEWIP) Block 3 [7].

11.6 DARPA Arrays of Commercial Timescales (ACT)

In around 2012–2013, DARPA initiated the ACT program, which aimed to move beyond the traditional antenna array design process, and to focus on new ways of developing electromagnetic arrays. This program aimed to solve this fundamental problem by developing common hardware module that can be applied to a wide range of array applications. The program also sought to develop reconfigurable electromagnetic interface for various polarizations, frequencies, and bandwidths [12]. A conceptual view of the aims of the program is shown in Figure 11.33.

One of the key thrusts in solving the issue of developing common modules that fit many applications was moving to digital beamforming. As shown in Figure 11.34

Table 11.1 Performance summary from [11].

	RX	TX
Performance Summary at 34–39 GHZ		
Gain @36 GHz (single)	–1 dB	2.0 dB
Gain* @36 GHz (array)	–2.5 dB	0 dB
Phase error (RMS)	< 5.6° @ 36-38 GHz	< 4.9° @ 36-38 GHz
	< 12° @ 34-39 GHz	< 12° @ 34-39 GHz
Gain error (RMS)	< 0.9 dB (w/o VGA)	< 0.7 dB (w/o VGA)
	< 0.5 dB (w VGA)	< 0.5 dB (w VGA)
Common port return loss	> 16 dB	> 16 dB
	> 14 dB (single)	> 14 dB (single)
Antenna port return loss	> 20 dB	> 8 dB
	> 20 dB (single)	> 8.6 dB (single)
Power consumption	142 mW /35.5 mW (single)	171 mW /43 mW (single)
OP1dB @ 36 GHz	–18 dBm (single)	4.7 dBm (single)
IP1dB @ 36 GHz	–16 dBm (single)	3.7 dBm (single)
IIP3 @ 36 GHz	–5.9 dBm (sinqle)	13 dBm (single)
Common port P1dB	–13.5 dBm (array)	11.3 dBm (array)
Gain control	7 dB (5-bit)	
Phase control	5-bit	
Noise figure	9 dB (single)	
Supply voltage	1.8 V (analog), 1.5 V (digital)	
Area	2.00 x 2.02 mm^2 / 1.7 x 0.75 mm^2(single)	
Technology	0.13-μm SiGe BiCMOS	

Source: Kang et al. [11] (C) 2009, IEEE.

from [13], analog beamforming phased arrays are frequency and bandwidth specific, and so they are not reusable across many systems. The cost to develop the antenna array and beamforming electronics leads to a high SWaP-C. However, moving to a digital beamforming approach, the beamforming hardware can be designed over wide frequencies and bandwidths, allowing for more flexible use and also allowing for reconfigurability. Going to an ACT-enabled system, the projected non-recurring engineering (NRE) is also reduced greatly since a common module can be used across many platforms (Figure 11.35). Much of the driving force behind the ability to move to digital beamforming is the evolution of data converter technology. Figure 11.36 shows the famous Walden curve showing the figure of merit for analog-to-digital converters (ADCs), and as monolithic

Figure 11.32   EW/IO/Comms aperture for InTop. *Source*: De Long [7].

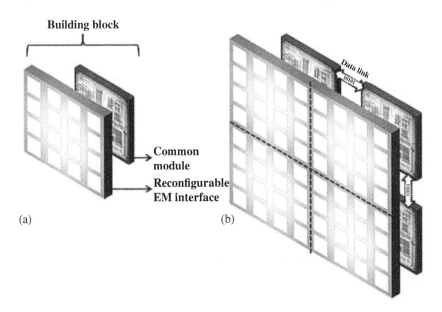

(a)   (b)

Figure 11.33   DARPA ACT module from [12]. (a) shows a common sub-array module (b) shows a full array using the sub-array modules. *Source*: ARLINGTON. "DARPA seeks to speed RF and microwave array development for radar, comms, and SIGINT", Military & Aerospace Electronics, 2013. https://www.militaryaerospace.com/trusted-computing/article/16715991/darpa-seeks-to-speed-rf-and-microwave-array-development-for-radar-comms-and-sigint.

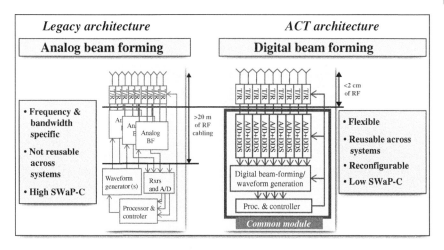

Figure 11.34 ACT digital beamforming [13]. *Source*: Epstein et al. [13]. © 2018, IEEE.

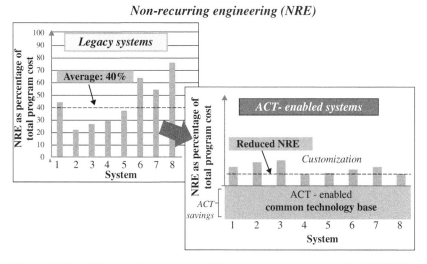

Figure 11.35 ACT digital beamforming NRE versus analog beamforming NRE [13]. *Source*: Epstein et al. [13]. © 2018, IEEE.

technologies continue to improve, the data rate and spurious free dynamic range (SFDR) of the ADCs increase. At the >1 Gbps data rate and >75 dB SFDR, extremely wideband digital beamforming is feasible [13]. Figure 11.37 shows a notional ACT common module developed by Rockwell Collins [13], and shows a hybrid analog and digital beamforming approach. The ACT program developed a number of key technologies and DARPA has initiated the Millimeter-Wave Digital Arrays (MIDAS) program which aims to further the frequency reach of digital beamformers to the

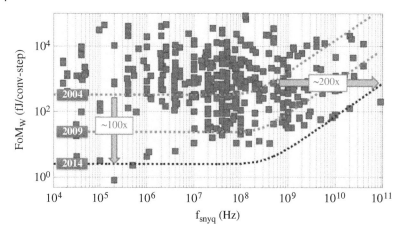

Figure 11.36 Walden curve from [13]. *Source*: Epstein et al. [13]. © 2018, IEEE.

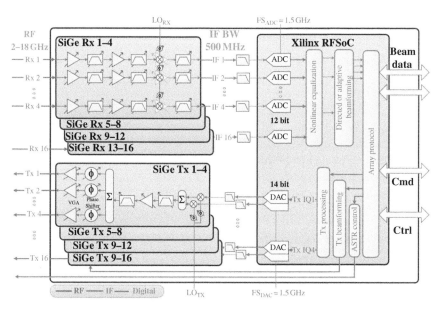

Figure 11.37 ACT common module developed by Rockwell Collins from [13]. *Source*: Epstein et al. [13]. © 2018, IEEE.

millimeter-wave regime. Figure 11.38 from [14] shows a conceptual rendition of a millimeter-wave digital beamformer developed by Jariet Technologies and Northrup Grumman Mission Systems. The digital beamformer is developed on CMOS 12LP with an ADC/DAC sampling rate of 12 Gbps [14]. These array antennas are examples of application and mission reconfigurable antennas.

TA1	TA2
Jariet Technologies	**Northrop Grumman mission systems**

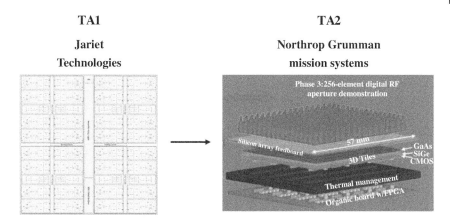

Figure 11.38 DARPA MIDAS concept from [14]. *Source*: Dr. Timothy Hancock, "Millimeter Wave Digital Arrays (MIDAS)", DARPA. https://mmwrcn.ece.wisc.edu/wp-content/uploads/sites/678/2019/02/Keynote_2_Hancock_012919.pdf

11.7 AFRL Transformational Element Level Array (TELA)

Around 2010, the Air Force Research Laboratory initiated research into extremely wideband antenna arrays. The approach was to develop monolithic true-time-delay chipsets which operated from 1 to 8 GHz. This chipset is shown in Figure 11.39 and has 8-bits of time delay control with 764 ps of total delay and 4-ps of LSB. The chipset has a 6-bit attenuator and 20-dB of total gain with a noise figure (NF) of 5 dB. The size of the plastic quad-flat-no (QFN) leads package is 13 mm × 9 mm. Figure 11.40 shows the measured response of the chipset from [15].

Figure 11.39 M/A-COM 1–8 GHz true time delay chipset from [15]. *Source*: http://d2xunoxnk3vwmv.cloudfront.net/uploads/03-Thomas-Dalrymple-2-.pdf. https://www.microwaves101.com/encyclopedias/time-delay-unit-tdu

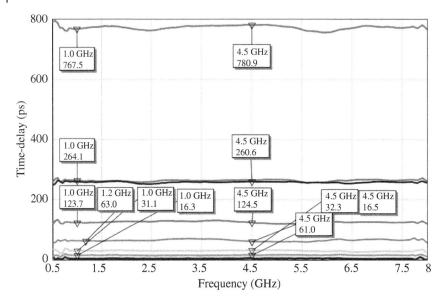

Figure 11.40 Measured response of the M/A-COM 1–8 GHz true time delay chipset from [15]. *Source*: Thomas W. Dalrymple. TELA Wideband Array Demonstration. 6 April 2010. http://d2xunoxnk3vwmv.cloudfront.net/uploads/03-Thomas-Dalrymple-2-.pdf. https://www.microwaves101.com/encyclopedias/time-delay-unit-tdu. © 2010, Air Force Research Laboratory

   The chipset was demonstrated with an antenna array. A 256-element wideband array was utilized, but only the center eight-columns were interfaced with the TTD chips (Figure 11.41). Figure 11.42 shows the measured radiation pattern of the eight-element linear array with the TTDs. Three patterns are shown for comparison. The first is the array steered to 45° using the TTDs, the second is the ideal pattern from MATLAB, and the third is when the array is steered to 45° using cables. As can be seen, there is no beam squint over the 2 GHz instantaneous bandwidth, as expected from utilizing TTDs. This is an example of an application and mission reconfigurable antenna.

11.8 Conclusion

This chapter presented various military reconfigurable antennas and related beamforming technologies utilizing analog, digital, and hybrid approaches. It also discusses various programs supporting these array antennas. The antennas and systems described demonstrate pattern reconfigurability as well as mission reconfigurability. Future defense antenna systems will leverage reconfigurability to support multi-functionality.

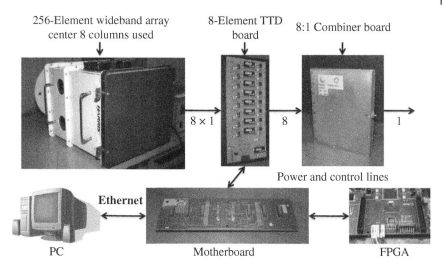

256-Element wideband array center 8 columns used

8-Element TTD board

8:1 Combiner board

8×1

8

1

Power and control lines

Ethernet

PC

Motherboard

FPGA

Figure 11.41 TTD array from [15]. *Source*: http://d2xunoxnk3vwmv.cloudfront.net/uploads/03-Thomas-Dalrymple-2-.pdf

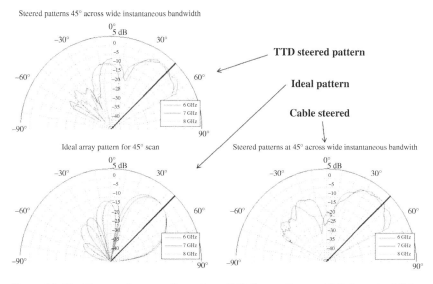

Figure 11.42 Measured array patterns from [15]. *Source*: Thomas W. Dalrymple. TELA Wideband Array Demonstration. 6 April 2010. http://d2xunoxnk3vwmv.cloudfront.net/uploads/03-Thomas-Dalrymple-2-.pdf. © 2010, Air Force Research Laboratory.

References

1 Colin, R. and Dodington, S. (1956). *Principles of TACAN*, vol. 33, 11–25. March: Electrical Communication.

2 Shestag, L. (January 1974). A cylindrical array for the TACAN system. *IEEE Transactions on Antennas and Propagation* 22 (1): 17–25.

3 Christopher, E. (January 1974). Electronically scanned TACAN antenna. *IEEE Transactions on Antennas and Propagation* 22 (1): 12–16.

4 Shipboard Electronics Materials Officer, *NAVEDTRA, 10478-A*, 1982.

5 A brief history of the sea-based X-Band Radar-1, Missile Defense Agency History Office, 2008.

6 Tavik, G.C., Hilterbrick, C.L., Evins, J.B. et al. (March 2005). The advanced multifunction RF concept. *IEEE Transactions on Microwave Theory and Techniques* 53 (3): 1009–1020.

7 De Long, B. (2017). An integrated approach to topside design. *2017 IEEE MTT-S International Microwave Symposium (IMS)*, Honolulu, HI, pp. 326–329.

8 Kindt, R.W., Kragalott, M., Parent, M.G., and Tavik, G.C. (December 2009). Preliminary investigations of a low-cost ultrawideband array concept. *IEEE Transactions on Antennas and Propagation* 57 (12): 3791–3799.

9 Koh, K. and Rebeiz, G.M. (November 2007). 0.13 μm CMOS phase shifters for X-, Ku-, and K-band phased arrays. *IEEE Journal of Solid-State Circuits* 42 (11): 2535–2546.

10 Koh, K. and Rebeiz, G.M. (June 2008). An X- and Ku-band 8-element phased-array receiver in 0.18 μm SiGe BiCMOS technology. *IEEE Journal of Solid-State Circuits* 43 (6): 1360–1371.

11 Kang, D., Kim, J., Min, B., and Rebeiz, G.M. (December 2009). Single and four-element Ka-band transmit/receive phased-array silicon RFICs with 5-bit amplitude and phase control. *IEEE Transactions on Microwave Theory and Techniques* 57 (12): 3534–3543.

12 https://www.militaryaerospace.com/trusted-computing/article/16715991/darpa-seeks-to-speed-rf-and-microwave-array-development-for-radar-comms-and-sigint (accessed May 2020)

13 Epstein, B., Olsson, R.H., and Bunch, K. (2018). Arrays at commercial timescales: addressing development and upgrade costs of phased arrays. *2018 IEEE Radar Conference (RadarConf18)*, Oklahoma City, OK, pp. 0327–0332.

14 https://mmwrcn.ece.wisc.edu/wp-content/uploads/sites/678/2019/02/Keynote_2_Hancock_012919.pdf (accessed July 2020).

15 http://d2xunoxnk3vwmv.cloudfront.net/uploads/03-Thomas-Dalrymple-2-.pdf (accessed July 2020).

12

5G Silicon RFICs-Based Phased Array Antennas

Jia-Chi S. Chieh and Satish K. Sharma

12.1   Introduction

Phased array antennas are a subset of reconfigurable antennas in that the pattern can be reconfigured and scanned. In the last five years, significant progress has been made in this area, especially since the advent of silicon beamformer chipsets for 5G applications.

12.2   Silicon Beamformer Technology

Phased array antennas have traditionally been very expensive and therefore adoption has been limited to the military applications. Several efforts were made to try to develop cost-effective methods to produce phased array antennas. The bottleneck for cost-effective solutions has always been the phase shifter, which allows the antenna to be scanned electronically. Historically, the phase shifter has always been implemented on compound semiconductors such as gallium arsenide (GaAs) or indium phosphide (InP). In the late 1990s, much effort was made to develop passive electronically scanned arrays (PESAs) through the use of radio frequency micro-electro-mechanical-systems (RF MEMS) technology. MEMS promised low-loss high-performance phase shifters. In [1], researchers developed a 4-bit RF MEMS phase shifter using single-pole four-throw (SP4T) MEMS switches. The developed MEMS phase shifter exhibited 1.2 dB of insertion loss and operated from DC to 10 GHz for a very broadband design. Figure 12.1 shows the developed 4-bit phase shifter from [1] and Figure 12.2 shows the measured

Multifunctional Antennas and Arrays for Wireless Communication Systems, First Edition.
Edited by Satish K. Sharma and Jia-Chi S. Chieh.
© 2021 John Wiley & Sons, Inc. Published 2021 by John Wiley & Sons, Inc.

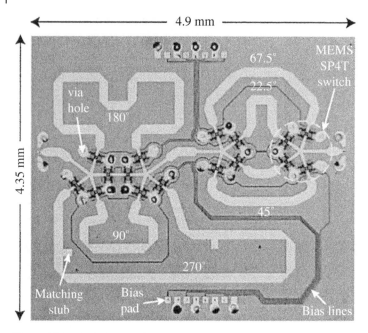

Figure 12.1 4-Bit RF MEMS-based phase shifter from [1]. *Source*: Tan et al. [1]. © 2003, IEEE.

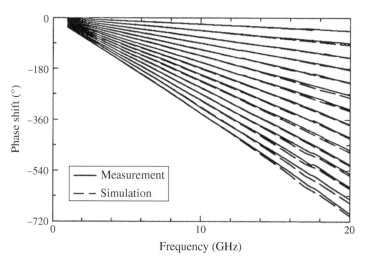

Figure 12.2 Measured and simulated phase response for all phase state from [1]. *Source*: Tan et al. [1]. © 2003, IEEE.

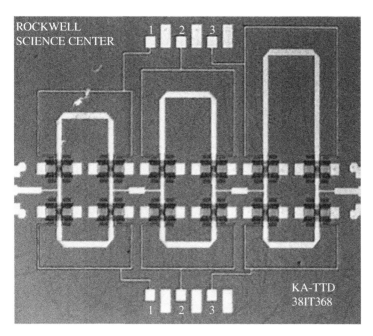

Figure 12.3 Fabricated 3-bit RF MEMS phase shifter from [2]. *Source*: Hacker et al. [3].

and simulated phase response over frequency for all of the phase states. In [2], researchers produced a 3-bit Ka-Band phase shifter based on single-pole double-throw (SPDT) RF MEMS switches. Figure 12.3 shows the fabricated prototype and Figure 12.4 shows the measured response. Although RF MEMS showed great promise for PESAs, ultimately RF MEMS technology was not compatible with silicon complementary-metal–oxide–semiconductor (CMOS) processes, and therefore could not produce cost-effective economies of scale. In around 2012, there was a resurgence of interest in PESAs utilizing holographic antenna technologies (Figure 12.5). Companies such as Kymeta emerged touting cost-effective PESAs [3].

In the early 2000s, researchers began thinking about how to realize phase shifters in silicon. Three architectures emerged. Local oscillator (LO)-based phase shifting from [4], RF-based phase shifting from [5], and intermediate frequency (IF)-based phase shifting from [6].

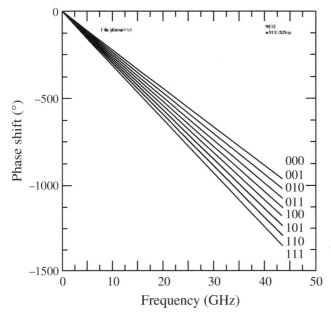

Figure 12.4 Measured phase response from [2]. *Source*: Hacker et al. [3]. © 2003, IEEE.

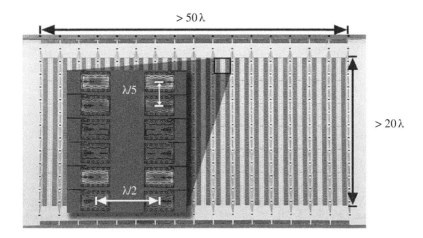

Figure 12.5 Kymeta holographic antenna from [3]. *Source*: Johnson et al. [2].

12.3 LO-Based Phase Shifting

This approach leveraged the fact that phase shift could be achieved in the RF mixer by adjusting the phase of the LO. The output phase of the mixer is a linear combination of its input phase, defined by the following equation:

$$Vout(t) = \cos(\omega_{RF}t + \phi_{RF}) \times \cos(\omega_{LO}t + \phi_{LO})$$

$$= \frac{1}{2}\cos\left[(\cos\omega_{RF} \pm \omega_{LO})t + (\phi_{RF} \pm \phi_{LO})\right].$$

Figure 12.6 shows a schematic view of this approach. It requires the LO signal to be the same frequency, but with phase offsets equal to the progressive phase shift required in a phased array. The challenge with this design is that the LO distribution and generation network is complex. This is illustrated in Figure 12.7. Researchers were successful in implementing this design in silicon germanium and an eight-element chipset was fabricated and proven. Although this approach was elegant, it suffers from linearity issues. The complex LO distribution also ultimately prevented this design from being truly scalable to create large phased array antennas. This architecture also requires the Up/Down conversion mixer to be

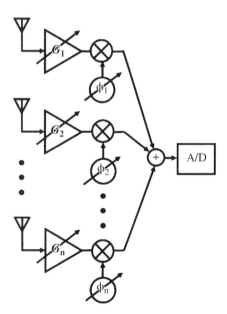

Figure 12.6 LO-based phase shift approach from [4]. *Source*: Hashemi et al. [4]. © 2005, IEEE.

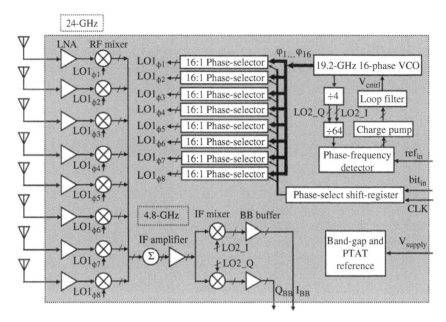

Figure 12.7 LO-based phase shift approach from [4] showing the LO generation and distribution. *Source*: Hashemi et al. [4]. © 2005, IEEE.

fully integrated onto the beamformer integrated circuit (IC), making this architecture limited in terms of scalability for large phased arrays (Figure 12.8).

12.4 IF-Based Phase Shifting

In the IF-based phase shifting approach, the complexity of designing high-frequency RF and microwave phase shifters are removed. The phase shift happens after the Up/Down conversion mixer at an IF, which is much lower. This moves the complexity of the phase shifter design from high frequency to low frequency, a much manageable trade-off. This architecture is shown in Figure 12.9 for an eight-element receive (Rx) and eight-element transmit (Tx) phased array chipset. The phase shifter has 5-bits of resolution. Figure 12.10 shows the fabricated prototype chipset. As can be seen, the major drawback of this approach is that the Up/Down conversion mixers and LO generators must all reside on the phased array chipset. This increases the complexity of what needs to be realized in silicon and tends to be a less scalable design. For example, this approach would not be practical for 1024-element phased arrays.

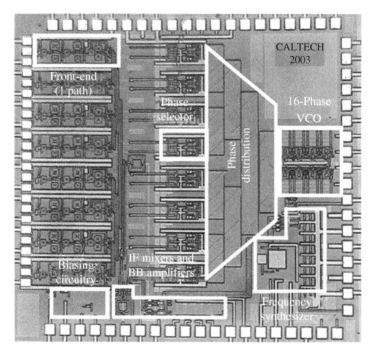

Figure 12.8 Die micrograph of an eight-element LO-based phase shifting phased array chipset at 24 GHz. *Source*: Hashemi et al. [4].

12.5 RF-Based Phase Shifting

Around the same time in the early 2000s, researchers were also starting to develop RF-based phase shifters in silicon (Si) and silicon germanium (SiGe). This architecture is preferred for large-scale arrays as the Up/Down conversion mixers do not need to be fully integrated onto the silicon beamformer, allowing for truly scalable phased arrays. This type of architecture also enhances the linearity of the phased array. In the LO- and IF-based architectures, an Up/Down conversion mixer is integrated into each beamforming channel. The antenna on each beamforming channel tends to be significantly less directional, allowing interferers to hit the RF front end and mixer. In the RF-based phase shifting architecture, the mixer is placed after the beamformer. In this way, the directionality of the antenna array acts like a spatial filter to reject interferers from unwanted directions, enhancing the linearity of the RF front end. Figure 12.11 shows block diagram model of how the RF-based phase shifting can enhance linearity from [7].

In this topology, the main enabler is how to realize highly integrated RF phase shifters in silicon, such that multichannel beamformer's can be achieved. Also,

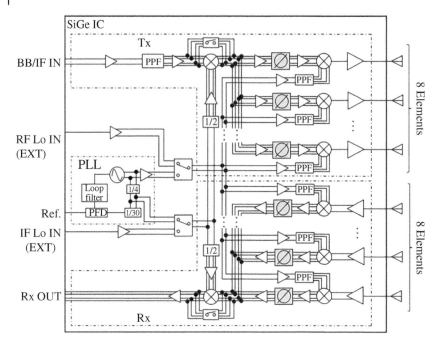

Figure 12.9 IF-based phase shift approach from [5]. *Source*: Oshima et al. [5].© 2015, IEEE.

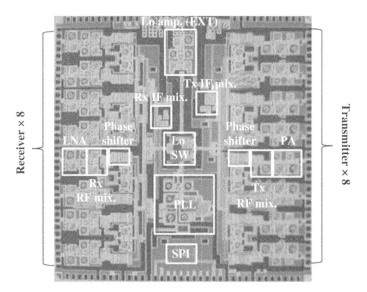

Figure 12.10 Micrograph of RFIC from [5] using IF based phase shifters. *Source*: Oshima et al. [5].

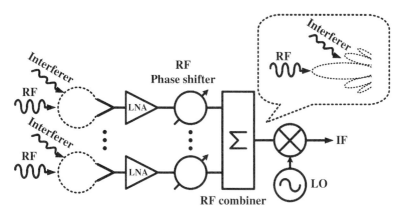

Figure 12.11 RF-based phase shifting from [7]. *Source*: Koh and Rebeiz [7]. © 2008, IEEE.

since the beamformer does not require an Up/Down conversion mixer and LO distribution network per beamformer chip, this architecture is much more scalable for large phased arrays. The first paper to introduce a highly integrated silicon phase shifter was from [6]. Up to this point, solid-state phase shifters were mostly implemented as switched delay line phase shifters, switched filter phase shifters, or high-pass/low-pass phase shifters. These types of phase shifter architectures utilized many passive lumped elements or transmission lines to realize, and as such their designs were large. As an example, in [8], a phase shifter is realized on GaAs (Figure 12.12). The total size of the phase shifter is 2.11 mm × 1.47 mm and operates from 15 to 19 GHz, at 0.10λ in the largest dimension. This type of phase shifter monolithic microwave integrated circuit (MMIC) is ideal for transmit/receive (T/R) multi-chip modules (MCMs) where each block in the T/R module is a discrete MMIC. This type of approach precludes the ability to realize a highly integrated beamformer chipset. This is a primary reason why beamformer core chipsets are not widely available commercially from compound semiconductor vendors.

In [6], researchers used a novel architecture to realize a highly integrated phase shifter in silicon. This type of phase shifter was named the vector modulator phase shifter. Figure 12.13 shows a schematic block diagram of how this topology works. Essentially, this block can be explained using the phasor diagram shown in Figure 12.14. If you have two vectors which are orthogonal to each other, a real vector and an imaginary vector, the two can be combined to form a vector of an arbitrary angle. The phase angle that is formed is determined by the magnitude of the two orthogonal vectors, defined as Eq. (12.1). As can be seen in Figure 12.13, the core of the phase shifter is composed of two variable gain amplifiers (VGAs) controlled by a digital-to-analog converter (DAC). A quadrature all-pass filter takes the differential signal and creates a quadrature

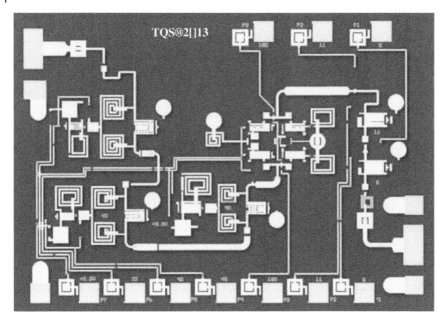

Figure 12.12 6-Bit phase shifter MMIC in GaAs from [8]. *Source*: Datasheet for TGP2615 [8].

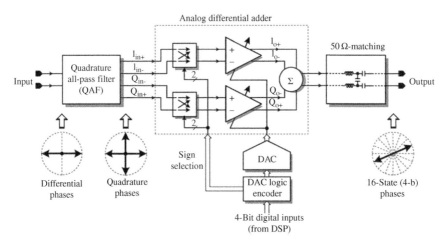

Figure 12.13 Silicon phase shifter utilizing vector modulator approach from [6]. *Source*: Koh and Rebeiz [6].

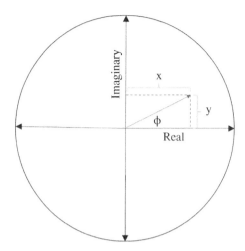

Figure 12.14 Unit circle phasor description.

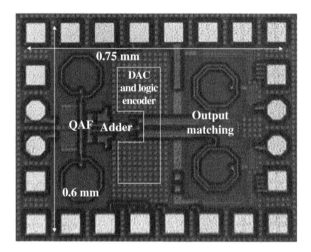

Figure 12.15 Vector modulator phase shifter in K-band from [6]. *Source*: Koh and Rebeiz [6].

pair. In order to traverse the entire unit circle, the quadrature signals are fully differential.

$$\phi = \tan^{-1}\left(\frac{x}{y}\right). \tag{12.1}$$

Figure 12.15 shows the realized 4-bit vector modulator phase shifter in silicon germanium design for the K-band from 10 to 16 GHz. The total size of the test chip is 0.75 mm × 0.6 mm, a significant size reduction when compared to [8].

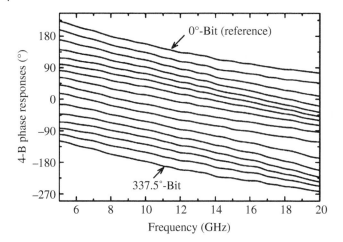

Figure 12.16 Vector modulator 4-bit phase shifter measured phase response in K-band from [6]. *Source*: Koh and Rebeiz [6].

This size reduction is what makes it possible to realize a multichannel beam-former chipset in silicon. Figure 12.16 shows the measured phase response of this chip. Using this building block, an eight-channel K-band phased array beamformer chipset was developed and realized in SiGe [7]. Figure 12.17 shows

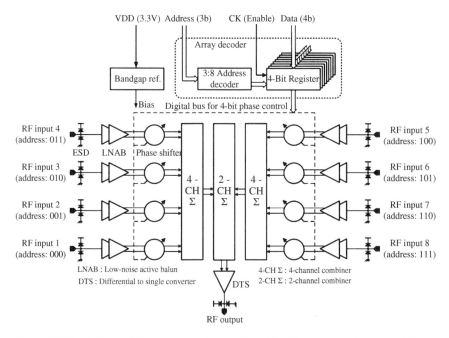

Figure 12.17 8-Channel K-band phased array chipset in silicon germanium from [7]. *Source*: Koh and Rebeiz [7]. © 2008, IEEE.

a schematic block diagram of the chipset and Figure 12.18 shows a micrograph of the chipset. Both [6, 7] marked a revolution in the development of phase arrays. No longer were MCMs necessary, where discrete MMICs representing each of the blocks within a T/R module were combined. Instead, each of these blocks is now realized in a single highly integrated silicon chip. The chip was developed on 0.18 μm SiGe BiCMOS (1P6M) and operates from 6 to 18 GHz. The chip is a receive-only beamformer with a noise figure (NF) of 4.2 dB and 8-channels. Each channel has a 4-bit phase shifter with a root mean square (RMS) phase mismatch of <2.7°. The total size of the chip is 2.2 mm × 2.45 mm. This represents an order of magnitude reduction in size but at the same time a magnitude order of improvement in performance. In addition, for the first time ever, phased arrays could finally reap the benefits of economies of scale that is possible because of silicon IC technology. The ability to miniaturize a multichannel beamformer allows the chip to be placed on the backside of an antenna array even with the constraints of the lattice spacing of the array. This allows for a truly low-profile high-performance design.

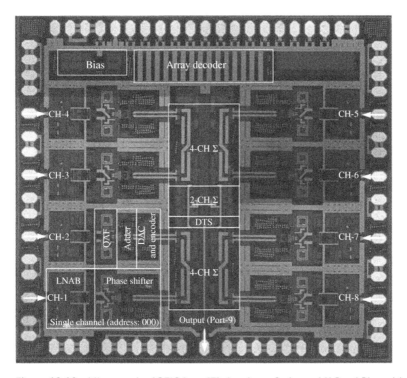

Figure 12.18 Micrograph of RFIC from [7] showing a 8-channel K-Band Phased Array Chipset in Silicon Germanium. *Source*: Koh and Rebeiz [7].

12.6 Ku-Band Phased Arrays Utilizing Silicon Beamforming Chipsets

In this section, we present a Ku-band phased array utilizing commercially available single-element T/R silicon chipset. We developed a small 8×8 array that can scan from $\pm 30°$ while maintaining a cross-polarization level below 15 dB. The frequency of operation is between 11 and 13 GHz. This work was presented in [9].

The fabricated prototype of the phased array is shown in Figure 12.19. The total size of the printed circuit board (PCB), including auxiliary connectors, is 4.2 in. × 5.5 in. The commercial beamformer chipset used was the Anokiwave AWMF-0117. This chipset is a single-channel T/R chip that can support dual polarizations and operates between 10.5 and 16 GHz. On receive, it has a NF of 3.5 dB, with an input third-order intercept point (IIP3) of −19 dBm. On transmit, it has a 1 dB compression point output power (OP1dB) of +12 dBm. Both transmit and receive paths have 6-bit control over both amplitude and phase. The chip comes in a wafer level chip scale (WLCS) flip-chip package, which allows for easy heat extraction.

In order to support low-cost solutions, a planar antenna integrated onto the circuit board is preferable. In order to enhance the bandwidth as well as support dual polarizations, a stacked-patch antenna was selected. The lower and upper patch elements are 193 mil × 193 mil and 178 mil × 178 mil, respectively. The antennas reside on a Rogers 4350 substrate, with 30 mils between the upper and lower patch elements. The antennas are probe fed with two vias, one for linear

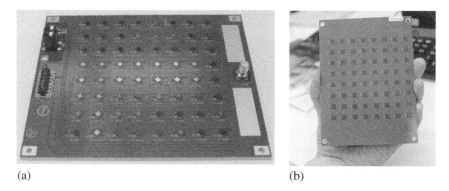

(a)   (b)

Figure 12.19 Developed prototype Ku-band phased array utilizing commercial silicon beamforming chipsets. (a) Backside view and (b) front view. *Source*: Chieh et al. [9].

horizontal polarization, and one for linear vertical polarization. The array has an inter-elemental spacing of approximately 0.6λ at 12 GHz.

The 8×8 array utilizes a stripline corporate feed network utilizing Wilkinson combiners. Embedded nickel-chrome (NiCr) thin film resistors were utilized as the feed network resides in between dielectric layers. Via fencing was implemented on the stripline layer to ensure proper characteristic impedance of transmission lines. The simulated excess feeder loss is approximately 9 dB. The AWMF-0117 chipset utilizes a five-wire serial peripheral interface (SPI) for controlling the array. A custom LabVIEW graphical user interface (GUI) was written to control the array along with a National Instruments USB-8452 providing the SPI outputs. Since both phase and amplitude can be controlled, we demonstrated a simple Taylor distribution weighting for sidelobe suppression (SLS). Measurements were made from 10 to 15 GHz for both polarizations and for both azimuth and elevation planes. For brevity, results from the vertical polarization at 13 GHz are presented. No phase or amplitude calibrations are applied. Figure 12.20 shows the measured scan patterns of the array on the azimuth plane, and as can be seen, the measured cross-polarization levels are well below 15 dB. The sweep angles in the elevation plane are limited to ±45° because of the limitation of the antenna positioner (Figure 12.21). The co-polarization patterns have symmetric sidelobe patterns, indicating that the standard deviation is tight from chip to chip for both amplitude and phase. Figure 12.22 shows the de-embedded measured

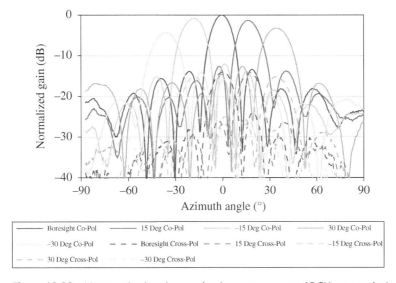

Figure 12.20 Measured azimuth scanning beam patterns at 13 GHz on vertical polarization. *Source*: Chieh et al. [9]. © 2019, IEEE.

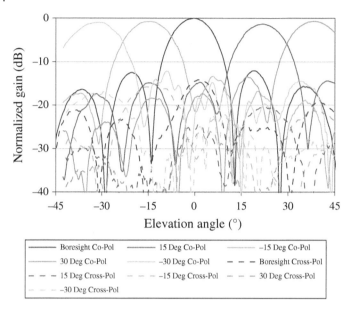

Figure 12.21 Measured elevation scanning beam patterns at 13 GHz on vertical polarization. *Source*: Chieh et al. [9]. © 2019, IEEE.

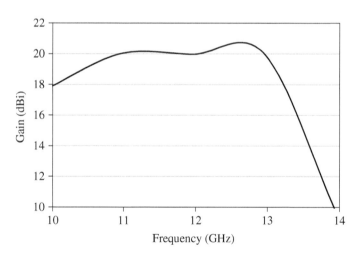

Figure 12.22 Measured de-embedded gain of the antenna array. *Source*: Chieh et al. [9]. © 2019, IEEE.

gain of the antenna array after subtraction of the feeder loss and frequency response of the AWMF-0117. Figure 12.23 shows the measured normalized gain pattern of the array when a Taylor amplitude weighting has been applied for 25 dB

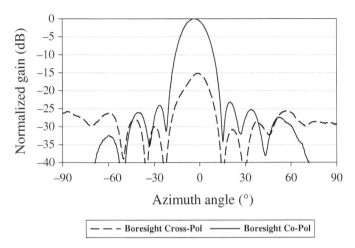

Figure 12.23 Measured beam pattern with Taylor weighting for 25 dB sidelobe level. *Source*: Chieh et al. [9]. © 2019, IEEE.

sidelobe levels. Finally, the array runs at 63.2 °C in receive mode without active cooling and in an ambient temperature of 22 °C. External considerations such as a cold plate could be used to reduce the operating temperature.

12.7 Ku-Band Phased Arrays on ROHACELL Utilizing Silicon Beamforming Chipsets

The bandwidth of the stacked patch is dependent on the relative permittivity of the dielectric spacer. That design sacrificed some bandwidth for production simplicity. In this section, we explore the use of ROHACELL as an air foam spacer between the stacked patch elements in order to broaden the bandwidth of operation. This work was presented in [10].

The fabricated prototype array is shown in Figure 12.24. The total size of the PCB, including auxiliary connectors, is 4.2 in. × 5.5 in. The commercial beamformer chipset we used was the Anokiwave AWMF-0117. The chip comes in a WLCS flip-chip package, which allows for easy heat extraction. In this design, we wanted to explore the use of air spacer foam, namely ROHACELL, to enhance gain and bandwidth. ROHACELL is known to have a relative permittivity of close to 1.041 and a loss tangent of 0.0017 at 10 GHz [11]. A simulation study was performed to determine the optimal height of the air spacer. We determined that if the height of the spacer was too much, the coupling between antenna elements spaced at $\lambda/2$ was strong and the impedance match was degraded both at

Figure 12.24 Ku-band ROHACELL phased array under test. *Source*: ROHACELL HF [11].

broadside and when the beam is scanned. For single antenna elements, this is not a big issue and greater air spacing is preferred and results in wider bandwidth as well as higher gain. For phased array applications, the mutual coupling between elements needs to be considered. In our study, we determined that a 40 mil ROHACELL HF spacer would be optimum resulting in higher gain and broader bandwidth compared to an all-PCB design. Figure 12.25 shows the designed stackup for the phased array. Figure 12.26 shows a comparison between the simulated broadside gain of an array utilizing all PCB and an array utilizing ROHACELL. As can be seen, the ROHACELL design has a greater bandwidth with a more graceful roll-off.

Figure 12.27 shows the fabricated array cross-section and as can be seen, the ROHACELL foam is roughly 40 mils high with a cut tolerance of ±2 mils. We do

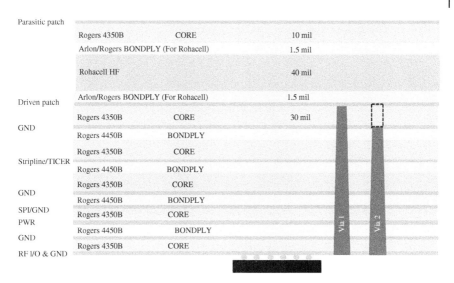

Figure 12.25 Phased array board stackup [11]. *Source*: Chieh et al. [9]. © 2019, IEEE.

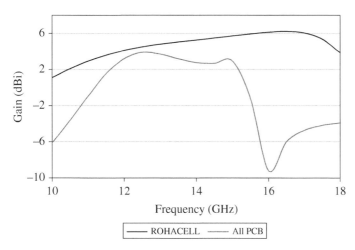

Figure 12.26 Simulated broadside gain comparison between all PCB and ROHACELL antenna element [11]. *Source*: Chieh et al. [9]. © 2019, IEEE.

note that the fabricated cross-section shown in Figure 12.27 differs from the notional proposed cross-section shown in Figure 12.25 where bondply was replaced by very high bond (VHB) tape in the fabricated prototype. The replacement of the bondply with the VHB tape had significant performance implications. There are a total of nine metal layers, with a stripline distribution layer

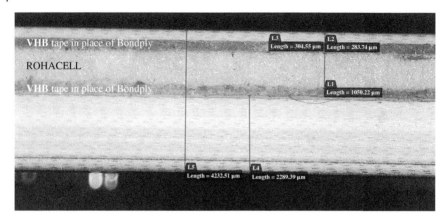

Figure 12.27 Cross-section of PCB with ROHACELL. *Source*: ROHACELL HF [11].

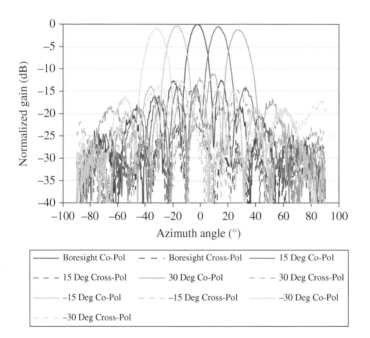

Figure 12.28 Measured azimuth scanning beam pattern at 15 GHz on vertical polarization [11]. *Source*: Chieh et al. [9]. © 2019, IEEE.

sandwiched in the middle of the circuit board. The stripline layer also has embedded TICER thin film resistors. Figure 12.28 shows the measured beam patterns at 15 GHz on the vertical polarization for azimuth scan. Figure 12.29 shows the measured beam pattern at 15 GHz on the horizontal polarization for vertical scan. As can be seen, the beam can easily be scanned ±30° with cross-polarization

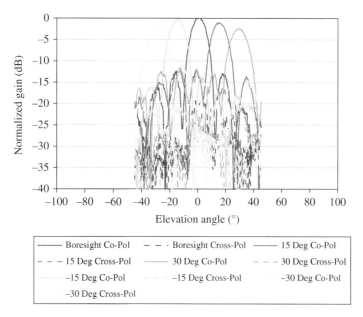

Figure 12.29 Measured elevation scanning beam pattern at 15 GHz on horizontal polarization [11]. *Source*: Chieh et al. [9]. © 2019, IEEE.

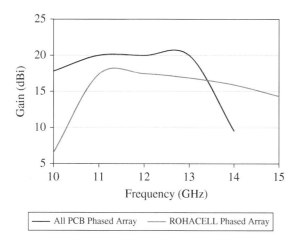

Figure 12.30 Measured broadside gain comparison [11]. *Source*: Chieh et al. [9]. © 2019, IEEE.

levels below 15 dB even at wide scan angles. In the elevation scan, the measurement was limited to ±45° because of the positioner rocker. Scan measurements in the elevation and azimuth planes were limited to ±30°. Figure 12.30 shows the measured broadside gain after de-embedding the loss from the feeder network and the gain from the AWMF-0117 chipset and compares between results from the all-PCB design and the ROHACELL design. Unfortunately, we see that the overall gain is lower across frequency compared to the all-PCB design. In addition, as the frequency increases, the gain degrades which is contradictory as the aperture becomes electrically larger at higher frequencies. The inconsistent result is attributed to the assembly of the ROHACELL. In Figure 12.25, we show that two bondply films, each 1.5 mil thick, were to be used to laminate the ROHACELL to the two individual PCBs. In Figure 12.27, it is apparent that a bondply was not used to adhere the ROHACELL to the PCB.

Instead, it was determined to use two-sided VHB adhesive transfer tape in place of a bondply. From Figure 12.27, it can be seen that the tape has a thickness of approximately 12 mils (each) with a low-frequency relative permittivity of approximately 4.7 and a loss tangent of approximately 0.1 [11]; a poor substitution for an air core. Since two pieces of the tape adhesive were needed, a total of 24 mils of the VHB adhesive were used. It is postulated that the VHB tape is highly lossy at microwave frequencies, which explains why the gain of the array degrades as the frequency increases. This also explains why the overall gain is lower than that of the all-PCB design. To validate this theory, we re-ran post fabrication simulations of the single antenna element replacing the 1.5 mil bondply with 12 mil VHB tape on each side of the ROHACELL. Figure 12.31 shows the simulated results, and as can be seen, the gain and bandwidth are significantly degraded. At 14 GHz, the degradation in peak gain is around 2.84 dB, very significant. To further add fidelity to the model, the relative permittivity and loss tangent of the VHB tape across frequency needs to be known, which is not known beyond 1 GHz.

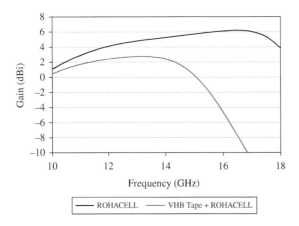

Figure 12.31 Post simulation comparison on single antenna element [11]. *Source*: Chieh et al. [9]. © 2019, IEEE.

12.8 Ku-Band Phased Arrays with Wide Axial Ratios Utilizing Silicon Beamforming Chipsets

Circular polarized active electronically scanned arrays (AESAs) are of great interest for satellite communications (SATCOM). Phased arrays which are capable of maintaining circular polarization over wide scan angles are of great interest especially for new satellite constellation that are being deployed in Low-Earth Orbit (LEO) and Medium-Earth Orbit (MEO).

One method that has been used to enhance the AR bandwidth in passive arrays has been to use sequential rotation (SQR). This technique was first discussed in [12–14]. Since then, several iterations of this technique have been used primarily for fixed beam passive arrays. In order to further improve the AR bandwidth, architectures utilize a "nested" SQR. In this approach, radiator elements within a sub-array utilize SQR, and then SQR is applied at the sub-array level. This concept is presented in [15–17]. All of these examples are variations of using SQR to realize a passive fixed beam array.

A prototype 16-element phased array using the nested SQR concept was designed and fabricated [18]. Figure 12.32 shows the fabricated antenna

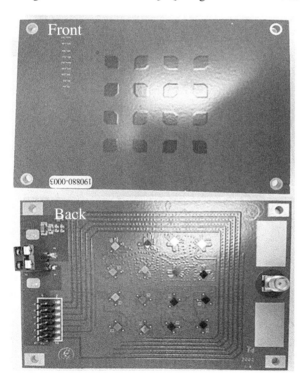

Figure 12.32 Photograph of the fabricated prototype array.

prototype. Anokiwave AWMF-0117 single channel silicon beamforming chipsets were used. This chipset supports both transmit and receive (T/R) functionality as well as double-pull double-throw (DPDT) switch to support dual polarizations. The PCB has 10 metal layers with an internal stripline layer for the feed distribution. Phase compensation due to the nested SQR is applied using this chipset. The measured beam patterns have been normalized since the phased array has gain on receive. The effects of the low-noise amplifier (LNA) and feed distribution network need to be de-embedded to obtain a calibrated gain measurement. Figure 12.33 shows the simulated and measured azimuth beam scan patterns for the LHCP polarization and they correlate very well. Measurements were made up to θ = ±80° due to constraints on the test setup in the anechoic chamber. Figure 12.34 shows the measured normalized co- and cross-polarization gain over frequency for beam angles 0, 15, 30, and 45°. As can be seen, the difference between LHCP and RCHP over frequency and scan angle is approximately better than 20 dB. The chipset allows for support of dual polarization through a SPDT T/R switch. When switching between LHCP and RHCP, since nested SQR is used,

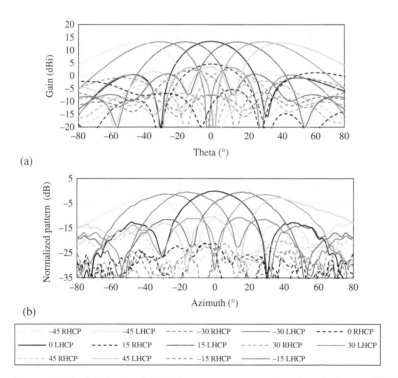

Figure 12.33 (a) Simulated LHCP beam scan patterns for azimuth plane at 12.5 GHz and (b) measured LHCP beam scan patterns for azimuth plane at 12.5 GHz.

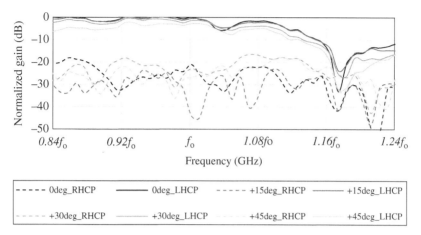

Figure 12.34 Measured co- and cross-polarization gain across frequency for LHCP.

the phase compensation applied through the phase shifter is reversed. Figure 12.35 shows the measured axial ratio over frequency for both polarizations (LHCP and RHCP) and for both azimuth and elevation beam scans. For both LHCP and RHCP polarizations on azimuth scan, the axial ratio remains below 3 dB for scan angles up to θ = ±45°. However, for both LHCP and RHCP in the elevation scan, the axial ratio for +45° and −45° are degraded. Since the RF currents are localized near the edge of the element, there is a stronger coupling at large scan angles in the elevation axis for either polarization.

12.9 28 GHz Phased Arrays Utilizing Silicon Beamforming Chipsets

The developed 64-element T/R array operates from 26 to 30 GHz. Comparison between the simulation and measurement results is presented in [19]. The beamformer board draws 14 W DC power on receive (Rx) and 23 W on transmit (Tx). Rogers 4350 (ε_r = 3.48 and tan δ = 0.004) is used through the board. The stacked patch elements reside on two 10 mil layers. The feed distribution layer resides on stripline utilizing Wilkinson power combiners. This feed distribution layer also utilizes embedded foil resistors from Ticer™. Various power, ground, and digital routing planes reside underneath the stripline feed layer to provide biasing and control to the chipsets. Finally, finite ground plane coplanar waveguide (FGCPW) are used to route RF signal from the chip to the antennas and from the chip to the corporate feed network.

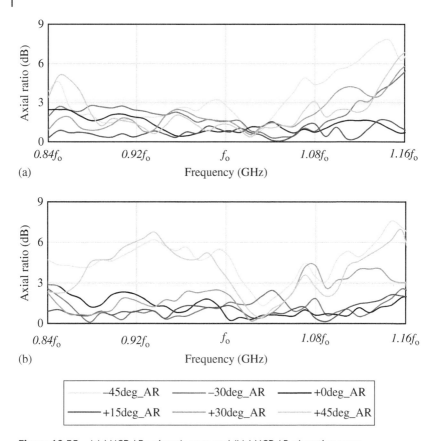

Figure 12.35 (a) LHCP AR azimuth scan and (b) LHCP AR elevation scan.

The developed phased array is shown in Figure 12.36. Figure 12.37 shows the antenna array under test at the Antenna and Microwave Lab (AML) at San Diego State University. The phased array antenna utilizes a vertical-mount 2.92 mm coaxial connector for RF input and output. We measured the antenna array inside an anechoic chamber with calibration done by standard gain horn antennas. A custom LabVIEW user interface was developed to control the array via SPI with the help of suitable beamforming algorithm. This software allows for beam scanning in both planes as well as the application of various amplitude tapers.

Figure 12.38 shows the simulated and measured array beam scan patterns for azimuth scanning under uniform excitation when selected linear polarization is $\phi = 45°$ at 28 GHz. As can be seen, the simulated and measured results correspond very well with the beam scanning up to $\pm55°$ with a 5 dB drop from broadside. As

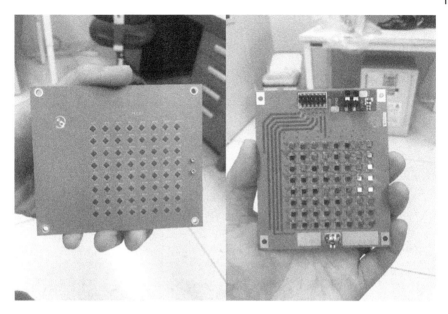

Figure 12.36 Developed 64-element T/R phased array at 28 GHz.

Figure 12.37 Phased array under test in anechoic chamber.

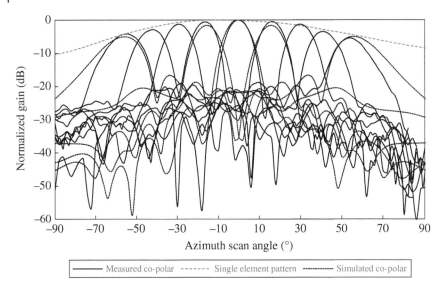

Figure 12.38 Simulation and measurement of azimuth scan at 45°-Pol, 28 GHz.

can be seen, the array patterns follow the pattern envelope of the single antenna element. Since the inter-elemental spacing is fixed, and the phase weights are static, as the frequency increases, the scan angle reduces. In theory, the phase weights could be recalculated at each frequency to ensure the scan angle performance. Measured cross-polarization levels are at least 15 dB below the main lobe. Sidelobe levels are less than 20 dB from the main lobe as measurements were taken in the $\phi = 45°$ cut plane, while $\phi = 0°$ and $\phi = 90°$ are the array planes.

Figure 12.39 shows the measured and simulated broadside gain of the array over frequency. In measurement, the antenna array gain is calculated by removing the receive gain of the AWMF-0116 and by removing the feeder loss of the corporate feed network. The feeder loss is not easy to de-embed as we did not fabricate a dedicated structure. This was done mainly through estimation through simulation. For this reason, Figure 12.39 shows some measurement uncertainty bars (± 3 dB) that account for this.

Finally, since the array has 6-bits of amplitude control per element, amplitude tapers can be performed. Amplitude tapers are extremely interesting from a tracking perspective as the phased array antenna can be used to generate monopulse patterns. This is especially applicable for SATCOM on the move or other platforms that require agile tracking. Figure 12.40a demonstrates a Taylor weighting [16] for 30 dB SLS. When implemented with the array elements, the achieved SLS is around 25 dB across all frequencies. This type of pattern can be used as the summation beam in a monopulse tracker. Figure 12.40b demonstrates a Bayliss [17]

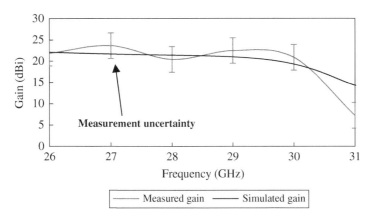

Figure 12.39 Measured and simulated peak gain.

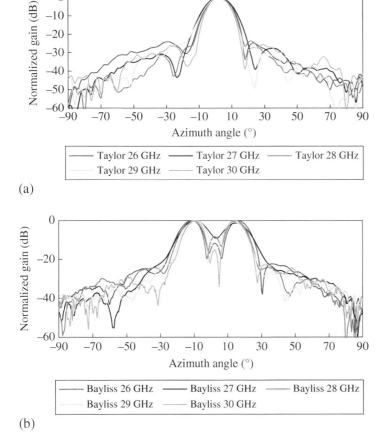

Figure 12.40 Measured (a) Taylor and (b) Bayliss weighted pattern.

weighting for creating a null at broadside. This type of pattern can be used as a difference beam in a monopulse tracker. As can be seen, at 30 GHz, the Bayliss weighting achieves around 35 dB of rejection at 0°. Transmit performance of this array was not verified.

12.10 Phased Array Reflectors Utilizing Silicon Beamforming Chipsets

The hybrid reflector-phased array antenna consists of a parabolic-cylindrical reflector and an 8×4 planar microstrip patch array [17]. The planar microstrip patch array is used as the feed of the parabolic-cylindrical reflector antenna and is located along the focal line of the reflector. The geometry and the coordinate system of the parabolic-cylindrical reflector fed by the phased array antenna are shown in Figure 12.41.

A MATLAB routine is written to evaluate the far-field components of the parabolic-cylindrical reflector fed by the phased array antenna. An offset parabolic-cylindrical reflector of size $A \times B = 50\,cm \times 50\,cm$ and $f/D = 0.4$ is chosen. The results for the current distribution, broadside radiation pattern, and the beam steering patterns of the symmetric parabolic-cylindrical reflector are shown in Figures 12.42 and 12.43.

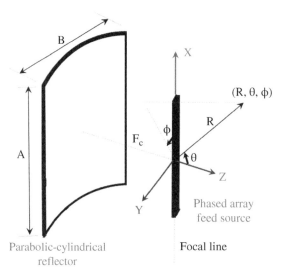

Figure 12.41 Geometry and coordinate system of the parabolic-cylindrical reflector fed by the phased array antenna [17]. *Source*: Mishra et al. [20].

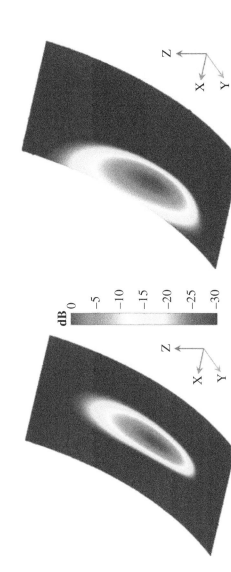

Figure 12.42 PO-computed current density using MATLAB for the offset parabolic-cylindrical reflector fed by 8 × 4 phased array antenna at 13 GHz: (a) broadside angle and (b) 30° beam scan angle [17].
Source: Mishra et al. [20].

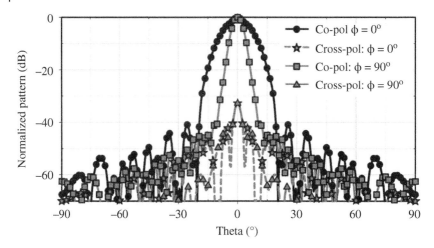

Figure 12.43 PO-computed broadside 2D normalized radiation pattern of the offset parabolic-cylindrical reflector fed by an 8×4 phased array antenna at 13 GHz [17]. *Source*: Mishra et al. [20].

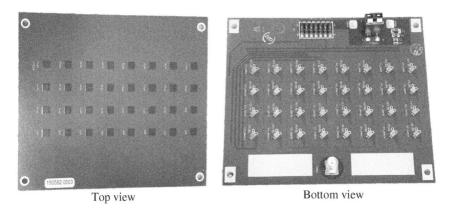

Top view   Bottom view

Figure 12.44 Photograph of fabricated dual linear-polarized phased array antenna integrated with the Anokiwave AWMF-0117 silicon RFIC chips-based beamforming network [17]. *Source*: Mishra et al. [20].

A dual linear-polarized stacked patch antenna is used as an element of the 8×4 phased array antenna with the inter-element spacing of dx = dy = 12 mm, as presented in Figure 12.44. The proposed dual-linear polarized stacked patch antenna with the beamforming network is prototyped at *Ku*-band to provide the beam steering range of ±45° for a 3 dB reduction in the gain. Anokiwave

AWMF-0117 integrated silicon core chips are used in the beamforming network for achieving the beam steering.

The parabolic-cylindrical reflector is manufactured using the aluminum sheet metal technique, and Figure 12.45 shows the photograph of the integrated reflector-phased array prototype. A 3D printed strut support is used to integrate the phased array with the reflector. The strut is fastened to the reflector using flathead surface flushed bolts. The strut and the bolts effects have been analyzed in the modeling of the reflector. The hybrid reflector-phased array beam steering radiation pattern is characterized in the compact antenna test range (CATR) facility at the Naval Information Warfare Center Pacific. The measurement of the parabolic-cylindrical reflector also exhibited the beam steering range of $\pm 30°$ for a 3 dB reduction in the gain for both the X-polarization and Y-polarization, as presented in Figure 12.46. The measured broadside 3 dB beamwidth at 13 GHz for the parabolic-cylindrical reflector antenna in the $\phi = 0°$ plane is 14°, and 14.5° for the X-polarization and Y-polarization, respectively, and the measured 3 dB beamwidth in the $\phi = 90°$ plane is 4.2° for both the X-polarization and Y-polarization. The peak simulated directivity of the proposed parabolic-cylindrical reflector fed by the phased array antenna is 26.85 and 27.2 dBi for the X-polarization and Y-polarization, respectively. The simulated directivity for the reflector antenna and the simulated and measured 3 dB beamwidth for different scan angles for the X-polarization and Y-polarization are shown in Figure 12.47.

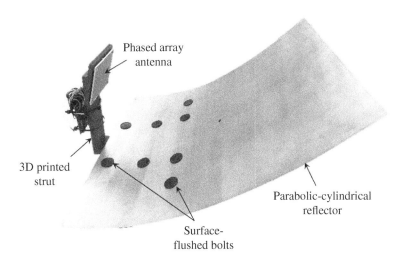

Phased array antenna

3D printed strut

Parabolic-cylindrical reflector

Surface-flushed bolts

Figure 12.45 Photograph of the fabricated parabolic-cylindrical reflector using sheet metal technique and integrated with the phased array antenna [17]. *Source*: Mishra et al. [20].

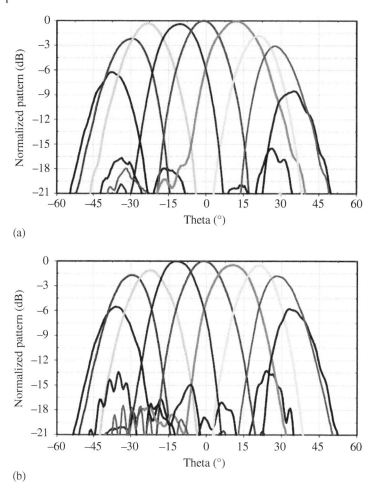

Figure 12.46 Measured normalized co-polarization beam steering radiation pattern of the parabolic-cylindrical reflector along the cylindrical axis: (a) X-polarization and (b) Y-polarization at 13 GHz [17]. *Source*: Mishra et al. [20].

12.11 Conclusion

In this chapter, we presented phased array antenna discussion by employing emerging 5G silicon RFICs. The silicon RFICs are small footprint chips which include Tx/Rx functions. The Tx/Rx paths include digital phase shifters, VGAs, LNAs, high power amplifiers, power divider, and SPI control. The phased array antennas using these RFICs need multilayer PCBs which needs unique circuit layout knowledge and skill. These phased array antennas are all-flat panel, low profile, and low weight. This technology has a very bright future and as the paradigm of economies of scale

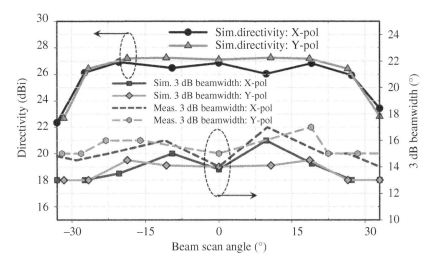

Figure 12.47 The simulated peak directivity of the proposed parabolic-cylindrical reflector antenna and the simulated and measured 3 dB beamwidth for different scan angles for X-polarization and Y-polarization along $\phi = 0°$ plane [17]. *Source*: Mishra et al. [20].

reduces the price point of phased arrays, they will be more ubiquitous. These phased arrays can find applications in 5G communications and satellite communications.

References

1   Tan, G.-L., Mihailovich, R.E., Hacker, J.B. et al. (2003). Low-loss 2- and 4-bit TTD MEMS phase shifters based on SP4T switches. *IEEE Transactions on Microwave Theory and Techniques* 51 (1): 297–304.
2   Johnson, M.C., Brunton, S.L., Kundtz, N.B., and Kutz, J.N. (2015). Sidelobe canceling for reconfigurable holographic metamaterial antenna. *IEEE Transactions on Antennas and Propagation* 63 (4): 1881–1886.
3   Hacker, J.B., Mihailovich, R.E., Kim, M., and DeNatale, J.F. (2003). A Ka-band 3-bit RF MEMS true-time-delay network. *IEEE Transactions on Microwave Theory and Techniques* 51 (1): 305–308.
4   Hashemi, H., Xiang, G., Komijani, A., and Hajimiri, A. (2005). A 24-GHz SiGe phased-array receiver-LO phase-shifting approach. *IEEE Transactions on Microwave Theory and Techniques* 53 (2): 614–626.
5   Oshima, N. I. Ando, Y. Kitagishi et al. (2015). A X-band reconfigurable phased array antenna system using 0.13-μm SiGe BiCMOS IC with 5-bit IF phase shifters. *Proceedings of the 2015 IEEE Compound Semiconductor Integrated Circuit Symposium (CSICS)*, New Orleans, LA (11–14 October 2015), 1–4.
6   Koh, K. and Rebeiz, G.M. (2007). 0.13-μm CMOS phase shifters for X-, Ku-, and K-band phased arrays. *IEEE Journal of Solid-State Circuits* 42 (11): 2535–2546.

7 Koh, K. and Rebeiz, G.M. (2008). An X- and Ku-band 8-element phased-array receiver in 0.18-μm SiGe BiCMOS technology. *IEEE Journal of Solid-State Circuits* 43 (6): 1360–1371.

8 Qorvo. Datasheet for TGP2615. https://www.qorvo.com/products/p/TGP2615 (accessed July 20)

9 Chieh, J.S., Yeo, E., Kerber, M., and Olsen, R. (2019). Ku-band dual polarized phased array utilizing silicon beamforming chipsets. *Proceedings of the 2019 IEEE Topical Workshop on Internet of Space (TWIOS)*, Orlando, FL (20–23 January 2019), 1–3.

10 Chieh, J.S., Yeo, E., Kerber, M., and Olsen, R. (2019). Dual polarized Ku-band phased array on ROHACELL utilizing silicon beamforming chipsets. *Proceedings of the 2019 IEEE International Symposium on Phased Array System & Technology (PAST)*, Waltham, MA (15–18 October 2019), 1–5.

11 Datasheet for ROHACELL HF. *Product Information*, 2018.

12 Hall, P.S. (1989). Application of sequential feeding to wide bandwidth, circularly polarised microstrip patch arrays. *IEE Proceedings H - Microwaves, Antennas and Propagation* 136 (5): 390–398.

13 Hall, P.S. (1992). Dual polarisation antenna arrays with sequentially rotated feeding. *IEE Proceedings H - Microwaves, Antennas and Propagation* 139 (5): 465–471.

14 Teshirogi, T., Tanaka, M., and Chujo, W. (1985). Wideband circularly polarised array antennas with sequential rotations and phase shift of elements. *Proceedings of the International Symposium on Antennas and Propagation, ISAP85*, Tokyo (20 August 1985), 117–120.

15 Hall, P.S. (1987). Feed radiation effects in sequentially rotated microstrip patch arrays. *Electronics Letters* 23 (17): 877–878.

16 George, R.R., Castro, A.T., and Sharma, S.K. (2017). Comparison of a four stage sequentially rotated wideband circularly polarized high gain microstrip patch array antennas at Ku-band. *Proceedings of the 2017 11th European Conference on Antennas and Propagation (EUCAP)*, Paris (9–24 March 2017), 2307–2311.

17 Chen, A., Zhang, Y., Chen, Z., and Yang, C. (2011). Development of a Ka-band wideband circularly polarized 64-element microstrip antenna array with double application of the sequential rotation feeding technique. *IEEE Antennas and Wireless Propagation Letters* 10: 1270–1273.

18 J-C. Chieh, E. Yeo, R. Farkouh, A. Castro, M. Kerber, R. Olsen, E. Merulla, and S. K. Sharma, "Development of Flat Panel Active Phased Array Antennas using 5G Silicon RFICs at Ku- and Ka-Bands", IEEE ACCESS, Volume 8, 2020, Pages: 192669-192681.

19 Chieh, J.S., Yeo, E., Kerber, M. et al. (2019). A 28 GHz dual slant polarized phased array using silicon beamforming chipsets. *Proceedings of the 2019 IEEE International Symposium on Phased Array System & Technology (PAST)*, Waltham, MA (15–18 October 2019), 1–5.

20 Mishra, G., Sharma, S.K., Chieh, J.S., and Olsen, R.B. (2020). Ku-band dual linear-polarized 1-D beam steering antenna using parabolic-cylindrical reflector fed by a phased array antenna. *IEEE Open Journal of Antennas and Propagation* 1: 57–70.

Index

Multifunctional Antennas and Arrays for Wireless Communication Systems, First Edition.
Edited by Satish K. Sharma and Jia-Chi S. Chieh.
© 2021 John Wiley & Sons, Inc. Published 2021 by John Wiley & Sons, Inc.